Lecture Notes of the Unione Matematica Italiana

21

More information about this series at http://www.springer.com/series/7172

Editorial Board

The Editorial Policy can be found
at the back of the volume.

Leticia Brambila Paz • Ciro Ciliberto •
Eduardo Esteves • Margarida Melo • Claire Voisin
Editors

Moduli of Curves

CIMAT Guanajuato, Mexico 2016

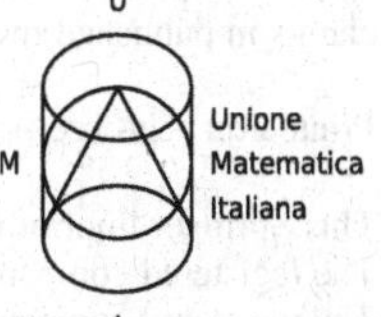

Editors
Leticia Brambila Paz
CIMAT
Guanajuato, Mexico

Ciro Ciliberto
Dipartimento di Matematica
Università di Roma Tor Vergata
Roma, Italy

Eduardo Esteves
Instituto Nacional de Matemática Pura
Rio de Janeiro, Brazil

Margarida Melo
Dipartimento di Matematica
Università di Roma Tre
Roma, Italy

Claire Voisin
Collège de France
Paris, France

ISSN 1862-9113 ISSN 1862-9121 (electronic)
Lecture Notes of the Unione Matematica Italiana
ISBN 978-3-319-59485-9 ISBN 978-3-319-59486-6 (eBook)
DOI 10.1007/978-3-319-59486-6

Library of Congress Control Number: 2017953005

Mathematics Subject Classification (2010): 14D20, 14D22, 14H10, 14E30, 14N05, 14F05, 14T05

Printed on acid-free paper

This Springer imprint is published by Springer Nature
The registered company is Springer International Publishing AG
The registered company address is: Gewerbestrasse 11, 6330 Cham, Switzerland

Preface

The present volume collects some of the lecture notes of a CIMPA-ICTP school on "Moduli of Curves," which took place at CIMAT, Guanajuato, Mexico, in the 2 weeks from February 22 to March 4, 2016.

Algebraic geometry is a classical key area of mathematical research. The guiding problem in algebraic geometry is the classification of algebraic varieties, which depends on discrete as well as on continuous invariants. The latter were named "moduli" by B. Riemann in the mid-nineteenth century. Moduli theory is the study of the way in which algebraic varieties, or more generally, geometric objects attached to them, vary in families. Moduli spaces have been studied since Riemann, but they were first rigorously constructed only in the 1960s by Deligne, Mumford, and others and the theory has continued to develop since then, with the infusion of ideas from physics after 1980. In the last decades, the theory has experienced an extraordinary development, finding an increasing number of connections with many other areas of mathematics (differential geometry, topology, number theory, representation theory, discrete mathematics, etc.) and also with other disciplines (in the present context, particularly theoretical physics).

The aim of the CIMPA-ICTP school in Guanajuato has been to give an introduction to subjects of current interest in the study of moduli spaces M_g of curves of genus g. More specifically, the main objective was to introduce young mathematicians to the classical foundations of the theory, to some of its major developments and to the main tools used in it, and also to provide glimpses on some of the more advanced topics, such as recent work on the so-called Hassett–Keel program, on the topology of M_g, on Bridgeland stability, on the tropical viewpoint, etc. As such, the school has been of interest not only for young persons but also for more advanced researchers in mathematics and physics interested in the subject.

There were 106 participants, coming from Argentina, Brazil, Chile, Colombia, Cuba, Costa Rica, El Salvador, France, Germany, Great Britain, Italy, Japan, Mexico, Spain, and the USA, including the main speakers. The school developed with a mixture of courses, lectures, poster sessions, informal discussions, tutorials, and/or problem sessions. This is the list of the main courses:

- Melody Chan, Moduli and degenerations of algebraic varieties via tropical geometry
- Paolo Cascini, Higher dimensional varieties and their moduli spaces
- Olivier Debarre, Higher dimensional varieties and their moduli spaces
- Gavril Farkas, Syzygies of algebraic curves
- Sam Grushevsky, Minimal model program, birational geometry and topology of M_g
- Radu Laza, Geometric invariant theory and Bridgeland stability
- Emanuele Macrì, Geometric invariant theory and Bridgeland stability
- Edoardo Sernesi, Algebraic curves and their moduli spaces

The main speakers are leading algebraic geometers and, in particular, specialists in the field of moduli spaces of curves and their geometry. Each course consisted of five 1-h lectures. The topics that appear in the title of two courses were thus covered in ten lectures. Each course started with basic material and finished with the presentation of the state of the art. Videos of the lectures, as well as more information, are available on the web site of the school:

http://moduli2016.eventos.cimat.mx/node/304

This book contains six of the eight courses. These lecture notes provide a timely description of the present state of the art of moduli spaces of curves and their geometry, and they are written in a way which will make their reading extremely useful not only for young people who want to approach this important field but also for well-established researchers, who will find references, problems, original expositions, new viewpoints, etc.

We thank all the speakers and especially the authors of the notes: all of them did a great job delivering spectacular lectures. We thank all the participants for the lively atmosphere they created during the school. We also thank the following institutions for their support:

- CIMAT (Centro de Investigaciòn en Matemàticas A.C.)
- CIMPA (Centre International de Mathématiques Pures et Appliquées)
- CCM (Centro de Ciencias Matemàticas, UNAM, Morelia)
- FCM (Foundation Compositio Mathematica)
- ICTP (International Centre for Theoretical Physics)
- SMM (Sociedad Matemàtica Mexicana)

Last but not least, we thank the local organizers: Abel Castorena, Osbaldo Mata, Hugo Torres, with special thanks to Lupita Hernandez and Guillermo Villanueva.

The Scientific Committee and Editors of the Volume:

Leticia Brambila Paz, CIMAT, Guanajuato, Mèxico
Ciro Ciliberto, Università di Roma "Tor Vergata," Roma, Italia
Eduardo Esteves, IMPA, Rio de Janeiro, Brasil
Margarida Melo, Università di Roma Tre, Roma, Italia
Claire Voisin, Collége de France, Paris, France

Contents

Lectures on Tropical Curves and Their Moduli Spaces

Melody Chan

1 Introduction

These are notes for five lectures on moduli and degenerations of algebraic curves via tropical geometry. What do I mean by degenerations of algebraic curves? The basic idea is that one can get information about the behavior of a smooth curve by studying one-parameter families of smooth curves, which degenerate in the limit to a singular curve, instead. The singular curve typically has many irreducible components, giving rise to a rich combinatorial structure. This technique obviously relies on having a robust notion of family of curves, that is, a moduli space. Thus moduli spaces immediately come to the fore.

Tropical geometry is a modern degeneration technique. You can think of it, to begin with, as a very drastic degeneration in which the limiting object is entirely combinatorial. We will flesh out this picture over the course of the lectures. It is also a developing field: exactly *what* tropical geometry encompasses is a work in progress, developing rapidly.

The powerful idea of using degenerations to study algebraic curves is at least several decades old and has already been very successful. But recent developments in tropical geometry make it timely to return to and expand upon these ideas. The focus of these lectures will be on one beautiful recent meeting point of algebraic and tropical geometry: the *tropical moduli space of curves*, its relationship with the Deligne-Mumford compactification by stable curves, and its implications for the topology of $\mathcal{M}_{g,n}$. I am not assuming any prior background in tropical geometry.

M. Chan (✉)
Department of Mathematics, Brown University, Providence, RI 02912, USA
e-mail: mtchan@math.brown.edu

L. Brambila Paz et al. (eds.), *Moduli of Curves*, Lecture Notes of the Unione Matematica Italiana 21, DOI 10.1007/978-3-319-59486-6_1

2 From the Beginning: Tropical Plane Curves

Let's start from scratch. Centuries ago, some of the very first objects considered in algebraic geometry were just *plane curves*: the zero locus in $\mathbb{P}^2$ of a homogeneous equation in three variables. Only later was the perspective of studying curves abstractly, free from a particular embedding in projective space, developed. We'll do the same, starting with some plane tropical curves, which are a special case of *embedded tropicalizations.*

2.1 Embedded Tropicalizations

The natural setting for tropical geometry is over *nonarchimedean fields.* Let K be a field, and write $K^* = K \setminus \{0\}$ as usual.

Definition 2.1 A *nonarchimedean valuation* on K is a map $v: K^* \to \mathbb{R}$ satisfying:

(1) $v(ab) = v(a) + v(b)$, and
(2) $v(a+b) \geq \min(v(a), v(b))$

for all $a, b \in K^*$. By convention we may extend v to K by declaring $v(0) = \infty$.

The ring R of elements with nonnegative valuation is the *valuation ring* of K. Note that R is a local ring; write $k = R/\mathfrak{m}$ for its residue field.

If you are arithmetically minded, you might immediately think of the p-adic field $K = \mathbb{Q}_p$, or $\mathbb{C}_p$, the completion of its algebraic closure. For another example, take your favorite field, perhaps $\mathbb{C}$, and equip it with the all-zero valuation, also known as the *trivial valuation.* This example sounds unimportant, but it is theoretically important, because it permits a unified theory of tropicalization.

Another good example to keep in mind is $K = \mathbb{C}((t))$, the field of *Laurent series,* with valuation $v(\sum_{i\in\mathbb{Z}} a_i t^i) = \min\{i : a_i \neq 0\}$. In this example, K is discretely valued, in that $v(K^*) \cong \mathbb{Z}$. Note that the algebraic closure of $\mathbb{C}((t))$ is the field of *Puiseux series* $\mathbb{C}\{\{t\}\} = \bigcup_n \mathbb{C}((t^{1/n}))$, whose elements are power series with bounded-denominator fractional exponents. This field is a favorite of tropical geometers.

Exercise 2.2 Let $a, b \in K$. If $v(a) \neq v(b)$, then in fact

$$v(a+b) = \min(v(a), v(b)).$$

In other words, for all $a, b \in K$, the minimum of $v(a), v(b), v(a+b)$ occurs at least twice.

Exercise 2.3 Suppose K is a nonarchimedean field. If K is algebraically closed then its residue field k is also algebraically closed.

Remark 2.4 Already, you can get a glimpse of *why* tropical geometry is performed over nonarchimedean fields. For example, a curve X over $K = \mathbb{C}((t))$ can be regarded as a family of complex curves over an infinitesimal punctured complex disc Spec K around $t = 0$. The eventual *tropicalization* of this curve will be a metric graph (decorated with a little bit of extra stuff). It records data about how the special fiber $t = 0$ must be filled in according to the properness of $\overline{\mathcal{M}}_g$.

A **valued extension** of valued fields is an extension L/K in which the valuation on L extends the valuation on K.

Definition 2.5 (Embedded Tropicalization) Fix a nonarchimedean field K. Let X be a subvariety of the algebraic torus $(\mathbb{G}_m)^n$. So X is defined by an ideal of the Laurent polynomial ring $K[x_1^{\pm}, \ldots, x_n^{\pm}]$.

The *tropicalization* of X is the subset of $\mathbb{R}^n$

$$\{(v(x_1), \ldots, v(x_n)) : (x_1, \ldots, x_n) \in X(L) \text{ for } L/K \text{ a valued extension}\}.$$

In particular, if K is algebraically closed and nontrivially valued, then Trop(X) is the closure, in the usual topology on $\mathbb{R}^n$, of the set

$$\{(v(x_1), \ldots, v(x_n)) : (x_1, \ldots, x_n) \in X(K)\}$$

of coordinatewise valuations of K-points of X.

I am skimping a little bit on notation, and denoting all valuations by v, even over different fields.

Let's immediately practice Definition 2.5 in the case of a line.

Example 2.6 (A Line in the Plane) Let $f(x, y) = x + y - 1$ and let $X = V(f) \subset \mathbb{G}_m^2$. So X is $\mathbb{P}^1$ minus three points. What is Trop(X)?

Answer: Suppose x and y are such that $x + y - 1 = 0$. By Exercise 2.2,

the minimum of $v(x)$, $v(y)$, and $v(1)$ is attained at least twice.

In other words,

$$\text{Trop}(X) \subseteq \{(z, w) \in \mathbb{R}^2 : \text{ the minimum of } z, w, \text{ and } 0 \text{ occurs at least twice}\}.$$

You can draw this latter set: it is polyhedral, consisting of three rays from the origin in the directions of the standard basis vectors e_1, e_2, as well as $-e_1 - e_2$. See Fig. 1. The content of the next theorem is that the containment above is actually an equality.

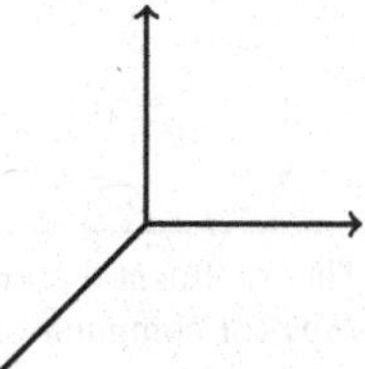

Fig. 1 The tropical line from Example 2.6

We state the theorem below in a baby case, the case of curves in the plane. But it holds verbatim for arbitrary hypersurfaces. See [24, Theorem 3.1.3] for the full statement and its proof.[1]

Theorem 2.7 (Kapranov's Theorem) *Let*

$$f = \sum_{(i,j)\in\mathbb{Z}^2} c_{ij}x^i y^j \quad \in K[x^{\pm}, y^{\pm}].$$

Then

$$\mathrm{Trop}(X) = \{(z, w) \in \mathbb{R}^2 : \min_{(i,j)\in\mathbb{Z}^2} v(c_{ij}) + iz + jw \textit{ occurs at least twice.}\}$$

Remark 2.8 Notice that the expression $\min\big(v(c_{ij}) + iz + jw\big)$ is obtained from f by

- replacing addition by minimum,
- replacing multiplication by $+$, and
- replacing scalars c_{ij} by their valuations.

This explains the slogan you may hear that *tropical geometry is the algebraic geometry of the min-plus semiring* $(\mathbb{R} \cup \{\infty\}, \min, +)$. This may also help explain the naming of the field of tropical geometry, which was in honor of the Brazilian mathematician and computer scientist Imre Simon, a pioneer in the study of the min-plus semiring.

Remark 2.9 It is not nearly as straightforward to compute tropicalizations of subvarieties of $\mathbb{G}_m^n$ that are not hypersurfaces. It can however be done by A.N. Jensen's software `gfan`, using Gröbner methods [19]. See the `gfan` manual for details.

Exercise 2.10 What are all possible tropical lines in the plane, i.e., subsets of $\mathbb{R}^2$ of the form Trop(X) where $X = V(ax + by + c)$?

2.2 *A Very, Very Short Treatment of Berkovich Analytifications*

All of this could be said much more elegantly using the language of Berkovich spaces [5]. We now assume that K is a *complete* valued field. This means that K is complete as a metric space, with respect to its nonarchimedean valuation.[2]

[1]The statement there also gives an equivalent formulation in terms of Gröbner initial ideals, which is key for computations, and which I won't talk about at all.

[2]The valuation v on K defines a norm on K by setting $|a| = \exp(-v(a))$.

Conceptually, the assumption that K is complete is not such a big deal, because we can always base change from a given field K to its completion $\hat{K}$.

Example 2.11 $\mathbb{Q}_p$ is complete by construction. Any trivially valued field is of course complete. On the other hand, the field of Puiseux series $\mathbb{C}\{\{t\}\}$ is not complete. What is its completion?

Let X be finite type scheme over K. We shall define the *Berkovich analytification* X^{an}, a locally ringed topological space associated to X. Actually, my plan is to entirely ignore the structure sheaf of analytic functions of X^{an}. So we will just regard X^{an} as a topological space for the duration.

I'll do things in an unconventional order, starting with a very quick way to say what the points of X^{an} are. I find this simple description very useful—especially when X is some kind of moduli space.

Definition 2.12 (Points of the Berkovich Space) Let X be a finite type scheme over a complete nonarchimedean valued field K. The points of the *Berkovich analytification* X^{an} are in bijection with maps $\operatorname{Spec} L \to X$ for all valued field extensions L/K, modulo identifying $\operatorname{Spec} L \to X$ with $\operatorname{Spec} L' \to \operatorname{Spec} L \to X$, where L'/L is again a valued field extension.

Does that sound strange? It should be compared with the more familiar situation of a scheme Y over any field K, with no valuations in sight. Then you can see for yourself the following way to name the points of Y: points of Y correspond to maps $\operatorname{Spec} L \to Y$ for all extensions L/K, modulo identifying $\operatorname{Spec} L \to Y$ with $\operatorname{Spec} L' \to \operatorname{Spec} L \to Y$ for all further extensions L'/L.

Next we will define the topology on X^{an}.

Definition 2.13 (Topology on the Berkovich Space, Affine Case) We continue to assume that K is a complete nonarchimedean valued field. Let $X = \operatorname{Spec} A$ be an affine scheme of finite type over K.

We take X^{an} to have the coarsest topology such that for all $f \in A$, the function

$$\begin{aligned} v_f \colon X^{\mathrm{an}} &\longrightarrow \mathbb{R} \\ (\operatorname{Spec} L \overset{p}{\to} X) &\longmapsto v(p^{\#} f) \end{aligned} \tag{1}$$

is continuous. Here v denotes the valuation on the valued field L and $p^{\#} \colon A \to L$ is the map of rings coming from p.

Now for an arbitrary finite type scheme X over K that is not necessarily affine, the topological space X^{an} is obtained by taking an affine open cover of X, analytifying everything separately, and then gluing.[3]

[3]Usually, the points of $(\operatorname{Spec} A)^{\mathrm{an}}$ are described as multiplicative seminorms $\|\cdot\|_p$ on A extending the norm on K, equipped with the coarsest topology such that for every $f \in A$, the map $X^{\mathrm{an}} \to \mathbb{R}$

Now let's try Definition 2.5 over again:

Redefinition 2.14 (Embedded Tropicalization, Again) Let $X \subseteq (\mathbb{G}_m)^n$, given explicitly as $X = \operatorname{Spec} K[x_1^{\pm}, \ldots, x_n^{\pm}]/I$. The **tropicalization** of X, denoted $\operatorname{Trop}(X)$, is the image of the map $X^{\mathrm{an}} \to \mathbb{R}^n$ that sends, for $p\colon \operatorname{Spec} L \to X$ a point of X^{an},

$$p \quad \longmapsto \quad (\nu_{x_1}(p), \ldots, \nu_{x_n}(p)).$$

The maps ν_{x_i} were defined in (1), and we set up the topology of X^{an} precisely so that each ν_{x_i} is continuous. Thus $\operatorname{Trop}(X)$ is, by Definition 2.13, a *continuous* image of X^{an}. This is helpful! For example, Berkovich tells us that X^{an} is connected if X is connected [5]. Therefore, in this situation, $\operatorname{Trop}(X)$ is connected too.

We've just hinted at the fact that passing to analytifications can be a helpful perspective for viewing tropicalizations. But actually, one could just as well say the reverse. Namely, one of the reasons tropicalizations are useful is that they can provide a faithful "snapshot" of a piece of the much hairier[4] and more complicated space X^{an}. See [4, 27].

3 Abstract Algebraic and Tropical Curves

Next, what is an *abstract* tropical curve, and how does such a gadget arise from an algebraic curve over a valued field K? That is the subject of this lecture. The relationship with the previous lecture is as follows. In the last section, we concerned ourselves with *embedded tropicalizations*, i.e. tropicalizations of a subvariety of a torus or toric variety. In this section, we are fast-forwarding many decades in the parallel story in the history of algebraic geometry, and treating curves now *in the abstract*, free from a particular embedding in projective space, say. Also, just as in algebraic geometry, once this bifurcation between abstract and embedded tropicalization happens, it then becomes interesting to study the relationship between the two. This is also a very interesting story (see [4, 17]), but I won't have

---

sending

$$\|\cdot\|_p \mapsto \|f\|_p$$

is continuous. It's not hard to describe the correspondence between Definition 2.12 and this definition. A seminorm $\|\cdot\|_p$ corresponds to the map

$$\operatorname{Spec} \operatorname{Frac}(A/\ker(\|\cdot\|_p)) \to \operatorname{Spec} A.$$

[4] Almost literally.

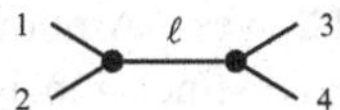

Fig. 2 A picture of the abstract tropicalization of the curve X/K of Example 3.1. We have drawn marked points as marked rays attached at vertices

time for it. I also highly recommend [3] for a survey of the state of the art in tropical linear series and Brill-Noether theory.

Before launching into all the definitions, let me give one example in full. It's such a tiny example that I can guarantee that it's not that interesting on its own. But it will serve as a little laboratory in which we can see all the definitions in action at once.

Example 3.1 (A Preview) Let $K = \mathbb{C}((t))$, with $R = \mathbb{C}[\![t]\!]$ its valuation ring. Let ℓ be a positive integer, and let X/K be the projective plane curve with equation

$$xy = t^\ell z^2. \tag{2}$$

So, X is just a smooth conic over K, but we regard X as defining a germ of a family, with base parameter $t \neq 0$, of smooth plane conics in the complex projective plane. Let's consider the four marked points

$$p_1, p_2 = (\pm t^\ell : \pm 1 : 1) \qquad p_3, p_4 = (\pm 1 : \pm t^\ell : 1)$$

on X.

Now, Eq. (2) also defines a scheme $\mathfrak{X}/R$, in which the special fiber $\mathfrak{X}_k = \mathfrak{X} \times_R k$ has equation $xy = 0$ in $\mathbb{P}^2_{\mathbb{C}}$. That is, the special fiber is a union of two rational curves meeting at a node. Furthermore, the horizontal closures $\overline{p_i}$ of the four marked points of X do indeed meet $\mathfrak{X}_k$ in four regular points, namely $(0 : \pm 1 : 1)$ and $(\pm 1 : 0 : 1)$. Note that $\mathfrak{X}$ will qualify as a *stable model* for X, as defined in Definition 3.6.

The *abstract tropical curve* associated to $\mathfrak{X}$ will be the vertex-unweighted metric graph with two vertices and an edge between them of length ℓ, with marked points $1, 2$ on one vertex and $3, 4$ on the other. See Fig. 2.

Now we'll launch into the definitions of stable curves, their dual graphs, and abstract tropicalizations.

3.1 Stable Curves

Fix k an algebraically closed field. By a *curve* we shall mean a reduced, proper, connected scheme X of dimension 1 over k. The arithmetic genus of the curve is $h^1(X, \mathcal{O}_X)$. A *node* of X is a point $p \in X(k)$ with the property that $\widehat{\mathcal{O}}_{X,p} \cong k[\![x, y]\!]/(xy)$. A *nodal* curve is a curve whose only singularities, if any, are nodes.

Definition 3.2 (Stable n-Pointed Curves) A nodal, n-marked curve of genus g is $(X, p_1, \ldots, p_n)$, where $p_i \in X(k)$ are distinct nonsingular points of a genus g nodal curve X.

We say that a nodal, marked curve $(X, p_1, \ldots, p_n)$ is **stable** if $\mathrm{Aut}(X, p_1, \ldots, p_n)$ is finite, that is, there are only finitely many automorphisms of the curve X that fix each $p_1, \ldots, p_n$ pointwise. This is often equivalently stated as follows: $(X, p_1, \ldots, p_n)$ is stable if the restriction of $\omega_X(p_1 + \cdots + p_n)$ to every irreducible component of X is a line bundle of positive degree. Here ω_X denotes the dualizing sheaf of X.

Notice that all smooth curves of genus $g \geq 2$ already have only finitely many automorphisms. A smooth curve of genus 1 has finitely many automorphisms once one fixes one marked point; and a smooth curve of genus 0, also known as $\mathbb{P}^1$, has finitely many automorphisms once one fixes three marked points. So we could equally phrase the stability condition as follows:

Observation 3.3 For every irreducible component C of X, let $\phi \colon C^\nu \to C$ denote the normalization of C. An n-marked nodal curve $(X, p_1, \ldots, p_n)$ is stable if and only if

(1) for every component C of geometric genus 0,

$$|C \cap \{p_1, \ldots, p_n\}| + |\{q \in C^\nu : \phi(q) \in X^{\mathrm{sing}}\}| \geq 3;$$

(2) for every component C of geometric genus 1,

$$|C \cap \{p_1, \ldots, p_n\}| + |\{q \in C^\nu : \phi(q) \in X^{\mathrm{sing}}\}| \geq 1.$$

(The second condition sounds misleadingly general. You can trace through the definition yourself to see that it excludes only one additional case, the case that the whole of X is just a smooth curve of genus 1 with no marked points.)

Exercise 3.4 Let $g, n \geq 0$. Check that stable n-marked curves of genus g exist if and only if $2g - 2 + n > 0$.

3.2 *Stable Models*

Let K be an algebraically closed field that is complete with respect to a nonarchimedean valuation. Good examples include the completion of the field of Puiseux series $\mathbb{C}\{\{t\}\}$ or the completion $\mathbb{C}_p$ of the algebraic closure of the field $\mathbb{Q}_p$.[5]

---

[5]By the way, you might complain that some of this theory can be developed with weaker hypotheses on the field K. That is true. For example, the stable reduction theorem holds for arbitrary complete nonarchimedean fields, up to *passing to a finite, separable field extension.* I am taking this approach partly for expository ease, especially for one's first exposure to this material. It's kind

As before, let R denote the valuation ring of K and let $k = R/\mathfrak{m}$ be the residue field. Recall that Spec R has two points η and s, corresponding to the ideals (0) and $\mathfrak{m}$ respectively. If $\mathfrak{X}$ is a scheme over Spec R, then the *generic fiber* of $\mathfrak{X}$ is the fiber over η; the *special fiber* is the fiber over s.

Definition 3.5 (Models) If X is any finite type scheme over K, then by a *model* for X we mean a flat and finite type scheme $\mathfrak{X}$ over R whose generic fiber is isomorphic to X.

Now let's define stable models. First, let me forget about marked points, and just suppose that X is a smooth curve over K.

Definition 3.6 (Stable Models) Suppose X is a smooth, proper, geometrically connected curve over K of genus $g \geq 2$. A *stable model* for X is a proper model $\mathfrak{X}/R$ whose special fiber $\mathfrak{X}_k = \mathfrak{X} \times_R k$ is a stable curve over k.

Definition 3.7 (Stable Models, Allowing Marked Points) Now say $2g-2+n > 0$ and suppose $(X, p_1, \ldots, p_n)$ is a smooth, n-marked, genus g curve. Then a stable model for X is a proper model $\mathfrak{X}/R$ with n sections $\overline{p}_1, \ldots, \overline{p}_n$: Spec $R \to \mathfrak{X}$ restricting to the marked points p_i on the general fiber, making the special fiber a stable n-marked curve of genus g over k.

Let $2g - 2 + n > 0$. When does an n-marked genus g curve X/K admit a stable model? The answer is: *always*. This is the content of the Stable Reduction Theorem of Deligne-Mumford-Knudsen, which also gives that the stable model is essentially unique. More precisely, the version we are using here, for fields whose valuations are not necessarily discrete, goes back to [6]; see also [4, 29, 31].[6] You can also see Harris and Morrison's book [18, §3.C] for a relatively explicit, algorithmic explanation of the Stable Reduction Theorem, at least in characteristic 0.

3.3 *Dual Graphs of Stable Curves*

We are working towards the goal of associating a graph, with some vertex decorations and some edge lengths, to a smooth curve X/K. The graph we are going to associate to X is the *dual graph* of the special fiber of a stable model

of like learning algebraic geometry over $\mathbb{C}$ first. My other defense is that in the tropical context it is often not a big deal to pass to a possibly huge field extension, at least in theory. See Definition 2.12, for example.

[6] Again, the typical formulation of the stable reduction theorem says that if X/K is a smooth curve, then there exists a finite separable field extension K'/K such that $X \times_K K'$ admits a stable model. Here, we've folded the need to pass to a finite field extension into the assumption that K itself is algebraically closed.

for X. Basically, the dual graph of a stable curve Y is a combinatorial gadget that records:

- how many irreducible components Y has, and what their geometric genera are;
- how the irreducible components of Y intersect; and
- the way in which the n marked points are distributed on Y.

Now we will explain this completely, starting with the graph theory.

Conventions on Graphs All graphs will be finite and connected, with loops and parallel edges allowed. (Graph theorists would call such objects finite, connected *multigraphs*.) Remember that a graph G consists of a set of vertices $V(G)$ and a set of edges $E(G)$. Each edge is regarded as having two endpoints which are each identified with vertices of G, possibly the same.

Definition 3.8 (Vertex-Weighted Marked Graphs) A *vertex-weighted, n-marked graph* is a triple (G, m, w) where:

- G is a graph;
- $w\colon V(G) \to \mathbb{Z}_{\geq 0}$ is any function, called a *weight function*, and
- $m\colon \{1, \ldots, n\} \to V(G)$ is any function, called an n-marking.[7]

The *genus* of (G, m, w) is

$$g(G) + \sum_{v \in V(G)} w(v)$$

where

$$g(G) = |E| - |V| + 1$$

is the first Betti number of G, considered as a 1-dimensional CW complex, say.

Definition 3.9 (Stability for Vertex-Weighted Marked Graphs) With (G, m, w) as above, we'll say that (G, m, w) is *stable* if for every $v \in V(G)$,

$$2w(v) - 2 + \mathrm{val}(v) + |m^{-1}(v)| > 0.$$

Here $\mathrm{val}(v)$ denotes the graph-theoretic *valence* of the vertex v, which is defined as the number of half-edges incident to it.

Figure 3 shows the seven distinct stable vertex-weighted graphs of type $(g, n) = (2, 0)$.

Exercise 3.10 Find the five stable, 2-marked weighted graphs of genus 1.

---

[7]Another common setup for marking a tropical curve is to attach infinite rays to a graph, labeled $\{1, \ldots, n\}$. Our marking function m is obviously combinatorially equivalent.

Fig. 3 The seven genus 2 stable vertex-weighted graphs with no marked points. The vertices have weight zero unless otherwise indicated

Exercise 3.11 Prove that there are only finitely many stable marked, weighted graphs for fixed g and n, up to isomorphism. It may be helpful to consider the partial order of *contraction* that we will define in Sect. 4.

Definition 3.12 (Dual Graph of a Stable Curve) Let k be an algebraically closed field, and let $(Y, p_1, \ldots, p_n)$ be a stable, n-marked curve over k.

The *dual graph* of $(Y, p_1, \ldots, p_n)$ is the vertex-weighted, marked graph (G, m, w) obtained as follows.

- The vertices v_i of G are in correspondence with the irreducible components C_i of Y, with weights $w(v_i)$ recording the geometric genera of the components.
- For every node p of Y, say lying on components C_i and C_j, there is an edge e_p between v_i and v_j.
- The marking function $m: \{1, \ldots, n\} \to V(G)$ sends j to the vertex of G corresponding to the component of Y supporting p_j.

Note that by Observation 3.3, (G, m, w) is stable since (Y, p_i) is stable.

3.4 Abstract Tropical Curves

A *metric graph* is a pair (G, ℓ), where G is a graph, and ℓ is a function $\ell: E(G) \to \mathbb{R}_{>0}$ on the edges of G. We imagine ℓ as recording real lengths on the edges of G.

An abstract tropical curve is just a vertex-weighted, marked *metric* graph:

Definition 3.13 ([7, 8, 26] Abstract Tropical Curves) An **abstract tropical curve** with n marked points is a quadruple $\Gamma = (G, \ell, m, w)$ where:

- G is a graph,
- $\ell: E(G) \to \mathbb{R}_{>0}$ is any function, called a *length function*, on the edges,
- $m: \{1, \ldots, n\} \to V(G)$ is any function, called a *marking function*, and
- $w: V(G) \to \mathbb{Z}_{\geq 0}$ is any function.

The **combinatorial type** of Γ is the triple (G, m, w), in other words, all of the data of Γ except for the edge lengths. We say that Γ is **stable** if its combinatorial type is stable. The **volume** of Γ is the sum of its edge lengths.
From now on, I will mean "stable abstract tropical curve" when I say "abstract tropical curve," even if I forget to say so.

Hints of a Tropical Moduli Space Informally, we view a weight of $w(v)$ at a vertex v as $w(v)$ loops, based at v, of infinitesimally small length. Each infinitesimal loop contributes 1 to the genus of C. Permitting vertex weights will ensure that the moduli space of tropical curves, once it is constructed, is complete. That is, a sequence of genus g tropical curves obtained by sending the length of a loop to zero will still converge to a genus g curve.

Of course, the real reason to permit vertex weights is so that the combinatorial types of genus g tropical curves correspond precisely to dual graphs of stable curves in $\overline{\mathcal{M}}_{g,n}$, and that the eventual moduli space will indeed be the boundary complex of $\mathcal{M}_{g,n} \subset \overline{\mathcal{M}}_{g,n}$.

3.5 *From Algebraic to Tropical Curves: Abstract Tropicalization*

Now let's put everything together. We continue to let K be an algebraically closed field, complete with respect to a nonarchimedean valuation. Let $2g - 2 + n > 0$. Suppose $(X, p_1, \ldots, p_n)$ is a smooth, proper, n-marked curve over K of genus g. Let us extract a tropical curve from the data of (X, p_i).

The procedure will go like this. First we will extend X to a family $\mathfrak{X}$ over $\operatorname{Spec} R$ along with n sections $\operatorname{Spec} R \to \mathfrak{X}$, filling in a stable, n-marked curve of genus g over k in the special fiber. The fact that this is possible is the Stable Reduction Theorem.

Then we will associate to $\mathfrak{X}$ the *vertex-weighted dual graph* of the special fiber $\mathfrak{X}_k$. It only remains to equip the edges of the dual graph with real lengths. We do this as follows: for every node q of $\mathfrak{X}_k$, say lying on components C_i and C_j, the completion of the local ring $\mathcal{O}_{\mathfrak{X},q}$ is isomorphic to $R[\![x, y]\!]/(xy - \alpha)$ for some $\alpha \in R$, and $v(\alpha) > 0$ is independent of all choices. (See e.g [31], and see Remark 3.16 below, for more on this independence.) Then we put an edge e_q between v_i and v_j of length $v(\alpha)$. The result is a stable vertex-weighed, marked metric graph. See again Example 3.1. Summarizing:

Definition 3.14 (Abstract Tropicalization) Let K be an algebraically closed field, complete with respect to a nonarchimedean valuation. Suppose $(X, p_1, \ldots, p_n)$ is a smooth, proper, n-marked curve over K of genus g. The **abstract tropicalization** of (X, p_i) is the dual graph of the special fiber of a stable model $(\mathfrak{X}, \overline{p}_i)$ for (X, p_i), declaring an edge corresponding to a node q to have length $v(\alpha)$ if the local equation of q in $\mathfrak{X}$ is $xy - \alpha$.

Remark 3.15 We can now take Definition 3.14 and extend it quite painlessly, to tropicalize *stable* curves, not just smooth ones. In this situation, the local equation of a node in the special fiber may be of the form $xy = 0$; in other words, the node may have simply persisted from the general fiber. Thus the natural result of tropicalization is a stable *extended* tropical curve: just like a tropical curve, but with edge lengths taking values in $\mathbb{R}_{>0} \cup \{\infty\}$.

Remark 3.16 Another way to say this whole story is that abstract tropicalization sends X/K to its *Berkovich skeleton*: the minimal skeleton of X^{an}, with respect to the n marked points $p_1, \ldots, p_n$, equipped with the skeleton metric. See [4, §5]. (Actually, the Berkovich skeleton has n infinite rays attached to the vertices, for the n marked points). There is a lot to be said here, but the main point at the moment is that the interpretation of the abstract tropicalization of X as its Berkovich skeleton shows that it's canonically associated with X: the construction is in fact independent of all choices.

4 Definition of the Moduli Space of Tropical Curves

It is time to construct the moduli space of tropical curves. This construction is due to Brannetti-Melo-Viviani [7] and subsequently Caporaso [8], building on work of Mikhalkin [26] and with antecedents in related constructions of Gathmann-Markwig [15, 25]. Actually, many of the ideas can be traced back even further to the work of Culler-Vogtmann [11].

Fix g and n with $2g - 2 + n > 0$. Suppose we fix a single combinatorial type (G, m, w) of type (g, n), and allow the edge lengths l to vary over all positive real numbers. Then we clearly obtain all tropical curves of that type. This motivates our construction of the moduli space of tropical curves below. We will first group together curves of the same combinatorial type, obtaining one cell for each combinatorial type. Then, we will glue our cells appropriately to obtain the moduli space.

To make this construction, for the moment we will just follow our noses combinatorially. But the whole point of the next lectures will be that the space we get out the other side is a good one algebro-geometrically: it can be identified with the *boundary complex* of the Deligne-Mumford compactification $\overline{\mathcal{M}}_{g,n} \supset \mathcal{M}_{g,n}$.

Let's begin. First, fix a combinatorial type (G, m, w) of type (g, n). What is a parameter space for all tropical curves of this type? Our first guess might be a positive orthant $\mathbb{R}_{>0}^{|E(G)|}$, that is, a choice of positive length for each edge of G. But we have overcounted by symmetries of the combinatorial type (G, m, w). For example, in the "figure 8" depicted leftmost in Fig. 3, the edge lengths $(2, 5)$ and $(5, 2)$ give the same tropical curve.

Furthermore, with foresight, we will allow zero length edges as well, with the understanding that such a curve will soon be identified with one obtained by contracting those edges. This suggests the following definition:

Definition 4.1 Given a combinatorial type (G, m, w), let the **automorphism group** $\mathrm{Aut}(G, m, w)$ be the set of all permutations $\varphi : E(G) \to E(G)$ that arise from automorphisms of G that preserve m and w. The group $\mathrm{Aut}(G, m, w)$ acts on the set $E(G)$, and hence on the orthant $\mathbb{R}_{\geq 0}^{E(G)}$, with the latter action given by permuting coordinates. We define $\overline{C(G,m,w)}$ to be the quotient space

$$\overline{C(G,m,w)} = \mathbb{R}_{\geq 0}^{E(G)}/\mathrm{Aut}(G,m,w).$$

Next, we define an equivalence relation on the points in the union

$$\coprod \overline{C(G,m,w)},$$

as (G,m,w) ranges over all combinatorial types of type (g,n). Regard a point $x \in \overline{C(G,m,w)}$ as an assignment of lengths to the edges of G. Now, given two points $x \in \overline{C(G,m,w)}$ and $x' \in \overline{C(G',m',w')}$, identify x and x' if one of them is obtained from the other by contracting all edges of length zero. By *contraction*, we mean the following. Contracting a loop, say based at vertex v, means deleting that loop and adding 1 to $w(v)$. Contracting a nonloop edge, say with endpoints v_1 and v_2, means deleting that edge and identifying v_1 and v_2 to obtain a new vertex whose weight we set to $w(v_1)+w(v_2)$.

Let $\sim$ denote the equivalence relation generated by the identification we have just defined. Now we glue the cells $\overline{C(G,m,w)}$ along $\sim$ to obtain our moduli space:

Definition 4.2 The **moduli space** $M_{g,n}^{\mathrm{trop}}$ is the topological space

$$M_{g,n}^{\mathrm{trop}} := \coprod \overline{C(G,m,w)}/\sim,$$

where the disjoint union ranges over all combinatorial types of genus (g,n), and $\sim$ is the equivalence relation defined above.

A picture of $M_{1,2}^{\mathrm{trop}}$ is shown in Fig. 4. The picture is not entirely accurate, in that there is a 2-dimensional cone with a nontrivial symmetry which is drawn with a dotted line through it, which is supposed to remind us of the self-gluing of this cone induced by the symmetry.

Exercise 4.3 Label the other cones of $M_{1,2}^{\mathrm{trop}}$ according to Exercise 3.10.

Exercise 4.4 Verify that $M_{g,n}^{\mathrm{trop}}$ is pure $(3g-3+n)$-dimensional, i.e., that the Euclidean dimension of every maximal cone $\overline{C(G,m,w)}$ of $M_{g,n}^{\mathrm{trop}}$ is $3g-3+n$.

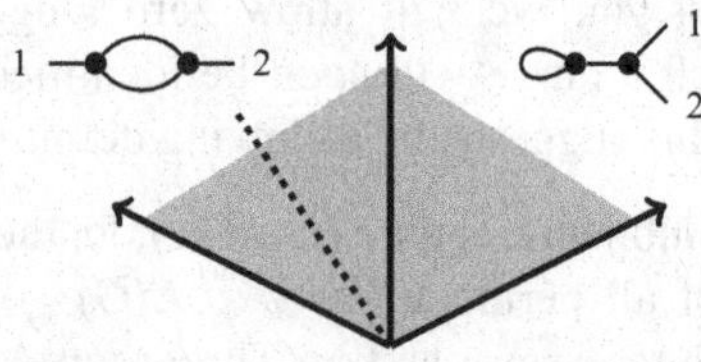

Fig. 4 A picture of the tropical moduli space $M_{1,2}^{\mathrm{trop}}$. Only the two top-dimensional strata are labeled. We have drawn marked points as marked rays attached at vertices

Of course $M_{g,n}^{\mathrm{trop}}$, being built out of cones, is contractible: e.g., it retracts onto its cone point, corresponding to the tropical curve, denoted $\bullet_{g,n}$, consisting of a single vertex with weight g, n marked points, and no edges. But the *link* of $M_{g,n}^{\mathrm{trop}}$, meaning a cross-section of $M_{g,n}^{\mathrm{trop}}$, is topologically very interesting and is a main character of these lectures.

Definition 4.5 (Link of Tropical Moduli Space) The *link* $\Delta_{g,n}$ of $M_{g,n}^{\mathrm{trop}}$ at the tropical curve $\bullet_{g,n}$ is the quotient of $M_{g,n}^{\mathrm{trop}} \setminus \{\bullet_{g,n}\}$ induced by uniform scaling of edge lengths. It can also be identified with the subspace of $M_{g,n}^{\mathrm{trop}}$ parametrizing tropical curves of volume 1.[8]

5 Boundary Complexes of Toroidal Embeddings

We are working towards the fundamental statement that *the link of the tropical moduli space of curves is the boundary complex of the stable curves compactification of* $\mathcal{M}_{g,n}$. This is one of the main results of [1]. It is this identification that allows us to re-examine the boundary complex of $\mathcal{M}_{g,n} \subset \overline{\mathcal{M}_{g,n}}$ as a worthy combinatorial moduli space in its own right, and to obtain new results about the topology of $\mathcal{M}_{g,n}$ using tropical geometry techniques. I will describe some of those applications in the last lecture. Right now, we will answer the question: what is a *toroidal embedding*, and what is its *boundary complex*?

5.1 Étale Morphisms

Let's recall the definition of an étale morphism of schemes, which provides a more flexible notion of neighborhoods than do Zariski open neighborhoods. Let $f: X \to Y$ be a morphism of schemes of finite type over a field k. Then f is called *unramified* if for all $x \in X$, letting $y = f(x)$, we have that $\mathfrak{m}_y \mathcal{O}_{X,x} = \mathfrak{m}_x$, and furthermore $k(y)$ is a separable field extension of $k(x)$. Then f is étale if it is both flat and unramified.

Étale morphisms are, roughly, the algebro-geometric local analogue of finite covering spaces. For example:

Example 5.1 Let $k = \mathbb{C}$. Then the map $\mathbb{A}^1 \to \mathbb{A}^1$ sending $z \mapsto z^n$ is étale away from $z = 0$.

[8]Having said all of that, you can find a slightly different way of describing $M_{g,n}^{\mathrm{trop}}$, as a colimit of a diagram of rational polyhedral cones over the appropriate category of graphs, in [1, §4], [10, §2]. This colimit presentation equips $M_{g,n}^{\mathrm{trop}}$ with the structure of a generalized cone complex, and its link has the structure of a smooth generalized Δ-complex in the sense of [10, §3].

5.2 Toroidal Embeddings

The theory of toroidal embeddings is due to Kempf-Knudsen-Mumford-Saint-Donat [20]. Let k be an algebraically closed field. Let U be an open subvariety of a normal variety X over k. We say that $U \subset X$ is a *toroidal embedding* if it is locally modeled by toric varieties. More precisely, it is a toroidal embedding if for every $x \in X$,

$$\widehat{\mathcal{O}}_{X,x} \cong \widehat{\mathcal{O}}_{Y_\sigma,y}$$

where $y \in Y_\sigma$ is a point in an affine toric variety Y_σ with torus T, and furthermore, the ideals of $X \setminus U$ and $Y_\sigma \setminus T$ correspond in the respective completed local rings.

We'll write $D_1, \ldots, D_r$ for the irreducible components of $X - U$. There are two cases, one of which makes the theory more intricate: we say that a toroidal embedding has *self-intersections* if the components D_i are not all normal.

It is actually not a problem if you aren't familiar with toric varieties, because the most relevant example for our purposes is one you definitely know: the usual embedding of the torus $\mathbb{G}_m^n$ into $\mathbb{A}^n$. Note that the complement $\mathbb{A}^n \setminus \mathbb{G}_m^n$ is the union of n coordinate hyperplanes, intersecting transversely.

Indeed, toroidal embeddings whose local toric charts are all affine spaces are called normal crossings divisors, and this is the case we'll be most interested in.

Definition 5.2 (Normal Crossings and Simple Normal Crossings) Let X be a normal variety, and D a divisor. We say D is a *normal crossings divisor* if for every $x \in X$, we have $\widehat{\mathcal{O}}_{X,x} \cong k[\![x_1, \ldots, x_n]\!]$ and the equation of D in $\widehat{\mathcal{O}}_{X,x}$ is $x_1 \cdots x_i$ for some i. Equivalently, $U = X - D \subset X$ is a toroidal embedding locally modeled by affine spaces.

We say D is *simple normal crossings* if in addition D has no self-intersections.

Example 5.3 The nodal cubic $V(y^2 = x^2 + x^3)$ in $\mathbb{A}^2$ is a normal crossings divisor, but not a simple normal crossings divisor.

5.3 Boundary Complexes of Toroidal Embeddings

The theory of boundary complexes for toroidal embeddings without self-intersection is due again to Kempf-Knudsen-Mumford-Saint-Donat [20]. For simplicity, we will state this theory in the case of simple normal crossings divisors, while emphasizing that both the work [20] and the work of Thuillier [28] takes place in the more general case of toroidal embeddings.[9]

---

[9]Indeed, to a toroidal embedding $U \subset X$ without self-intersections, one associates a *rational polyhedral cone complex*, whose cones correspond to the toroidal strata of $U \subset X$ [20]. Next, if $U \subset X$ is toroidal with self-intersections, one may associate a *generalized cone complex*, a more general object in which self-gluings of cones are permitted [1, 28]. How do these definitions

Definition 5.4 (Boundary Complex, No Self-Intersections) Suppose $U \subset X$ is an open inclusion whose boundary is simple normal crossings. Let $D_1, \dots, D_r$ be the irreducible components of $\partial X = X - U$. The *boundary complex* $\Delta(U \subset X)$, or just $\Delta(X)$, is the Δ-complex on vertices $D_1, \dots, D_r$ with a d-face for every irreducible component of an intersection $D_{i_1} \cap \cdots \cap D_{i_{d+1}}$.

Example 5.5 The boundary complex $\Delta(\mathbb{G}_m^n \subset \mathbb{A}^n)$ is the simplex Δ^{n-1}.

Next, Thuillier recently extended the theory of boundary complexes in a way that is important for our applications, dropping the assumption that the D_i are normal [28].

Definition 5.6 (Boundary Complex, Self-Intersections) Now let $U \subset X$ have normal crossings boundary. Let $V \to X$ be an étale surjective morphism to X, such that $U_V = U \times_X V \subset V$ is simple normal crossings, and let $V_2 = V \times_X V$, with $U_2 = U_V \times_X U_V$. The boundary complex of $U \subset X$ is the coequalizer, in the category of topological spaces, of the diagram[10]

$$\Delta(U_2 \subset V_2) \rightrightarrows \Delta(U_V \subset V).$$

Example 5.7 Let $k = \mathbb{C}$. Consider, as in [1, Example 6.1.7], the *Whitney umbrella*

$$D = \{x^2y = z^2\} \quad \subset \quad X = \mathbb{A}^3 \setminus \{y = 0\},$$

drawn in Fig. 5. Let $U = X - D$. We will explain why $\Delta(U \subset X)$ is a "half-segment," meaning the quotient of a line segment by a $\mathbb{Z}/2\mathbb{Z}$ reflection.[11]

Let $V \cong \mathbb{A}^2 \times \mathbb{G}_m \to X$ be the étale cover of degree 2 given by a base change $y = u^2$. Then

$$D_V = D \times_X V = \{x^2u^2 - z^2 = 0\}$$

is simple normal crossings, and $D_2 = D_V \times_X D_V \cong D_V \times \mathbb{Z}/2\mathbb{Z}$, since D_V is degree 2 over D. Explicitly, one component of D_2 parametrizes pairs (p, p) of points in D_V, and the other parametrizes pairs (p, q) with $p \neq q$ lying over the same point of D.

---

specialize to Definitions 5.4 and 5.6, in the special cases of simple normal crossings and normal crossings, respectively? In these cases, the cone complexes associated to $U \subset X$ are glued from *smooth* cones. The operation of replacing each smooth d-dimensional cone with a $(d-1)$-simplex produces the desired correspondence between these two pairs of definitions. This is explained more precisely in [10, §3], in terms of an equivalence of categories between smooth generalized cone complexes and *generalized Δ-complexes*.

[10]Thuillier actually shows that this construction is independent of all choices, because in fact it is intrinsic to the Berkovich analytification of the pair $U \subset X$. See [28] for the precise description.

[11]Of course, a line segment modulo a reflection is, topologically, just another segment. There is a more abstract definition of a boundary complex in which the half-segment and segment are nonisomorphic, and only their *geometric realizations* as topological spaces are homeomorphic. See [10, §3].

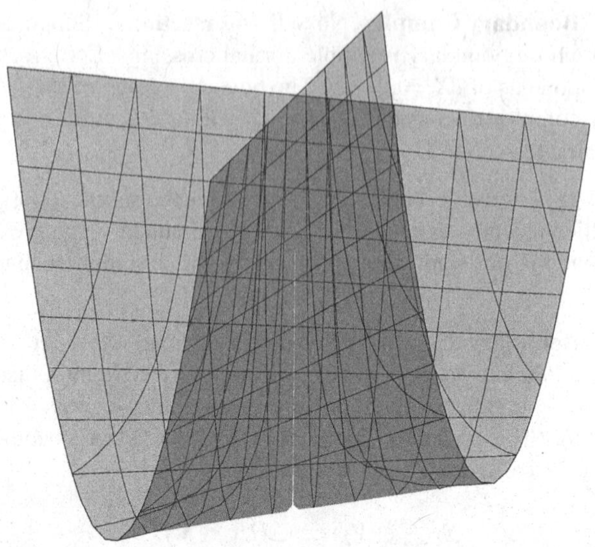

Fig. 5 The Whitney umbrella of Example 5.7

That means that $\Delta(V)$ is a segment and $\Delta(V_2)$ is two segments, and the two maps $\Delta(V_2) \rightrightarrows \Delta(V)$ differ by one flip. So the coequalizer is a segment modulo a flip.

What is going on complex-analytically? Let $Y = \{(0, y, 0) : y \neq 0\}$ be the umbrella pole. It is a punctured affine line $\mathbb{A}^1 - \{0\}$ over $\mathbb{C}$, to be visualized complex-analytically as a punctured plane. The points $(0, u, 0)$ and $(0, -u, 0)$ in D_0 correspond to the two analytic branches of D along Y at the point $(0, y, 0)$, where $y = u^2$. The equations of the branches are $z = xu$ and $z = -xu$. So taking y around a loop around the puncture precisely interchanges the branches.

Exercise 5.8 Compute the boundary complex of the complement of the nodal cubic $V(y^2 = x^2 + x^3)$ in $\mathbb{A}^2$.

6 Toroidal Deligne-Mumford Stacks

Boundary complexes can be defined for toroidal Deligne-Mumford stacks as well by following Thuillier's construction [1]. The punchline will be that toroidal Deligne-Mumford stacks admit étale covers by toroidal schemes, and so Definition 5.6 can be repeated with respect to toroidal étale covers with no changes. Here we'll give a quick-start guide to toroidal DM stacks. We will have to skip many details in order to get anywhere in the allotted time. But I'll try to indicate exactly what I'm skipping.

6.1 Categories Fibered in Groupoids

Let us fix an algebraically closed field k, without the structure of a valuation. Let Sch_k denote the category of schemes over k. Let me recall the following "negative result": it's impossible to define a scheme that deserves to be called a *fine moduli space* for genus g, n-marked curves. By a *fine moduli space* I mean a space $\mathcal{M}_{g,n}$ such that maps S to $\mathcal{M}_{g,n}$ correspond, functorially, to families of genus g, n-marked curves over S. The obstruction is that some curves have nontrivial automorphisms.

Nevertheless, we can start by axiomatizing the desired property of a fine moduli space into a category whose objects are families of genus g, n-marked smooth, respectively stable, curves. More precisely:

Definition 6.1 (The Category $\mathcal{M}_{g,n}$) We denote by $\mathcal{M}_{g,n}$ the category whose objects are flat, proper morphisms $f: X \to B$ of k-schemes, together with n sections $p_1, \ldots, p_n: B \to X$, such that the geometric fibers, with their n marked points induced by the p_i, are smooth curves of type (g, n). The morphisms in $\mathcal{M}_{g,n}$ are Cartesian diagrams

$$\begin{array}{ccc} X' & \longrightarrow & X \\ \downarrow & & \downarrow \\ B' & \longrightarrow & B. \end{array}$$

Definition 6.2 (The Category $\overline{\mathcal{M}}_{g,n}$) The definition of $\overline{\mathcal{M}}_{g,n}$ is the same as above, but with families of *stable* curves instead.

These two categories come equipped with obvious functors to Sch_k: take a family of curves $X \to B$, and remember only its base B. In fact, both $\mathcal{M}_{g,n}$ and $\overline{\mathcal{M}}_{g,n}$, along with their functors to Sch_k, are examples of *categories fibered in groupoids*, or CFGs for short. I won't state the condition that a category is *fibered in groupoids* over Sch_k, but I will state that in our case it is the condition that pullbacks of families of curves exist and are unique up to unique isomorphism.

Here is another CFG, verifying that CFGs encompass k-schemes:

Definition 6.3 (The Category $\underline{S}$) Let S be any k-scheme. The objects of the category $\underline{S}$ are morphisms $X \to S$ of k-schemes. The morphisms in $\underline{S}$ are commuting triangles $X' \to X \to S$. The functor $\underline{S} \to \mathrm{Sch}_k$ sends $(X \to S)$ to X.
In addition, the category $\underline{S}$ determines the scheme S, in the sense made precise by Yoneda's Lemma. In this case, the fact that $\underline{S}$ is a CFG boils down to the fact that the *composition* of two morphisms $X' \to X \to S$ exists and is unique—which is obvious.

A *morphism* of CFGs $\mathcal{C}$ and $\mathcal{D}$ is just what you think: it is a functor $F: \mathcal{C} \to \mathcal{D}$ making a commuting triangle with the functors $\mathcal{C} \to \mathrm{Sch}_k$ and $\mathcal{D} \to \mathrm{Sch}_k$. Using the Yoneda correspondence, you can check (and make more precise):

Exercise 6.4 To give a morphism $\underline{S} \to \mathcal{M}_{g,n}$ is precisely to give a family of genus g, n-marked curves over S.

6.2 Fast Forward: Deligne-Mumford Stacks, and Toroidal Embeddings

Now let's chat a little about Deligne-Mumford stacks. Not all categories fibered in groupoids are schemes. (In other words, not all CFGs are of the form $\underline{S}$ for some k-scheme S.) Stacks, and, even more restrictively, *Deligne-Mumford stacks*, are CFGs satisfying some extra conditions that make them behave a little more geometrically, even if they aren't exactly schemes.

One of these requirements is the following, which we state as a fact about Deligne-Mumford stacks:

Fact 6.5 *If $\mathcal{M}$ is a Deligne-Mumford stack, then there is an étale, surjective morphism from a scheme U to $\mathcal{M}$.*[12]
A very rough rephrasing is that locally, everywhere in a Deligne-Mumford stack there is a "scheme covering space."

Given Fact 6.5, we can define toroidal Deligne-Mumford stacks by looking on étale atlases, and define their boundary complexes in exactly the same way as in Definition 5.6. We continue to let k be an algebraically closed field without valuation. All our stacks are separated and connected over k.

Definition 6.6 (Toroidal Deligne-Mumford Stacks) An open substack $\mathcal{U} \subset \mathcal{X}$ of a Deligne-Mumford stack $\mathcal{X}$ is *toroidal* if for every étale morphism $V \to \mathcal{X}$ from a scheme, the induced map of schemes $\mathcal{U}_V := \mathcal{U} \times_{\mathcal{X}} V \to V$ is a toroidal embedding (of schemes).

Definition 6.7 (Boundary Complexes of Toroidal Deligne-Mumford Stacks) This is a reprise of Definition 5.6. Let $\mathcal{U} \subset \mathcal{X}$ be a toroidal Deligne-Mumford stack. Let $V \to \mathcal{X}$ be an étale cover by a scheme such that $\mathcal{U}_V \to V$ is a toroidal embedding of schemes without self-intersections. Then the *boundary complex* of $\mathcal{X}$ is the coequalizer, in the category of topological spaces, of

$$\Delta(V \times_{\mathcal{X}} V) \rightrightarrows \Delta(V).$$

That's it. Abramovich-Caporaso-Payne show that Thuillier's work can be extended to the setting of DM stacks. In particular, when $\mathcal{X}$ is proper, the boundary complex of $\mathcal{U} \subset \mathcal{X}$ can be found intrinsically inside the Berkovich analytification of the coarse moduli space of $\mathcal{X}$.[13]

---

[12] What do I even mean by saying that the morphism $U \to \mathcal{M}$ has a property like, say, étale? The other condition for a stack $\mathcal{M}$ to be Deligne-Mumford is a representability condition: it says that for any morphisms $f\colon S \to \mathcal{M}$ and $g\colon T \to \mathcal{M}$ from a scheme, the *fiber product* $S \times_{\mathcal{M}} T$ (which I will not define) is again a scheme. Then for any property P of morphisms that is preserved by base change, we say that f has property P if $S \times_{\mathcal{M}} T \to T$ is a map of schemes with property P.

[13] In fact, it can be found intrinsically inside the analytification of the stack $\mathcal{X}$ itself [30, IV].

7 Compactification of $\mathcal{M}_{g,n}$ by Stable Curves

We continue to let k be an algebraically closed field, with no valuation. At this point, I want to recall some of the essential facts about the Deligne-Mumford-Knudsen moduli stacks $\mathcal{M}_{g,n}$ and $\overline{\mathcal{M}}_{g,n}$ [14, 21–23].

Fact 7.1 *The category $\overline{\mathcal{M}}_{g,n}$ of smooth genus g, n-marked curves over k, defined in Definition 6.2, is a smooth, proper Deligne-Mumford stack containing $\mathcal{M}_{g,n}$ an open substack. The inclusion $\mathcal{M}_{g,n} \subset \overline{\mathcal{M}}_{g,n}$ is toroidal, indeed normal crossings (though far from simple normal crossings).*

7.1 The Boundary Strata $\mathcal{M}_{\mathbf{G}}$ of $\overline{\mathcal{M}}_{g,n}$

Moreover, the strata of the boundary $\overline{\mathcal{M}}_{g,n} \setminus \mathcal{M}_{g,n}$ are naturally indexed by genus g, n-marked combinatorial types $\mathbf{G} = (G, m, w)$, according to the dual graphs of the stable curves that they parametrize. I would now like to describe these strata, which we'll denote $\mathcal{M}_{\mathbf{G}}$. This description follows [1, §3.4] and the correctness of this description is proved in [2, §12.10].

Fix a combinatorial type $\mathbf{G} = (G, m, w)$. For each vertex v, let $n_v = \mathrm{val}(v) + |m^{-1}(v)|$ where $\mathrm{val}(v)$ is the valence of v. Let

$$\widetilde{\mathcal{M}_{\mathbf{G}}} = \prod_{v \in V(G)} \mathcal{M}_{w(v),n_v}.$$

If you think about it, $\widetilde{\mathcal{M}_{\mathbf{G}}}$ can be identified with the moduli space of n-marked genus g stable curves, *together with a chosen isomorphism of the dual graph with* $\mathbf{G}$. To get rid of that choice of isomorphism, we take the stack quotient $[\widetilde{\mathcal{M}_{\mathbf{G}}}/\,\mathrm{Aut}(\mathbf{G})]$. The theorem is then that there is a canonical isomorphism

$$\mathcal{M}_{\mathbf{G}} \cong [\widetilde{\mathcal{M}_{\mathbf{G}}}/\,\mathrm{Aut}(\mathbf{G})].$$

An explicit example is given in Example 7.2 below.

(It is worth noting that this stratification is *inclusion-reversing* with respect to the corresponding stratification of $M_{g,n}^{\mathrm{trop}}$ by combinatorial type. The more edges there are in $\mathbf{G}$, the smaller the stratum $\mathcal{M}_{\mathbf{G}}$ is, and the larger $\overline{C}(\mathbf{G})$ is in the tropical moduli space.)

Now, given a point $p \in \mathcal{M}_{\mathbf{G}}$ corresponding to a stable curve C, we may describe an étale neighborhood V_p of p in $\overline{\mathcal{M}}_{g,n}$ in which the boundary can be identified with the d coordinate hyperplanes inside $\mathbb{A}^d$. The boundary of $\overline{\mathcal{M}}_{g,n}$ in this neighborhood V_p is a union of irreducible divisors D with simple normal crossings; each D_i corresponds to an edge of $\mathbf{G}$ and parametrizes local smoothings of the corresponding node. So the boundary complex of V_p is just a simplex $\Delta^{E(\mathbf{G})-1}$.

But there is in fact *monodromy* manifested in the coequalizer

$$\Delta(V_p \times V_p) \rightrightarrows \Delta(V_p),$$

and it turns out that this monodromy identifies the coequalizer of the diagram above with

$$\Delta^{E(\mathbf{G})-1}/\mathrm{Aut}(\mathbf{G}),$$

where $\mathrm{Aut}(\mathbf{G})$ acts by permutation on $E(\mathbf{G})$. So there is concordance on the level of strata with $M_{g,n}^{\mathrm{trop}}$!

Example 7.2 Let's see everything at work in the following specific example of a stratum in $\overline{\mathcal{M}}_{1,3}$.

Let $\mathbf{G}$ be the combinatorial type below.

1
2
3

Consider the boundary stratum $\mathcal{M}_{\mathbf{G}}$ of $\overline{\mathcal{M}}_{1,3}$. Locally, it is a self-intersection of the boundary component whose dual graph is obtained from $\mathbf{G}$ by contracting either edge.

Let's describe $\mathcal{M}_{\mathbf{G}}$. I'll assume $\mathrm{char}\, k \neq 2$ in this example. According to the discussion above we have $\widetilde{\mathcal{M}_{\mathbf{G}}} \cong \mathcal{M}_{0,4}$. Essentially, to give a stable curve C with dual graph $\mathbf{G}$ *along with a fixed identification of the two nodes of C with the two edges of* $\mathbf{G}$, we choose (up to projective equivalence) four distinct points p_1, p_2, q_1, q_2 on a $\mathbb{P}^1$, with the understanding that p_1 will be marked 1, p_2 marked 2, and q_1 and q_2 will be the two points of attachment of the other rational curve. Of course $\mathcal{M}_{0,4}$ is an honest variety: for example, fixing $p_1 = 0, p_2 = 1$, and $q_1 = \infty$ identifies $\mathcal{M}_{0,4}$ with $\mathbb{A}^1 - \{0, 1\}$.

Now $\mathcal{M}_{\mathbf{G}}$ is then the stack quotient $[\mathcal{M}_{0,4}/(\mathbb{Z}/2\mathbb{Z})]$, where the action is the one that exchanges q_1 and q_2. You can work out that with the identification $\widetilde{\mathcal{M}_{\mathbf{G}}} = \mathbb{A}^1 - \{0, 1\}$ above, the action sends a to $1 - a$.

Thus the quotient $\mathcal{M}_{\mathbf{G}}$ is a once-punctured plane with a $\mathbb{Z}/2\mathbb{Z}$-stacky point, corresponding to the fixed point $(0, 1, \infty, 1/2)$ of $\mathcal{M}_{0,4}$ under $\mathbb{Z}/2\mathbb{Z}$. It is the stacky point that produces monodromy: walking around it interchanges the analytic branches of the boundary divisor that meet along it. This example is just like Example 5.7, except that the punctured complex plane in that example is now filled in with a $\mathbb{Z}/2\mathbb{Z}$-stacky point, and there is another (inconsequential) puncture elsewhere.

The result is that for a point p in $\mathcal{M}_{\mathbf{G}}$, the neighborhood V_p has boundary complex a segment modulo a flip, just as in Example 5.7. And this is indeed a slice of the cell $\overline{C(\mathbf{G})}$ corresponding to $\mathbf{G}$ in the tropical moduli space $M_{1,3}^{\mathrm{trop}}$.

I omitted many details here, but this can all be patched together to show:

Theorem 7.3 ([1]) *There is a canonical identification of the link $\Delta_{g,n}$ of $M_{g,n}^{\mathrm{trop}}$ with the boundary complex of the toroidal embedding $\mathcal{M}_{g,n} \subset \overline{\mathcal{M}}_{g,n}$.*

7.2 Applications to the Topology of $\mathcal{M}_{g,n}$

What good is all of that? Here is one application to the cohomology of $\mathcal{M}_{g,n}$.

Suppose U is a smooth variety over k, or even a smooth Deligne-Mumford stack. Let X be a normal crossings compactification of U. It is a fact that the homotopy type of the boundary complex $\Delta(X)$ is independent of the choice of compactification X; that is, any other normal crossings compactification produces a homotopy equivalent complex [12]. This means that all topological invariants of $\Delta(X)$ are actually invariants of U itself. One very interesting such invariant is the *rational homology* of $\Delta(X)$. The reason that it is particularly interesting is as follows.

Set $k = \mathbb{C}$. Then there is a *weight filtration*, due to Deligne, on the cohomology of U

$$W_0 H^k(U,\mathbb{Q}) \subset \cdots \subset W_{2k} H^k(U,\mathbb{Q}) = H^k(U,\mathbb{Q}).$$

Let's write $\mathrm{Gr}_k^W H^j$ for the quotient $W_k H^j / W_{k-1} H^j$. Letting $d = \dim U$, I'll refer to $\mathrm{Gr}_{2d}^W H^*$ as the *top-weight cohomology*, since cohomology never appears in weights above $2d$. The point is that there is a canonical identification

$$\widetilde{H}_{i-1}(\Delta(U \subset X),\mathbb{Q}) \cong \mathrm{Gr}_{2d}^W H^{2d-i}(U,\mathbb{Q}) \tag{3}$$

of reduced, rational homology of the boundary complex, up to shifting degrees, with the top-weight rational cohomology of U. These facts all follow from Deligne's work on mixed Hodge structures [13] in the case of varieties; in the case of stacks, they can be proved following Deligne's ideas as in [10, Appendix]. In short: the top-weight slice of cohomology is combinatorially encoded in the boundary complex of any normal crossings compactification.

The identification (3), along with Theorem 7.3, allows us to study the rational cohomology of $\mathcal{M}_{g,n}$ appearing in top weight exactly by studying the reduced rational homology of $M_{g,n}^{\mathrm{trop}}$. This is a useful shift in perspective, because it allows arguments from tropical geometry and metric graph theory to be employed to study these complexes. Along these lines, S. Galatius, S. Payne, and I have shown some general results on $\Delta(\mathcal{M}_{g,n})$, including a lower bound on the connectivity of those spaces.[14] When $g = 1$, our results allow us to describe the whole situation pretty thoroughly:

[14] We say that a space is n-connected if the homotopy groups $\pi_1, \ldots, \pi_n$ all vanish.

Theorem 7.4 ([10])

(1) $\Delta(\mathcal{M}_{1,n})$ *is homotopy equivalent to a wedge of* $(n-1)!/2$ *top-dimensional spheres, for* $n \geq 3$. *(It is contractible when* $n = 1$ *and 2.) Therefore:*
(2) *For each* $n \geq 1$, *the top weight cohomology of* $\mathcal{M}_{1,n}$ *is*

$$\mathrm{Gr}^W_{2n} H^i(\mathcal{M}_{1,n}, \mathbb{Q}) \cong \begin{cases} \mathbb{Q}^{(n-1)!/2} & \text{for } n \geq 3 \text{ and } i = n, \\ 0 & \text{otherwise.} \end{cases}$$

Moreover, for each $n \geq 3$, *the representation of* S_n *on* $\mathrm{Gr}^W_{2n} H^n(\mathcal{M}_{1,n}, \mathbb{Q})$ *induced by permuting marked points can be described explicitly, as in [10].*
(3) *When* $n \geq 3$, *a dual basis for* $\mathrm{Gr}^W_{2n}(\mathcal{M}_{1,n}, \mathbb{Q})$ *is given by the* torus classes *associated to the* $(n-1)!/2$ *terminal curves whose dual graph is a loop of* n *once-marked* $\mathbb{P}^1$*s.*

I should remark that when $n \geq 5$, Theorem 7.4(2) also follows in principle from an earlier calculation by Getzler [16], as we explain in [10]. There is also a companion paper for the case $g = 2$ [9], in which I again use tropical techniques to show vanishing of *integral* homology of $\Delta(\mathcal{M}_{2,n})$ outside the top two degrees, and compute the top-weight Euler characteristic of $\mathcal{M}_{2,n}$ for every n. Furthermore, using tropical geometry and *using a computer* one can fully compute the top-weight $\mathbb{Q}$-cohomology of $\mathcal{M}_{g,n}$ in a range of cases, as we discuss in [9, 10]. I reproduce the computations for $\mathcal{M}_{2,n}$ from [9] for your curiosity: the cohomology is concentrated in degrees $n+3$ and $n+4$ with ranks given in the table below.

n	0	1	2	3	4	5	6	7	8
$\dim \mathrm{Gr}^W_{2d} H^{n+3}(\mathcal{M}_{2,n}, \mathbb{Q})$	0	0	1	0	3	15	86	575	4426
$\dim \mathrm{Gr}^W_{2d} H^{n+4}(\mathcal{M}_{2,n}, \mathbb{Q})$	0	0	0	0	1	5	26	155	1066

Acknowledgements Thank you very much to the organizers of the Moduli of Curves School at CIMAT for inviting me to lecture and to write these notes, and to Dan Abramovich, Ethan Cotterill, Sam Payne, and the anonymous reviewer for giving me comments on them. I'm grateful to Dan Abramovich, Matt Baker, Lucia Caporaso, Joe Harris, Diane Maclagan, Sam Payne, and Bernd Sturmfels, and many others for many helpful conversations over the years.

References

1. D. Abramovich, L. Caporaso, S. Payne, The tropicalization of the moduli space of curves. Ann. Sci. Ec. Norm. Supér. **48**(4), 765–809 (2015)
2. E. Arbarello, M. Cornalba, P. Griffiths, *Geometry of Algebraic Curves. Volume II*. With a contribution by Joseph Daniel Harris. Grundlehren der Mathematischen Wissenschafte, vol. 268 (Springer, Heidelberg, 2011)
3. M. Baker, D. Jensen, Degeneration of linear series from the tropical point of view and applications. Preprint. `arXiv:1504.05544`

4. M. Baker, S. Payne, J. Rabinoff, Nonarchimedean geometry, tropicalization, and metrics on curves. Algebr. Geom. **3**(1), 63–105 (2016)
5. V. Berkovich, *Spectral Theory and Analytic Geometry over Non-Archimedean Fields*. Mathematical Surveys and Monographs, vol. 33 (American Mathematical Society, Providence, RI, 1990)
6. S. Bosch, W. Lütkebohmert, Stable reduction and uniformization of abelian varieties. I. Math. Ann. **270**(3), 349–379 (1985)
7. S. Brannetti, M. Melo, F. Viviani, On the tropical Torelli map. Adv. Math. **226**, 2546–2586 (2011)
8. L. Caporaso, Algebraic and tropical curves: comparing their moduli spaces, in *Handbook of Moduli, Volume I*, ed. by G. Farkas, I. Morrison. Advanced Lectures in Mathematics, vol. XXIV (Springer, Berlin, 2013), pp. 119–160
9. M. Chan, Topology of the tropical moduli spaces $M_{2,n}$. Preprint. `arxiv:1507.03878`
10. M. Chan, S. Galatius, S. Payne, The tropicalization of the moduli space of curves II: topology and applications. Preprint. `arxiv:1604.03176`
11. M. Culler, K. Vogtmann, Moduli of graphs and automorphisms of free groups. Invent. Math. **84**(1), 91–119 (1986)
12. V.I. Danilov, Polyhedra of schemes and algebraic varieties. Mat. Sb. (N.S.) **139**(1), 146–158, 160 (1975)
13. P. Deligne, Théorie de Hodge II, III, Inst. Hautes Études Sci. Publ. Math. **40**, 5–57 (1971) and **44**, 5–77 (1974)
14. P. Deligne, D. Mumford, The irreducibility of the space of curves of given genus. Inst. Hautes Études Sci. Publ. Math. **36**, 75–109 (1969)
15. A. Gathmann, H. Markwig, The numbers of tropical plane curves through points in general position. J. Reine Angew. Math. **602**, 155–177 (2007)
16. E. Getzler, Resolving mixed Hodge modules on configuration spaces. Duke Math. J. **96**(1), 175–203 (1999)
17. W. Gubler, J. Rabinoff, A. Werner, Skeletons and tropicalizations. Adv. Math. **294**, 150–215 (2016)
18. J. Harris, I. Morrison, *Moduli of Curves*. Graduate Texts in Mathematics, vol. 187 (Springer, New York, 1998), xiv + 366 pp.
19. A.N. Jensen, Gfan, a software system for Gröbner fans and tropical varieties. Available at http://home.imf.au.dk/jensen/software/gfan/gfan.html
20. G. Kempf, F. Knudsen, D. Mumford, B. Saint-Donat, *Toroidal Embeddings. I*. Lecture Notes in Mathematics, vol. 339 (Springer, Berlin, 1973)
21. F.F. Knudsen, The projectivity of the moduli space of stable curves. II. The stacks $\mathcal{M}_{g,n}$. Math. Scand. **52**(2), 161–199 (1983)
22. F.F. Knudsen, The projectivity of the moduli space of stable curves. III. The line bundles on $\mathcal{M}_{g,n}$, and a proof of the projectivity of $\overline{M}_{g,n}$ in characteristic 0. Math. Scand. **52**(2), 200–212 (1983)
23. F.F. Knudsen, D. Mumford, The projectivity of the moduli space of stable curves. I. Preliminaries on "det" and "Div". Math. Scand. **39**(1), 19–55 (1976)
24. D. Maclagan, B. Sturmfels, *Introduction to Tropical Geometry*. Graduate Studies in Mathematics, vol. 161 (American Mathematical Society, Providence, RI, 2014), vii + 359pp.
25. H. Markwig, The enumeration of plane tropical curves. Ph.D dissertation, Technischen Universität Kaiserslautern (2006)
26. G. Mikhalkin, Tropical geometry and its applications, in *International Congress of Mathematicians. Volume II* (European Mathematical Society, Zürich, 2006), pp. 827–852
27. S. Payne, Analytification is the limit of all tropicalizations. Math. Res. Lett. **16**(3), 543–556 (2009)
28. A. Thuillier, Géométrie toroïdale et géométrie analytique non archimédienne. Application au type d'homotopie de certains schémas formels. Manuscripta Math. **123**(4), 381–451 (2007)
29. I. Tyomkin, Tropical geometry and correspondence theorems via toric stacks. Math. Ann. **353**(3), 945–995 (2012)

30. M. Ulirsch, Tropical geometry of logarithmic schemes. Ph.D. Dissertation, Brown University (2015)
31. F. Viviani, Tropicalizing vs. compactifying the Torelli morphism, in *Tropical and Non-Archimedean Geometry*. Contemporary Mathematics, vol. 605 (American Mathematical Society, Providence, RI, 2013), pp. 181–210

Higher Dimensional Varieties and their Moduli Spaces

Guanajuato, Mexico

Paolo Cascini

1 Introduction

To explain some of the main ideas of the Minimal Model Program and some of the tools used, we use some basic facts from graph theory. In particular, we describe a directed graph associated to the category of projective varieties. For this reason, we recall some of the basic definitions in graph theory.

Recall that a *directed graph* is a set of vertices connected by oriented edges, i.e. ordered pairs of vertices. Two edges are said to be *consecutive* if the ending vertex of one coincides with the starting vertex of the other. A *chain* in a directed graph is a sequence of distinct vertices connected by consecutive edges. A *cycle* in a directed graph is a sequence of consecutive ordered edges, starting and ending at the same vertex. A *tree* is a directed graph which does not contain any cycle. Note that the topological space underlying the tree is not necessarily simply connected, e.g. it could contain two distinct vertices and two edges connecting the two vertices with the same orientation.

Given two vertices X and Y of a directed graph, we say that Y is *below* X if we can find a chain starting from X and ending in Y. Clearly, If Y is below X, then we say that X is *above* Y. An *end-point* for a directed graph, is a vertex which does not admit any other vertex below it.

P. Cascini (✉)
Imperial College London, London, UK
e-mail: p.cascini@imperial.ac.uk

L. Brambila Paz et al. (eds.), *Moduli of Curves*, Lecture Notes of the Unione Matematica Italiana 21, DOI 10.1007/978-3-319-59486-6_2

1.1 Projective Surfaces

The easiest example in the study of directed graphs associated to the birational geometry of projective varieties is given by the category of smooth projective surfaces. To this end, we consider the *directed graph* whose vertices are smooth projective surfaces defined over an algebraically closed field k and whose edges are proper birational morphisms. The connected component containing a projective surface X corresponds to the birational class of X. We now look at some easy properties of this component. First, it is easy to check that this graph is a tree. Indeed, if X, Y are non-isomorphic projective surfaces connected by an edge, i.e. if there exists a non-trivial projective morphism $f: X \to Y$, then the second Betti number of X is greater than the one of Y. Thus, the claim follows easily.

Note that there are always infinitely many vertices above a vertex associated to a projective surface X, as it is always possible to blow-up an infinite sequence of points to obtain an infinite chain above X. On the other hand, using the inequality on the second Betti number described above, it is easy to check that starting from a vertex X, it is always possible to find an end-point below X. More specifically, there exists no infinite chain starting from X. Thus, we can think of the end-point Y to be a good representative of the connected class of X. We will see that also in higher dimension, one the main goals of the minimal model programme is to find the end-point of a connected component associated to a projective variety X.

We now show that projective surfaces can be divided into two large classes. The same dichotomy is expected to hold also in higher dimension.

First, we assume that X is a smooth projective surface such that $h^0(X, mK_X) > 0$ for some positive integer m. Then the subgraph obtained by considering the vertices below X and the corresponding edges is finite. In addition, there exists a unique vertex which is an end-point for the connected component containing X. Such a vertex Y is called the *minimal model* of X and, by Castelnuovo theorem, it is characterised by the fact that it does not admit any smooth rational curve E of self-intersection -1. Alternatively, Y is the only surface in the connected component of X such that K_Y is nef, i.e. $K_Y \cdot C \geq 0$ for any curve C in Y.

We now assume that X is a smooth projective surface such that $h^0(X, mK_X) = 0$ for all positive integer m. In this case, X is *uniruled*, i.e. it is covered by rational curves. It is possible to show that although the graph below X might be finite, there are always infinitely many end-points for the connected component of X. For example, if $X = \mathbb{P}^2$ is the two-dimensional projective space over the field k, then the connected component containing the vertex associated to X, corresponds to the set of all the smooth *rational surfaces*. Clearly, X is an end-point of such a graph, but also each Hirzebruch surface $\mathbb{F}_n = \mathbb{P}(\mathcal{O}_{\mathbb{P}^1} \oplus \mathcal{O}_{\mathbb{P}^1}(-n))$, with $n \in \mathbb{N}, n \neq 1$, is such. Finally, note that not all the projective surfaces which admit a Mori fiber space is an end-point for the directed graph we have constructed (e.g. the blow-up of $\mathbb{P}^2$ at one point admits a Mori fiber space, but it corresponds to a vertex which is not an end-point).

1.2 Higher Dimensional Projective Varieties

The goal of the minimal model program is to generalise the study of the directed graph we have seen in the previous section, to projective varieties of any dimension. Although we expect a similar behaviour, we will see that there are many problems arising as we go from dimension 2 to dimension 3 or higher.

In order to define the directed graph associated to projective varieties over an algebraically closed field k, we first try to understand what the edges are. A simple analysis in birational geometry shows that it is not suitable to consider only proper birational morphism between projective varieties, as otherwise we might get end-points of the graph which do not admit any special property and it would be hard to find a suitable characterisation of a good representative of the birational class of a projective variety. Thus, to solve this issue, we consider birational map with some extra properties. First, a birational map between normal projective varieties $\varphi: X \dashrightarrow Y$ is called a *contraction* if φ^{-1} does not contract any divisor. In other word, we do not want to consider maps like the blow-up of a proper subvariety, as this would not improve the understanding of our projective variety. Unfortunately, it is not enough to consider these maps as edges of our directed graph. Indeed, even in dimension 3, it is possible to find pairs of non-isomorphic projective manifolds, which are isomorphic in codimension 1, i.e. there exists X and Y which are isomorphic only after removing finitely many curves from both of them. In this case, X and Y would be part of a cycle, as the birational moprhism connecting X and Y and its inverse are both contraction. Therefore, we need to be more restrictive in the choice of those maps that define the edges of our directed graph.

Let $\varphi: X \dashrightarrow Y$ be a birational contraction between normal varieties, whose canonical divisors K_X and K_Y are $\mathbb{Q}$-Cartier (i.e. there exist positive integers m_1 and m_2 such that $m_1 K_X$ and $m_2 K_Y$ are Cartier). Then φ is said to be *K-negative* if there exist proper birational maps $p: W \to X$ and $q: W \to Y$ such that

$$p^* K_X = q^* K_Y + E$$

where E is an effective $\mathbb{Q}$-divisor and the support of E is equal to the union of all the exceptional divisors contracted by q.[1] It is then easy to show that if $\varphi: X \dashrightarrow Y$ is a K-negative birational contraction then its inverse is not K-negative. More in general, it is possible to check that if we define an edge of our graph to be a K-negative birational contraction, then the graph is a tree as it does not admit any cycle.

We now define the vertices of our new directed graph. Although, it would be tempting to consider only smooth projective varieties, as above it is possible to show that the end-points of our graph will not satisfy any useful property. On the other

[1]Note that this definition is slightly different than the one given in [2].

hand, we want to show that if the canonical divisor of a projective variety is nef then the vertex associated to X is an end-point of the graph. Indeed, we have:

Proposition 1.1 *Let X be a smooth projective variety such that K_X is nef. Then any K-negative birational contraction $\varphi: X \dashrightarrow Y$ is trivial, i.e. $\varphi = id_X$.*

Proof Exercise. Hint: it follows from the Negativity Lemma below. □

Lemma 1.2 (Negativity of Contraction) *Let $p: W \to Z$ be a proper birational morphism of normal varieties. Let E be an effective p-exceptional $\mathbb{Q}$-Cartier $\mathbb{Q}$-divisor on W. Then there is a component F of E which is covered by curves Σ such that $E \cdot \Sigma < 0$.*

Proof Cutting by hyperplanes in W, we reduce to the case when W is a surface, in which case the claim reduces to the Hodge Index Theorem. E.g. see [2, Lemma 3.6.2] for more details. □

Thus, it is natural to ask under what condition, a projective variety corresponding to the end-point of the directed graph admits a nef canonical divisor. To this end, we need to consider projective varieties with some mild singularities. First, for simplicity, we assume that all the varieties we consider are $\mathbb{Q}$-*factorial*, i.e. we assume that X is normal and any Weil divisor S is such that mS is Cartier for some positive integer m. In particular, this implies that K_X is $\mathbb{Q}$-Cartier. Moreover, we assume that X is *terminal*, which means that X admits a smooth variety Y above X. We now show that this is equivalent to the more classical definition:

Proposition 1.3 *X is terminal if and only if there exists a resolution $f: Y \to X$ such that*

$$K_Y = f^*K_X + E$$

for some $\mathbb{Q}$-divisor $E \geq 0$ whose support coincides with the exceptional locus of f. It is easy to show that if X is terminal then the same property holds for any resolution of X.

Proof If $f: Y \to X$ is as in the proposition, then clearly X is terminal.

Let us assume now that there exists a smooth variety Z above X. Then there exist proper birational maps $p: W \to Z$ and $q: W \to X$ such that

$$p^*K_Z = q^*K_X + E$$

where E is an effective $\mathbb{Q}$-divisor and the support of E is equal to the union of all the exceptional divisors contracted by q. Let $h: \tilde{W} \to W$ be a resolution and let $p': \tilde{W} \to Z$ and $q': \tilde{W} \to X$ be the induced morphisms. Since Z is smooth, we may write

$$K_W = p'^*K_Z + G$$

for some $\mathbb{Q}$-divisor $G \geq 0$ whose support coincides with the exceptional locus of q'. Thus, we have

$$K_W = q'^* K_X + h^* E + G$$

and $h^* E + G \geq 0$. By assumption, the support of $h^* E$ contains the strict transform of the exceptional locus of q in $\tilde{W}$ and the support of G contains the exceptional locus of h. Thus, the support of $h^* E + G$ coincides with the exceptional locus of q' and we are done. □

Thus, terminal projective varieties appear naturally in the graph that we are considering. We can finally construct our directed graph: the vertices are terminal projective varieties defined over an algebraically closed field k and the edges are K-negative birational contractions.

Lemma 1.4 *Let X be a variety with terminal singularities. Then the singular locus of X has codimension at least* 3.

In particular, any terminal surface is smooth.

Proof The case of surfaces is easy to check by considering a minimal resolution $h: \tilde{X} \to X$. In higher dimension, as in Lemma 1.2, it is enough to cut by general hyperplanes. E.g. see [16, Corollary 5.18] for more details. □

At this point, it is natural to ask if the same properties described in the case of surfaces, would hold for this directed graph. In particular, if X is a terminal projective variety, we can ask if there might exist an infinite chain starting from X. It is possible to show that this coincides with the following famous open problem:

Conjecture 1.5 (Termination of Flips) Let X be a terminal projective variety. Then there exists no infinite sequence of flips

$$X = X_0 \dashrightarrow X_1 \dashrightarrow X_2 \dashrightarrow \ldots$$

starting from X

A *flip* is a special K-negative birational contraction which is an isomorphism in codimension 1 (see [16] for more details). Existence of flips for projective varieties defined over $\mathbb{C}$ was proven in [2, 11] and, later on, in [5, 8]. Termination of flips was proven by Shokurov in dimension 3 (cf. Theorem 3.1) and under some assumptions in higher dimension. In particular, one of the main achievements in this direction is obtained by combining the results in [1] and [13].

We now go back to the study of our directed graph and we can ask if the dichotomy within uniruled and non-uniruled surfaces extends to the case of higher dimensional projective varieties. More specifically, we expect that if $h^0(X, mK_X) > 0$ for some positive integer m, then there exists always an end-point below X, represented by a terminal projective variety Y with the property that K_Y is nef. As in the case of surfaces, Y will be called a *minimal model* of X.

Thus, we have the following:

Conjecture 1.6 (Existence of Minimal Models) Let X be a terminal projective variety with non-trivial canonical ring

$$R(X, K_X) = \bigoplus_{m \geq 0} H^0(X, mK_X).$$

Then X admits a minimal model, i.e. there exists a K-negative birational map

$$\varphi: X \dashrightarrow Y$$

into a terminal projective variety Y, such that K_Y is nef.

If X is a complex variety of general type, i.e. if K_X is big, then the conjecture holds in any dimension [2] (cf. Theorem 4.1). Note that even in this case, Conjecture 1.5 is still open. Thus, the sub-graph of all the varieties below a given terminal projective variety of general type could contain an infinite chain, such that each vertex of this chain is connected to an end-point after finitely many consecutive edges.

If X is uniruled, then a very similar picture as in the case of surfaces holds. Although the graph given by the terminal projective varieties below X might be infinite, we do have a good description of the end-points of the connected component containing X. Indeed, these vertices correspond to varieties Y which admit a *Mori fibre space* [2] (cf. Theorem 4.3), i.e. a non-trivial map $\eta: Y \to Z$ onto a lower dimensional projective variety, such that the general fiber F is a (possibly singular) Fano variety, i.e. the anti-canonical divisor $-K_F$ is ample. Note that Y might admit more than one structure as a Mori fiber space (e.g. the simplest example is $\mathbb{P}^1 \times \mathbb{P}^1$).

Although the picture for uniruled varieties is quite well-understood, we still need to understand whether there is no other connected component of our directed graph, corresponding to projective varieties which do not belong to the two classes of varieties described above (i.e. non uniruled varieties with trivial canonical ring). Currently, this is one of the most important open problem in the Minimal Model Programme. For simplicity, we will only discuss it in the case of complex projective varieties:

Conjecture 1.7 (Weak Abundance Conjecture) Let X be a terminal complex projective variety with trivial canonical ring, i.e.

$$R(X, K_X) = \mathbb{C}.$$

Then X is uniruled.

Note that in dimension 3, all the conjectures described above (i.e. Conjecture 1.5–1.7) hold in full generality for complex projective varieties (e.g. see [16, 17]).

2 Birational Geometry of Log Pairs

It is immediate to see from the arguments in the previous section, that the canonical divisor plays a very important role in the study of the birational geometry of a projective variety over any algebraically closed field k. In addition, if $k = \mathbb{C}$, several generalisations of Kodaira's vanishing theorem, such as Kawamata-Viehweg vanishing, had a huge impact in birational geometry (e.g. in [5], it was shown that finite generation of the canonical ring follows almost directly from these results). This is another evidence of the fact that the canonical divisor is an essential tool in birational geometry. On the other hand, there are at least two main problems if we work in this generality. First, if S is a normal hypersurface in a $\mathbb{Q}$-factorial projective variety X, then the adjunction formula [17] implies that

$$(K_X + S)|_S = K_S + \mathrm{Diff}_S$$

where Diff_S is an effective $\mathbb{Q}$-divisor on S, i.e. a linear combination of prime divisors of S with positive rational coefficients. Secondly, even in the simple case of a smooth minimal elliptic surface $\pi: X \to C$ over a smooth curve C, we have that

$$K_X = \pi^*(K_S + D)$$

for some effective $\mathbb{Q}$-divisor D on C. Thus, it is natural to consider a larger category, which includes not only varieties but pairs (X, Δ) where X is a $\mathbb{Q}$-factorial projective variety and Δ is an effective $\mathbb{Q}$-divisor. It is thus convenient to replace varieties by log pairs (X, Δ), and the canonical divisor by $K_X + \Delta$.

Therefore, our new goal is to define the right generalisation of the directed graph that we want to consider. To this end, we consider log pairs with Kawamata log terminal singularities: this is the most suitable condition in terms of the singularities of a log pair in the minimal model programme. A *log pair* (X, Δ) consists of a normal variety X with a $\mathbb{Q}$-divisor Δ whose coefficients are contained in the interval $(0, 1]$ and such that $K_X + \Delta$ is $\mathbb{Q}$-Cartier (for simplicity, we assume that the variety X is $\mathbb{Q}$-factorial so that this last condition is automatically satisfied). The log pair (X, Δ) is said to be *Kawamata log terminal* (resp. *log canonical*) if for any proper birational morphism $\varphi: Y \to X$, we may write

$$K_Y + \Delta_Y = \varphi^*(K_X + \Delta)$$

where Δ_Y is a (non-necessarily effective) $\mathbb{Q}$-divisor whose coefficients are strictly less than 1 (resp. not greater than 1). Note that this definition does not assume resolution of singularities.

Recall that a pair (X, Δ) is said to be *log smooth* if X is smooth and Δ is simple normal crossing, i.e. the components of Δ are smooth and they intersect everywhere transversally. Then, it is easy to check that a log smooth pair (X, Δ) is Kawamata log terminal (resp. log canonical) if and only if the coefficients of Δ are contained in the interval $(0, 1)$ (resp. $(0, 1]$).

We can now define the edges of our new directed graph. Let (X, Δ) and (Y, Δ) be two Kawamata log terminal pairs. An edge from (X, Δ) to (Y, Δ') is a birational contraction $f: X \dashrightarrow Y$ such that $f_*\Delta = \Delta'$ and f is $(K + \Delta)$-negative, i.e. there exist proper maps $p: W \to X$ and $q: W \to Y$ from a projective variety W which resolve the indeterminacy of f and such that

$$p^*(K_X + \Delta) = q^*(K_Y + \Delta') + E$$

where $E \geq 0$ is an effective $\mathbb{Q}$-divisor whose support coincides with the union of all the exceptional divisors of q. In particular, under these assumptions, if the log pair (Y, Δ') is below (X, Δ), then $H^0(X, m(K_X + \Delta)) \neq 0$ for some positive integer m if and only if the same property holds for (Y, Δ').

Thus, as in the absolute case (i.e. when $\Delta = 0$), given a Kawamata log terminal pair (X, Δ), we can investigate the graph given by pairs below (X, Δ). In particular, it is expected that there are no infinite chains below (X, Δ). As in Conjecture 1.5, this corresponds to *termination of log-flips*. In other words, it is expected that there exists always an end-point below (X, Δ). If $H^0(X, m(K_X+\Delta)) \neq 0$ for some positive integer m, then an end-point (Y, Γ) is characterised by the property that $K_Y + \Gamma$ is nef. The pair (Y, Γ) is called a *minimal model* of (X, Δ). In [2], it was proven that if Δ is big, then (X, Δ) always admits a minimal model (cf. Theorem 4.4).

We now consider the case of Kawamata log terminal pairs (X, Δ) such that $H^0(X, m(K_X + \Delta)) = 0$ for all positive integers m. Similarly as in the absolute case, it is expected that an end-point (Y, Δ') below (X, Δ) admits a *Mori fiber space*, which is a non-trivial morphism $\eta: Y \to Z$ such that $-(K_X + \Delta)|_F$ is ample for the general fiber F of η. In [2], it was proven that if $(K_X + \Delta)$ is not pseudo-effective (i.e. if there exists an ample $\mathbb{Q}$-divisor such that $K_X + \Delta + A$ is not big) then there exists an end-point below (X, Δ), which admits a Mori fiber space.

The main tools used in the study of log pairs are just generalisations of the results in the absolute case. E.g. Kodaira vanishing generalises to:

Theorem 2.1 (Kawamata-Viehweg Vanishing Theorem) *Let (X, Δ) be a complex $\mathbb{Q}$-factorial projective Kawamata log terminal pair. Let N be a Weil divisor such that*

$$N \equiv \Delta + A$$

where A is a big and nef $\mathbb{Q}$-divisor.

Then $H^i(X, \mathcal{O}_X.(K_X + N)) = 0$ for all $i > 0$.

Proof E.g. see [16, Theorem 2.70]. □

Furthermore, we have:

Theorem 2.2 (Base Point Free Theorem) *Let (X, Δ) be a complex $\mathbb{Q}$-factorial projective Kawamata log terminal pair. Let D be a nef $\mathbb{Q}$-divisor such that*

$$D = K_X + \Delta + A$$

where A is a big and nef $\mathbb{Q}$-divisor.

Then D is semi-ample, i.e. there exists a positive integer m such that mD is Cartier and $|mD|$ is base point free.

Proof E.g. see [16, Theorem 3.3]. □

Theorem 2.3 (Cone Theorem) *Let (X, Δ) be a complex projective Kawamata log terminal pair. Then there are countably many rational curves $C_1, C_2, \ldots$ such that*

$$0 < -(K_X + \Delta) \cdot C_i \leq 2 \dim X$$

and

$$\overline{NE}(X) = \overline{NE}(X)_{(K_X+\Delta)\geq 0} + \sum \mathbb{R}_{\geq 0}[C_j].$$

Proof E.g. see [16, Theorem 3.7]. □

Example 2.4 Assume that (X, Δ) is a complex log Fano pair (i.e. (X, Δ) is Kawamata log terminal and $-(K_X + \Delta)$ is big and nef). Let D be a nef divisor. Then, we may write

$$D = K_X + \Delta + (D - (K_X + \Delta))$$

and since $D - (K_X + \Delta)$ is big and nef, Theorem 2.2 implies that D is semi-ample.

Now assume that S is the blow-up of $\mathbb{P}^2$ at nine very general points and let E be the unique elliptic curve in $\mathbb{P}^2$ passing through these points. Then $-K_S = C$ where C is the strict transform of E in S and in particular, $-K_S$ is not semi-ample, as C is contained in the base locus of $|m(-K_S)|$ for any positive integer m (the problem is that $-K_S$ is not big and therefore S is not log Fano).

Finally, let H be an ample Cartier divisor and let $Z = \mathbb{P}(\mathcal{O}_S. \oplus \mathcal{O}_S.(-H))$ with projection map $p: Z \to S$. Let ξ the tautological class on Z and let $\Delta = \xi$.

Then, it is easy to check that

$$-(K_Z + \Delta) = -p^* K_S + \xi + p^* H$$

is big and nef. On the other hand, $D = -(K_Z + \Delta)$ is not semi-ample. Indeed

$$D|_S = -K_S$$

and we showed that $-K_S$ is semi-ample (see [10, Example 5.2] for more details). Note that (X, Δ) is log smooth but it is not Kawamata log terminal (the coefficient of Δ is equal to one).

We now consider a special case, which illustrates the fact that the point of view of directed graphs is useful to understand the birational geometry of a Kawamata log terminal pair (X, Δ). Assume that Δ is big and that $K_X + \Delta$ is not pseudo-effective. Since Δ is big, the non-vanishing theorem proven in [2] implies that the assumption that $K_X + \Delta$ is not pseudo-effective coincides with the a-priori stronger assumption that $H^0(X, m(K_X + \Delta)) = 0$ for any positive integer m. Thus, it is natural to ask if (X, Δ) admits infinitely many end-points below (X, Δ). Note that the picture in the absolute case (i.e. when $\Delta = 0$) is very different, as we have already showed that there might be infinitely many edges starting from the vertex associated to $(X, 0)$. On the other hand, assuming that Δ is big, it is possible to show, by using boundedness of the length of extremal rays [14], that there are only finitely many edges starting from (X, Δ). Thus, König's Lemma implies that the sub-graph given by all the vertices below (X, Δ) is finite if and only if there are no infinite chains starting from (X, Δ) (see [19, Lemma 6.7] for more details). Clearly, if the sub-graph below (X, Δ) is finite, there are only finitely many end-point below (X, Δ).

3 Termination of Threefold Flips

We now want to show how the study of singularities imply termination of threefold flips. In particular, we want to show:

Theorem 3.1 (Shokurov) *There exists no infinite sequence of three dimensional K-negative birational contractions.*

First, we need to define the log discrepancy of a variety with respect to a divisorial valuation. To this end, we consider the more general case of a log pair (X, Δ).

Definition 3.2 Let (X, Δ) be a log pair and let $p: Y \to X$ be a proper birational morphism. Then we may write

$$K_Y + \Delta_Y + \sum(1 - a_i)E_i = p^*(K_X + \Delta)$$

where Δ_Y is the strict transform of Δ on Y (i.e. $\Delta_Y = p_*^{-1}\Delta$) and the sum is taken over all the p-exceptional divisor E_i of p. The rational number $a_i = a(E_i, X, \Delta)$ denotes the *log discrepancy* of (X, Δ) with respect to E_i.

If $\Delta = 0$, then we denote $a(E, X, \Delta)$ simply by $a(E, X)$.

Remark 3.3 Note that, in many other references, such as in [16], the same symbol $a(E; X, \Delta)$ denotes the discrepancy of (X, Δ) with respect to E, which is, according to our notation, nothing but $a(E, X, \Delta) - 1$.

It is important to understand that the log discrepancy does not depend on the morphism p that we consider. In other words, using the same notation as above, if E is a p-exceptional divisor and $q: Z \to X$ is another proper birational morphism such that the induced birational map $\varphi: Y \dashrightarrow Z$ is an isomorphism at the general point of E then $a(E, X, \Delta)$ coincides with $a(\varphi_* E, X, \Delta)$ and therefore it can be computed on Z. It is a good exercise to check that the equality above holds. Thus, given a log pair (X, Δ), we can define $a(E, X, \Delta)$ with respect to any divisorial valuation E over X.

The idea is that the smaller is the minimum value of $a(E, X, \Delta)$ with respect to all the divisorial valuations E over X, and the more singular the log pair (X, Δ) is. In particular, we have:

Lemma 3.4 *A log pair* (X, Δ) *is Kawamata log terminal if and only if the coefficients of* Δ *are contained in the interval* $(0, 1)$ *and, for any proper birational morphism* $p: Y \to X$ *and for any* p*-exceptional divisor* E*, we have* $a(E, X, \Delta) > 0$.

A variety X *is terminal if and only if for any proper birational morphism* $p: Y \to X$ *and for any* p*-exceptional divisor* E*, we have* $a(E, X) > 1$.

Proof Exercise. □

Example 3.5 Let X be a smooth threefold and let $p: Y \to X$ be the blow-up of X along a smooth curve C, with exceptional divisor E. Then $a(E, X) = 2$.

Lemma 3.6 *Let* X *be a terminal variety. Then there are at most finitely many divisorial valuations* E *such that* $a(E, X) < 2$.

Proof Let $f: Y \to X$ be a resolution. Then, we may write

$$K_Y = f^* K_X + \Gamma \tag{1}$$

where $\Gamma \geq 0$ is a f-exceptional divisor. Let E be a divisorial valuation over X such that $a(E, X) < 2$ and let us assume that E is not a divisor over Y. Then, (1) implies that

$$a(E, Y) \leq a(E, X) < 2.$$

Since Y is smooth, we easily obtain a contradiction. Thus, E is a divisor on Y. Since there are only finitely many divisors on Y which are contracted by f, the claim follows immediately.

For a more general result, see [16, Proposition 2.36]. □

We define the *difficulty* of a terminal threefold as

$$d(X) = \#\{E \mid a(E, X) < 2\}.$$

By Lemma 3.6, $d(X)$ is a non-negative integer.

Example 3.7 It is easy to check that if X is a smooth or Gorenstein (i.e. K_X is Cartier) terminal variety, then $d(X) = 0$. The opposite is also true, at least over

$\mathbb{C}$, i.e. if $d(X) = 0$ then X is Gorenstein. But its proof relies on the classification of terminal singularities.

Lemma 3.8 *Let $\varphi: X \dashrightarrow Y$ be a K-negative birational contraction between terminal threefolds which is an isomorphism in codimension 1 and let E be a divisorial valuation over X such that $a(E, Y) < 2$.*

Then $a(E, X) < 2$ and, in particular,

$$d(X) \geq d(Y).$$

More in general, if $X \dashrightarrow Y$ is a K-negative birational contraction between terminal threefolds, then $d(X) \geq d(Y) - \rho(X/Y)$.

Proof By assumption, there exist proper birational maps $p: W \to X$ and $q: W \to Y$ such that

$$p^* K_X = q^* K_Y + F, \tag{2}$$

where F is an effective q-exceptional $\mathbb{Q}$-divisor. Let E be a divisorial valuation over Y such that $a(E, Y) < 2$. After possibly replacing W by a variety W' which admits a proper birational morphism $W' \to W$, we may assume that W is smooth and that E is a divisor on W (note that we cannot assume anymore that the support of F is equal to the union of all the exceptional divisors contracted by q as in our definition of K-negative birational contraction, but we do not need this fact here). There are two cases: either E is a divisor on X which is contracted by φ or it is contracted by p. If φ is an isomorphism in codimension 1, then only the second case can occur and (2) easily implies that $a(E, X) < 2$. Thus, the claim follows. □

We can finally prove Theorem 3.1.

Proof Assume that

$$X = X_0 \dashrightarrow X_1 \dashrightarrow X_2 \dashrightarrow \dots$$

is an infinite sequence of K-negative birational contraction in dimension 3. Then, for each i, we have

$$\rho(X_i) \geq \rho(X_{i+1})$$

where the equality holds if and only if $X_i \dashrightarrow X_{i+1}$ is an isomorphism in codimension 1. Since $\rho(X_i)$ is a positive integer, after possibly taking a subsequence, we may assume that each $X_i \dashrightarrow X_{i+1}$ is an isomorphism in codimension 1. It is enough to show that $d(X_i) > d(X_{i+1})$.

Lemma 3.8 implies that if E is a divisorial valuation over X_{i+1} such that $a(E, X_{i+1}) < 2$ then $a(E, X_i) < 2$. Thus, it is enough to construct a divisorial valuation E over X_i such that $a(E, X_i) < 2$ and $a(E, X_{i+1}) \geq 2$.

By assumption, there exist proper birational maps $p\colon W \to X_i$ and $q\colon W \to X_{i+1}$ such that

$$p^* K_{X_i} = q^* K_{X_{i+1}} + G, \tag{3}$$

where G is an effective $\mathbb{Q}$-divisor whose support is equal to the union of all the exceptional divisors contracted by q.

Lemma 1.4 implies that the singularities of X_{i+1} are isolated. On the other hand, one can check that there exists a divisor F on W such that $\xi = q(F)$ is a curve in X_{i+1}. Let $X_{i+1}^{prime} \to X_{i+1}$ be the blow-up of X_{i+1} along ξ and let E be the exceptional divisor. Then E is a divisorial valuation on X_{i+1} such that $a(E, X_{i+1}) = 2$ (see Example 3.5). On the other hand, (3) implies that $a(E, X_i) < 2$ and the claim follows.

Thus, $d(X_i) > d(X_{i+1})$ and the sequence must terminate. □

Note that the same proof, combined with Example 3.7, implies that if X is Gorenstein, then there are no non-trivial K-negative birational contraction $X \dashrightarrow Y$ in dimension 3. In particular, if X is Gorenstein of dimension 3, then there are no flips $X \dashrightarrow Y$.

4 Some Results

Some of the main results in [2] are:

Theorem 4.1 *Let X be a smooth complex projective variety of general type. Then X admits a minimal model $X \dashrightarrow Y$.*

Theorem 4.2 *Let X be a smooth complex projective variety. Then the canonical ring*

$$R(X, K_X) = \bigoplus_{m \geq 0} (X, mK_X)$$

is finitely generated.

Theorem 4.3 *Let X be a uniruled smooth complex projective variety. Then X admits a Mori fibre space $X \dashrightarrow Y$.*

We first want to show that the three results above follow from the following

Theorem 4.4 *Let (X, Δ) be a Kawamata log terminal pair over $\mathbb{C}$ such that Δ is a big $\mathbb{Q}$-divisor and $K_X + \Delta$ is pseudo-effective. Then (X, Δ) admits a minimal model $X \dashrightarrow Y$.*

Recall, that a $\mathbb{Q}$-divisor D is pseudo-effective if, given any ample $\mathbb{Q}$-divisor A, we have that $D + A$ is big. A minimal model for a pair (X, Δ) is a $(K_X + \Delta)$-negative birational contraction $\varphi\colon X \dashrightarrow Y$. such that $K_Y + \varphi_* \Delta$ is nef.

We first show how Theorem 4.4 implies the three results above:

Proof of Theorem 4.1 By assumption K_X is big. In particular, there exists a positive integer m such that $|mK_X| \neq \emptyset$. Let $D \in |mK_X|$. Since X is smooth, it is easy to check that if ε is a sufficiently small rational number and $\Delta = \varepsilon D$, then (X, Δ) is Kawamata log terminal. Note that, since

$$K_X + \Delta = (1 + \varepsilon m)K_X \tag{4}$$

it follows that $K_X + \Delta$ is big and, in particular, it is pseudo-effective. Thus, Theorem 4.4 implies that $(K_X + \Delta)$ admits a minimal model $\varphi\colon X \dashrightarrow Y$. It is then easy to check that (4) implies that φ is also a minimal model for X. □

Proof of Theorem 4.2 We first consider a slightly different set-up. We assume that (X, Δ) is a Kawamata log terminal pair such that Δ is big. We want to show that the ring

$$R(X, K_X + \Delta) = \bigoplus_{m \text{ suff. divisible}} H^0(X, m(K_X + \Delta))$$

is finitely generated. Clearly if $H^0(X, m(K_X + \Delta)) = 0$ for any m then $R(X, K_X + \Delta)$ is finitely generated. Thus, we may assume that $H^0(X, m(K_X + \Delta)) \neq 0$ for some m and in particular $K_X + \Delta$ is pseudo-effective. Theorem 4.4 implies that (X, Δ) admits a minimal model $\varphi\colon X \dashrightarrow Y$ and, since φ is $(K_X + \Delta)$-negative, it follows that $R(X, K_X + \Delta) \simeq R(Y, K_Y + \varphi_* \Delta)$. Thus, after replacing X by Y and Δ by $\varphi_* \Delta$, we may assume that $K_X + \Delta$ is nef. Since Δ is big, we can find an ample $\mathbb{Q}$-divisor A and an effective $\mathbb{Q}$-divisor E such that

$$\Delta \simeq A + E.$$

Let $\varepsilon > 0$. We may write

$$\Delta = (1 - \varepsilon)\Delta + \epsilon E + \epsilon A$$

and since (X, Δ) is Kawamata log terminal, it follows that if ε is sufficiently small, then

$$(X, (1 - \varepsilon)\Delta + \epsilon E)$$

is also Kawamata log terminal. We have,

$$K_X + \Delta = K_X + (1 - \varepsilon)\Delta + \epsilon E + \epsilon A$$

and since ϵA is ample, the base point free theorem (cf. Theorem 2.2) implies that $K_X+\Delta$ is semi-ample. Thus, there exists a morphism $\eta: X \to Z$ such that $K_X+\Delta = \eta^* H$ for some ample $\mathbb{Q}$-divisor H. It is easy to check, that up to truncation the ring $R(X, K_X + \Delta)$ is isomorphic to the ring

$$R(Z,H) = \bigoplus_{m \text{ suff. divisible}} H^0(X, mH)$$

which is finitely generated since it is the quotient of a polynomial ring.

We now go back to our set-up and we assume that X is a smooth projective variety. As above, we may assume that the Kodaira dimension of X is non-negative (otherwise there is nothing to prove). Let $\psi: X \dashrightarrow Y$ be the Iitaka fibration associated to K_X (e.g. see [18, §2.1.C]). Then, by a result of Fujino and Mori [9], it follows that, up to truncation, the canonical ring $R(X, K_X)$ of X is isomorphic to the ring associated to a Kawamata log terminal pair (Y, Γ) for some big $\mathbb{Q}$-divisor Γ on Y. Thus, the result follows from the claim above. □

Sketch of the Proof of Theorem 4.3 Since X is uniruled, it follows that K_X is not pseudo-effective. In particular, if A is an ample $\mathbb{Q}$-divisor on X then there exists $t > 0$ such that $K_X + tA$ is not big. Let

$$\lambda = \sup\{t > 0 \mid K_X + tA \text{ is not big}\}.$$

Then $K_X + \lambda A$ is pseudo-effective but not big. It is easy to show that there exists $\Delta \sim_{\mathbb{Q}} \lambda A$ such that (X, Δ) is Kawamata log terminal. Thus, by Theorem 4.4, we may find a proper birational contraction $\varphi: X \dashrightarrow Y$. such that $K_Y + \varphi_*\Delta$ is nef. Since $K_X + \lambda A$ is not big, it follows that $K_Y + \varphi_*\Delta$ is also not big. Since $\varphi_*\Delta$ is big, as in the proof of Theorem 4.2, it follows from the base point free theorem[2] that $K_Y + \varphi_*\Delta$ is semi-ample and in particular, there exists $\eta: Y \to Z$ such that

$$K_Y + \varphi_*\Delta = \eta^* H \tag{5}$$

for some ample $\mathbb{Q}$-divisor H on Z. In particular $\dim Z < \dim Y = \dim X$.

We now assume that the restriction of $\varphi_*\Delta$ to the general fibre of η is ample (e.g. this is true if $\rho(Y/Z) = 1$). Then, since the restriction of $\eta^* H$ to the general fibre of η is numerically trivial, it follows from (5), that the general fibre of Y is Fano. Thus Y is a Mori fibre space. The general case (without assuming that the restriction of $\varphi_*\Delta$ to the general fibre of η is ample) is slightly harder. □

[2]Note that a priori λ might be an irrational number. On the other hand, Theorem 2.2 holds in a more general context, assuming that Δ is a $\mathbb{R}$-divisor rather than a $\mathbb{Q}$-divisor.

5 Minimal Model Program with Scaling

We now illustrate some of the steps to prove Theorem 4.4. We begin by considering the classical minimal model program, due to Mori.

We start with a complex projective manifold X. We want to find a K-negative birational contraction $\varphi: X \dashrightarrow Y$ such that, either Y is minimal or it admits a Mori fibre space.

If K_X is nef, then we are done, as X is minimal and we can stop here. Thus, we may assume that K_X is not nef. By the cone theorem (cf. Theorem 2.3), it is easy to show that there exists an ample $\mathbb{Q}$-divisor A such that $K_X + A$ is nef but not ample, and there exists a rational curve ξ such that, $K_X \cdot \xi < 0$ and for any curve C in X, we have that

$$(K_X + A) \cdot C = 0 \qquad \text{if and only if } [C] \in \mathbb{R}_+[\xi],$$

where $[C]$ and $[\xi]$ denote the numerical equivalence class of the 1-cycles defined by C and ξ respectively. Since A is ample and $K_X + A$ is nef, the base point free theorem (cf. Theorem 2.2) implies that there exists $f: X \to Z$ with connnected fibers and such that $K_X \cdot C < 0$ for any curve C s.t. $f(C)$ is a point. More precisely, C is contracted if and only if $[C] \in \mathbb{R}_+[\xi]$.

We now distinguish three cases:

1) if $\dim Z < \dim X$ then, X is automatically a Mori fibre space, and we can stop here.
2) if f is a *divisorial contraction*, i.e. it is birational and the exceptional locus of f is a divisor, then we just replace X by Z and we start again. Note that f is automatically a K-negative birational contraction. On the other hand, Z is not anymore necessarily smooth, but it is always terminal and all the steps above would work without any change.
3) Otherwise, the morphism f is a *flipping contraction*, i.e. f is birational and the exceptional locus of f has codimension ≥ 2. In this case, Z is too singular and we cannot replace X by Z. Indeed, it is possible to show that K_Z is never $\mathbb{Q}$-Cartier and pretty much everything above would not hold true anymore. On the other hand, [2, 11] imply that there exists a flip $\varphi: X \dashrightarrow X^+$ of ξ. It follows again that X^+ is terminal and φ is a K-negative birational contraction. Thus, we can replace X by X^+ and start all over again.

At this point, it only remains to show that this process would terminate after finitely many steps, so that we obtain either a minimal model or a Mori fibre space which is birational to X. Note that 2) would only occur finitely many times, because, as we mentioned before, after each such step, the Picard number of X would decrease by one (and any time that case 3) occurs, the Picard number would not change). Thus, the only (big) problem is to show that 3) does not occur infinitely many times. Theorem 3.1 solves this problem in dimension 3, but in higher dimension, this is still an open question.

The idea of the minimal model program with scaling is that, instead of choosing a random sequence of flips, we show that there exist a special sequence which terminates. To this end, it turns out that it is more natural to work in the more general context of log pairs (X, Δ). We begin by assuming that (X, Δ) is log canonical and we pick a $\mathbb{Q}$-divisor H such that $K_X + \Delta + H$ is nef and the pair $(X, \Delta + H)$ is log canonical (e.g. we may choose A to be a sufficiently ample divisor, and for any sufficiently large, we define $H = \frac{1}{m}B$ where $B \in |mA|$ is a general element). We define

$$\lambda = \min\{t \in \mathbb{R}_{\geq 0} |\ K_X + \Delta + tH \text{ is nef}\}.$$

Note that if $\lambda = 0$ then $K_X + \Delta$ is nef and we can stop here. Thus, we may assume that $\lambda > 0$. Then by the cone theorem (cf. Theorem 2.3) there exists a rational curve ξ, such that $\mathbb{R}_+[\xi]$ is an extremal ray for $\overline{NE}(X)$ and such that

$$(K_X + \Delta) \cdot \xi < 0 \qquad \text{and}$$
$$(K_X + \Delta + \lambda H) \cdot \xi = 0.$$

As before, we may find a morphism $f: X \to Y$ which contracts any curve which is numerically equivalent to a multiple of ξ. If f defines a Mori fibre space, then we are done. Thus, we may assume that f is birational. If f is a divisorial contraction then we denote Y by X', otherwise f is a flipping contraction and we consider the flip $X \dashrightarrow X'$ associated to X. In both cases, if Δ' and H' denote the strict transform on X' of Δ and H respectively, then it is easy to check that (X', Δ') is log canonical and $K_{X'} + \Delta' + \lambda H'$ is nef. We now consider

$$\lambda' = \min\{t \in \mathbb{R}_{\geq 0} |\ K_{X'} + \Delta' + tH' \text{ is nef}\} \leq \lambda$$

and we proceed as above. Thus, we obtain a sequence of $(K+\Delta)$-negative birational contractions

$$X = X_0 \dashrightarrow X_1 \dashrightarrow X_2 \dashrightarrow \dots$$

but also a decreasing sequence of rational numbers

$$\lambda = \lambda_0 \geq \lambda_1 \geq \lambda_2 \geq \dots\dots$$

Note that the choice of contractions depends on the initial choice of the $\mathbb{Q}$-divisor H. For this reason, this process, is called a $(K_X + \Delta)$-*Minimal Model Program with scaling of* H.

We now present some of the main ingredients in the proof that, if (X, Δ) is Kawamata log terminal and Δ is big, then the minimal model program with scaling terminates. Some of the details are rather technical and we refer to [2].

We first need to enlarge the category of Kawamata log terminal pairs, to a slightly larger (and more suitable) one. A log pair (X, Δ) is *divisorially log terminal* if there exists a log resolution ϕ such that $a(E, X, \Delta) > 0$ for any ϕ-exceptional divisor E.

It is a good exercise to show that if (X, Δ) is divisorially log terminal and A is ample, there exists $\Delta' \sim_{\mathbb{Q}} \Delta + A$ such that (X, Δ') is Kawamata log terminal.

Before we proceed, we need the following result, due to Shokurov

Theorem 5.1 (Special Termination) *Assume that the main results of the Minimal Model Program hold in dimension $\leq n-1$. Assume that (X, Δ) is a divisorially log terminal pair of dimension n. Let*

$$X = X_0 \dashrightarrow X_1 \dashrightarrow \dots$$

be a sequence of $(K+\Delta)$-flips. Let $S = \llcorner \Delta \lrcorner$. and let S_i be the strict transform of S on X_i.

Then, if $i \gg 0$, the exceptional locus of $X_i \dashrightarrow X_{i+1}$ is disjoint from S_i.

Proof We omit the proof. It uses similar ideas as Theorem 3.1. See [7, Theorem 4.2.1] for more details. □

We begin by considering a very special case:

Lemma 5.2 (Key Lemma) *Let (X, Δ) be a log pair and let $D, H \geq 0$ be effective $\mathbb{Q}$-divisors such that*

(1) $\Delta = S + A + B$, where $\llcorner \Delta \lrcorner. = S$, $A \geq 0$ is an ample $\mathbb{Q}$-divisor and $B \geq 0$,
(2) $(X, \Delta + H)$ is divisorially log terminal (in particular S and Supp *H do not have any common component) and $K_X + \Delta + H$ is nef, and*
(3) the support of D is contained in S and there exists $\alpha \geq 0$ such that

$$K_X + \Delta \sim D + \alpha H.$$

Then the $(K_X + \Delta)$-Minimal model program with scaling of H terminates.

Sketch of the proof We define λ as above, and we consider a rational curve ξ such that

$$(K_X + \Delta) \cdot \xi < 0 \quad \text{and}$$
$$(K_X + \Delta + \lambda H) \cdot \xi = 0.$$

In particular, it follows immediately that $H \cdot \xi > 0$. Thus, (3) implies that $D \cdot \xi < 0$ and therefore ξ is contained in the support of D. Applying (3) again, it follows that ξ is contained in S. Thus, special termination (cf. Theorem 5.1) implies termination for the MMP with scaling and we are done. □

Note that special termination holds under the assumption that all the results of the minimal model program hold true in dimension lower than the dimension of X. On the other hand, it is possible to use a "special" version of special termination which holds unconditionally.

We now present the main ideas used in the proof of Theorem 4.4. Let (X, Δ) be a divisorially log terminal pair such that Δ is big and $K_X + \Delta$ is pseudo-effective. It is possible (but not easy) to show that there exists a $\mathbb{Q}$-divisor D such that $K_X + \Delta \sim_{\mathbb{Q}} D \geq 0$. We further assume that the pair $(X, \Delta + D)$ is log smooth.

We proceed by induction on the following positive integer

$$k := \# \text{ of irreducible components of } D \text{ which are not contained in } \llcorner \Delta \lrcorner.$$

We first consider the case $k = 0$. This means that all the support of D is contained in $\llcorner \Delta \lrcorner$. In this case, we take an ample divisor H such that $K_X + \Delta + H$ is ample. Thus, Lemma 5.2 immediately implies that the pair (X, Δ) admits a minimal model and we are done.

We now assume that $k > 0$. Then we may write

$$D = D_1 + D_2 \qquad \text{with } D_1, D_2 \geq 0$$

where the components of D_2 are exactly all the components of D which are not contained in $\llcorner \Delta \lrcorner$.. We define:

$$\lambda := \sup\{ t \in [0, 1] \mid (X, \Delta + tD_2) \text{ is log canonical} \}.$$

By assumption, we have $\lambda > 0$ and it follows that the number of irreducible components of $D + \lambda D_2$ which are not contained in $\llcorner \Delta + \lambda D_2 \lrcorner$ is less than k. Thus, by induction, it follows that the pair $(X, \Delta + \lambda D_2)$ admits a minimal model $X \dashrightarrow X'$. After replacing X by X', and with a bit of work,[3] it is possible to assume that if

$$\Theta = \Delta + \lambda D_2$$

then $K_X + \Theta$ is nef. Let $H = \lambda D_2$. We have

$$K_X + \Delta \sim_{\mathbb{Q}} D_1 + \frac{1}{\lambda} H$$

and the support of D_1 is contained in $\lfloor \Delta \rfloor$.. Thus, we may apply Lemma 5.2 again and we obtain that (X, Δ) admits a minimal model.

As a final remark, note that Theorem 4.4 can be proven as a consequence of finite generation [5, 8]. We refer to [4] for a short survey.

[3]The main idea, on why we can do this, relies on the fact that any divisor contracted by the birational contraction $X \dashrightarrow X'$ in contained in the stable base locus of $K_X + \Delta$.

6 Flops and Sarkisov Program

We have seen that in general projective varieties X [and Kawamata log terminal pairs (X, Δ)] are divided into two large families, depending on the existence of a global section of mK_X [respectively $m(K_X + \Delta)$] for some positive integer m. If there is no such a section, than the picture is more complicated, as the vertices might have infinitely many edges starting from it.

Now it is natural to investigate the relations within the end-points of the directed graphs we constructed. First, assume that X is an end-point in the absolute case (i.e. with $\Delta = 0$), represented by a terminal projective variety such that K_X is nef. Then by a result of Kawamata [15], any other end-point Y in the connected component of X is connected to X by a composition of flops (see [16] for a definition of flop). Note that a flop $\varphi: X \dashrightarrow Y$ is a special K-trivial isomorphism in codimension 1, i.e. if $p: W \to X$ and $q: W \to Y$ are proper morphisms which resolve the indeterminacy locus of φ, then

$$p^* K_X = q^* K_Y.$$

A similar picture holds for log pairs (see [15] for more details).

It is conjectured that any two terminal projective varieties which are isomorphic in codimension 1 and K-equivalent (i.e. they admit a K-trivial birational map which is an isomorphism in codimension 1) are connected by a finite sequence of flops (e.g. see [20]). At the moment, the conjecture is open even in the case of smooth projective varieties of dimension 3.

Finally, if X is a terminal projective variety of general type, then the number of end-points in the connected component of the graph containing X is always finite [2].

We now consider the case of uniruled projective varieties. We have seen that even in dimension 2, the number of end-points in each connected component of the graph which contains a uniruled projective variety is infinite. It is therefore natural to ask about the relation within birational pairs of projective varieties which admit a Mori fiber space. More specifically let $\eta: X \to Z$ and $\eta': X \to Z'$ be two Mori fiber spaces, with X and X' terminal projective varieties which admit a birational map $\psi: X \dashrightarrow Y$. Then the goal of the Sarkisov programme is to show that φ can be decomposed into a sequence of *Sarkisov links*, which are elementary transformations obtained as compositions of flops and divisorial contractions (e.g. see [12] for more details). The programme was successfully carried out in [3, 6] in dimension 3 and in [12] in full generality.

Acknowledgements These are the notes for the CIMPA-CIMAT-ICTP School "Moduli of Curves" in Guanajuato, México, 22 February–4 March 2016. I would like to thank the organisers and all the participants for the invitation and for giving me the opportunity to present this material at the school. I would also like to thank the referee for reading a preliminary version of these notes and providing many useful comments.

References

1. C. Birkar, Ascending chain condition for log canonical thresholds and termination of log flips. Duke Math. J. **136**(1), 173–180 (2007)
2. C. Birkar, P. Cascini, C.D. Hacon, J. McKernan, Existence of minimal models for varieties of log general type. J. Am. Math. Soc. **23**(2), 405–468 (2010)
3. A. Bruno, K. Matsuki, Log Sarkisov program. Int. J. Math. **8**(4), 451–494 (1997)
4. P. Cascini, V. Lazić, The minimal model program revisited, in *Contributions to Algebraic Geometry* (European Mathematical Society, Zurich, 2012), pp. 169–187
5. P. Cascini, V. Lazić, New outlook on the minimal model program, I. Duke Math. J. **161**(12), 2415–2467 (2012)
6. A. Corti, Factoring birational maps of threefolds after Sarkisov. J. Algebr. Geom. **4**(2), 223–254 (1995)
7. A. Corti (ed.), *Flips for 3-Folds and 4-Folds*. Oxford Lecture Series in Mathematics and its Applications, vol. 35 (Oxford University Press, Oxford, 2007)
8. A. Corti, V. Lazić, New outlook on the minimal model program, II. Math. Ann. **356**(2), 617–633 (2013)
9. O. Fujino, S. Mori, A canonical bundle formula. J. Differ. Geom. **56**(1), 167–188 (2000)
10. Y. Gongyo, On weak Fano varieties with log canonical singularities. J. Reine Angew. Math. **665**, 237–252 (2012)
11. C.D. Hacon, J. McKernan, Existence of minimal models for varieties of log general type II. J. Am. Math. Soc. **23**(2), 469–490 (2010)
12. C. Hacon, J. McKernan, The Sarkisov program. J. Algebr. Geom. **22**(2), 389–405 (2013)
13. C. Hacon, J. McKernan, C. Xu, Acc for log canonical thresholds (2012)
14. Y. Kawamata, On the length of an extremal rational curve. Invent. Math. **105**, 609–611 (1991)
15. Y. Kawamata, Flops connect minimal models. Publ. Res. Inst. Math. Sci. **44**(2), 419–423 (2008)
16. J. Kollár, S. Mori, *Birational Geometry of Algebraic Varieties*. Cambridge Tracts in Mathematics, vol. 134 (Cambridge University Press, Cambridge, 1998)
17. J. Kollár et al., Flips and abundance for algebraic threefolds, Société Mathématique de France, Paris, 1992
18. R. Lazarsfeld, *Positivity in Algebraic Geometry. I*. Ergebnisse der Mathematik und ihrer Grenzgebiete, vol. 48 (Springer, Berlin, 2004)
19. B. Lehmann, A cone theorem for NEF curves. J. Algebr. Geom. **21**(3), 473–493 (2012)
20. C.-L. Wang, *K*-equivalence in birational geometry and characterizations of complex elliptic genera. J. Algebr. Geom. **12**(2), 285–306 (2003)

Higher-Dimensional Varieties

Lecture Notes for the CIMPA-CIMAT-ICTP School on Moduli of Curves, February 22–March 4, 2016, Guanajuato, México

Olivier Debarre

1 Introduction

Mori's Minimal Model Program (MMP) is a classification program: given a (smooth) projective variety X, the aim is to find a "simple" birational model of X, ideally one whose canonical bundle is nef (although this is not possible for varieties covered by rational curves). Mori's original approach gave a prominent role to rational curves. Although it is not sufficient to complete his program (essentially because it cannot deal with the necessary evil of singular varieties), it contains beautiful geometric results which have their own interest and which are also much more accessible than the latest developments of the MMP.

Assuming that the reader is familiar with the basics of algebraic geometry (e.g., the contents of the book [8]), we present in these notes the necessary material (and a bit more) to understand Mori's cone theorem.

In Sect. 2, we review Weil and Cartier divisors and linear equivalence (this is covered in [8]). We explain the relation between Cartier divisors and invertible sheaves and define the Picard group. We define the intersection number between a Cartier divisor and a curve. This is a fundamental tool: it defines numerical equivalence, an equivalence relation on the group of Cartier divisors weaker than linear equivalence. The quotient space is therefore a quotient of the Picard group and, for proper varieties, it is a free abelian group of finite rank.

We explain the standard relation between linear systems and rational maps to projective spaces (also covered in [8]). We also explain global generation of coherent sheaves and define ample ($\mathbf{Q}$)-Cartier divisors on a scheme of finite type over a field. We prove Serre's theorems.

O. Debarre (✉)
École normale supérieure, 45 rue d'Ulm, 75230 Paris cedex 05, France
e-mail: olivier.debarre@ens.fr

L. Brambila Paz et al. (eds.), *Moduli of Curves*, Lecture Notes of the Unione Matematica Italiana 21, DOI 10.1007/978-3-319-59486-6_3

After proving a Riemann–Roch theorem on a smooth projective curve, we prove that ample divisors on a smooth projective curve are those of positive degree. Finally, we define, on any projective scheme X, nef divisors as those having non-negative intersection number with any curve. We then define, in the (finite-dimensional real) vector space $N^1(X)_{\mathbf{R}}$, the ample, nef, big, effective, and pseudo-effective cones. In order to shorten the exposition, we accept without proof that the sum of an ample divisor and a nef divisor is still ample. This is not very satisfactory since this result is usually obtained as a consequence of the material in the next section, but I did not have time to follow the standard (and logical) path.

Section 3 is devoted to asymptotic Riemann–Roch theorems: given a Cartier divisor D on a projective scheme X of dimension n, how fast does the dimension $h^0(X, mD)$ of the space of sections of its positive multiples mD grow? It is easier to deal with the Euler characteristic $\chi(X, mD)$ instead, which grows like $am^n/n! +$ lower order terms, where a is an integer which is by definition the self-intersection number (D^n). This is a fast way to define this product, and more generally the intersection number of n Cartier divisors $D_1, \ldots, D_n$ on X. This number turns out to count, when these divisors are effective and meet in only finitely many points, these intersection points with multiplicities. When D is nef, this is also the behavior of $h^0(X, mD)$. We end this section by defining and discussing the Kodaira dimension of a Cartier divisor. We define algebraic fibrations and explain what the Iitaka fibration is.

In Sect. 4, we discuss the moduli space of morphisms from a fixed (smooth projective) curve to a projective variety and construct it when the curve is $\mathbf{P}^1$. We explain without proof its local structure. We discuss free rational curves and uniruled varieties (i.e., varieties covered by rational curves). We state (without proof) Mori's bend-and-break lemmas and explain in more details Mori's beautiful proof of the fact that Fano varieties are uniruled (the nice part is the reduction to positive characteristics). In the last section, we explain an extension, based on a classical result of Miyaoka and Mori, of Mori's result to varieties with nef but not numerically trivial anticanonical bundle.

The fifth and last section is devoted to the proof of the cone theorem and its various consequences. The cone theorem describes the structure of the closed convex cone spanned by classes of irreducible curves (the "Mori cone") in the dual $N_1(X)_{\mathbf{R}}$ of the (finite-dimensional real) vector space $N^1(X)_{\mathbf{R}}$, for a smooth projective variety X. It is an elementary consequence of the Miyaoka–Mori theorem mentioned above and uses only elementary geometrical facts on the geometry of closed convex cones in finite-dimensional real vector spaces. We state without proof Kawamata's base-point-free theorem and explain how it allows us to construct, in characteristic 0, contractions of some extremal rays of the Mori cone: these are algebraic fibrations c from X to a projective variety which contract exactly the curves whose class is in the ray.

These fibrations are of three types: fiber-type (when all fibers of c have positive dimensions), divisorial (when c is birational with exceptional locus a divisor), small (when c is birational with exceptional locus of codimension ≥ 2). We remark that the image $c(X)$ of the contraction may be singular, but not too singular, except in the case of a small contraction. We end these notes with a two-page description of what the MMP is about and explain what the main problems are.

We prove some, but far from all, of the results we state. The bibliography provides a few references where the reader can find more detailed expositions. There are also a few exercises throughout this text.

2 Divisors

In this section and the rest of these notes, **k** is a field and a **k**-*variety* is an integral scheme of finite type over **k**.

2.1 Weil and Cartier Divisors

In this section, X is a scheme which is for simplicity assumed to be integral.[1]

A *Weil divisor* on X is a (finite) formal linear combination with integral coefficients of integral hypersurfaces in X. Its *support* is the union of the hypersurfaces which appear with non-zero coefficients. We say that the divisor is *effective* if the coefficients are all non-negative.

Assume moreover that X is *normal*. For each integral hypersurface Y of X with generic point η, the integral local ring $\mathcal{O}_{X,\eta}$ has dimension 1 and is regular, hence is a discrete valuation ring with valuation v_Y. For any non-zero rational function f on X, the integer $v_Y(f)$ (valuation of f along Y) is the order of vanishing of f along Y if it is non-negative, and the opposite of the order of the pole of f along Y otherwise. We define the divisor of f as

$$\operatorname{div}(f) = \sum_Y v_Y(f) Y.$$

The rational function f is regular if and only if its divisor is effective [8, Proposition II.6.3A].

2.1. Linearly Equivalent Weil Divisors Two Weil divisors D and D' on the normal scheme X are *linearly equivalent* if their difference is the divisor of a non-zero rational function on X; we write $D \underset{\text{lin}}{\equiv} D'$. Linear equivalence classes of Weil divisors form a group $\operatorname{Cl}(X)$ (the *divisor class group*) for the addition of divisors.

A *Cartier divisor* is a divisor which can be locally written as the divisor of a non-zero rational function. The formal definition is less enlightening.

Definition 2.2 (Cartier Divisors) A Cartier divisor on an integral scheme X is a global section of the sheaf $\underline{K(X)}^\times / \mathcal{O}_X^\times$, where $\underline{K(X)}$ is the constant sheaf of rational functions on X.

[1] The definitions can be given for any scheme, but they take a slightly more complicated form.

In other words, a Cartier divisor is given by a collection of pairs (U_i, f_i), where (U_i) is an affine open cover of X and f_i is a non-zero rational function on U_i such that f_i/f_j is a regular function on $U_i \cap U_j$ that does not vanish.

A Cartier divisor on X is *principal* if it can be defined by a global non-zero rational function on the whole of X.

2.3. Associated Weil Divisor Assume that X is normal. Given a Cartier divisor on X, defined by a collection (U_i, f_i), one can consider the associated Weil divisor $\sum_Y n_Y Y$ on X, where the integer n_Y is the valuation of f_i along $Y \cap U_i$ for any i such that $Y \cap U_i$ is non-empty (it does not depend on the choice of such an i).

A Weil divisor which is linearly equivalent to a Cartier divisor is itself a Cartier divisor.

When X is *locally factorial* (e.g., a smooth variety), i.e., its local rings are unique factorization domains, any hypersurface can be defined locally by one (regular) equation [8, Proposition II.6.11],[2] hence any divisor is locally the divisor of a rational function. In other words, there is no distinction between Cartier divisors and Weil divisors.

2.4. Effective Cartier Divisors A Cartier divisor D is *effective* if it can be defined by a collection (U_i, f_i), where f_i is in $\mathscr{O}_X(U_i)$. We write $D \geq 0$. When D is not zero, it defines a subscheme of codimension 1 by the "equation" f_i on each U_i. We still denote it by D.

2.5. Q-Divisors A Weil **Q**-divisor on a scheme X is a (finite) formal linear combination with *rational* coefficients of integral hypersurfaces in X. On a normal scheme X, one says that a **Q**-divisor is **Q**-*Cartier* if some multiple with integral coefficients is a Cartier divisor.

Example 2.6 Let X be the quadric cone defined in $\mathbf{A}^3_{\mathbf{k}}$ by the equation $xy = z^2$. It is integral and normal. The line L defined by $x = z = 0$ is contained in X; it defines a Weil divisor on X which cannot be defined near the origin by one equation (the ideal (x, z) is not principal in the local ring of X at the origin). It is therefore not a Cartier divisor. However, $2L$ is a principal Cartier divisor, defined by the regular function x, hence L is a **Q**-Cartier divisor. Similarly, the sum of L with the line defined by $y = z = 0$ is also a principal Cartier divisor, defined by the regular function z. So the "components" of a Cartier divisor need not be Cartier.

2.2 *Invertible Sheaves*

Definition 2.7 (Invertible Sheaves) An invertible sheaf on a scheme X is a locally free $\mathscr{O}_X$-module of rank 1.

[2]This is because in a unique factorization domain, prime ideals of height 1 are principal.

The terminology comes from the fact that the tensor product defines a group structure on the set of locally free sheaves of rank 1 on X, where the inverse of an invertible sheaf $\mathscr{L}$ is $\mathscr{H}om(\mathscr{L}, \mathscr{O}_X)$. This makes the set of isomorphism classes of invertible sheaves on X into an abelian group called the *Picard group* of X and denoted by $\operatorname{Pic}(X)$. For any $m \in \mathbf{Z}$, it is traditional to write $\mathscr{L}^m$ for the mth (tensor) power of $\mathscr{L}$ (so in particular, $\mathscr{L}^{-1}$ is the dual of $\mathscr{L}$).

Let $\mathscr{L}$ be an invertible sheaf on X. We can cover X with affine open subsets U_i on which $\mathscr{L}$ is trivial and we obtain changes of trivializations, or transition functions

$$g_{ij} \in \mathscr{O}_X^\times(U_i \cap U_j). \tag{1}$$

They satisfy the cocycle condition

$$g_{ij}g_{jk}g_{ki} = 1$$

hence define a Čech 1-cocycle for $\mathscr{O}_X^\times$. One checks that this induces an isomorphism

$$\operatorname{Pic}(X) \simeq H^1(X, \mathscr{O}_X^\times). \tag{2}$$

For any $m \in \mathbf{Z}$, the invertible sheaf $\mathscr{L}^m$ corresponds to the collection of transition functions $(g_{ij}^m)_{i,j}$.

2.8. Invertible Sheaf Associated with a Cartier Divisor Given a Cartier divisor D on an integral scheme X, given by a collection (U_i, f_i), one can construct an invertible subsheaf $\mathscr{O}_X(D)$ of $\underline{K(X)}$ by taking the sub-$\mathscr{O}_X$-module generated by $1/f_i$ on U_i. We have

$$\mathscr{O}_X(D_1) \otimes \mathscr{O}_X(D_2) \simeq \mathscr{O}_X(D_1 + D_2).$$

Every invertible subsheaf of $\underline{K(X)}$ is obtained in this way and two Cartier divisors are linearly equivalent if and only if their associated invertible sheaves are isomorphic [8, Proposition II.6.13]. Since X is integral, every invertible sheaf is a subsheaf of $\underline{K(X)}$ [8, Remark II.6.14.1 and Proposition II.6.15], so we get an isomorphism of groups

$$\{\text{Cartier divisors on } X, +\}/\underset{\text{lin}}{\equiv} \;\simeq \{\text{Invertible sheaves on } X, \otimes\}/\text{isom.} = \operatorname{Pic}(X).$$

In some sense, Cartier divisors and invertible sheaves are more or less the same thing. However, we will try to use as often as possible the (additive) language of divisors instead of that of invertible sheaves; this allows for example for $\mathbf{Q}$-divisors (which have no analogs in terms of sheaves).

We will write $H^i(X, D)$ instead of $H^i(X, \mathscr{O}_X(D))$ and $\mathscr{F}(D)$ instead of $\mathscr{F} \otimes_{\mathscr{O}_X} \mathscr{O}_X(D)$, if $\mathscr{F}$ is an $\mathscr{O}_X$-module.

Assume that X is moreover normal. One has

$$H^0(X, \mathscr{O}_X(D)) \simeq \{ f \in K(X) \mid f = 0 \text{ or } \operatorname{div}(f) + D \geq 0\}. \quad (3)$$

Indeed, if (U_i, f_i) represents D, and f is a non-zero rational function on X such that $\operatorname{div}(f) + D$ is effective, ff_i is regular on U_i (because X is normal!), and $f|_{U_i} = (ff_i)\frac{1}{f_i}$ defines a section of $\mathscr{O}_X(D)$ over U_i. Conversely, any global section of $\mathscr{O}_X(D)$ is a rational function f on X such that, on each U_i, the product $f|_{U_i} f_i$ is regular. Hence $\operatorname{div}(f) + D$ effective.

Remark 2.9 Let D be a non-zero effective Cartier divisor on X. If we still denote by D the subscheme of X that it defines (see 2.4), we have an exact sequence of sheaves[3]

$$0 \to \mathscr{O}_X(-D) \to \mathscr{O}_X \to \mathscr{O}_D \to 0.$$

Remark 2.10 Going back to Definition 2.2 of Cartier divisors, one checks that the morphism

$$\begin{aligned} H^0(X, \underline{K(X)}^\times / \mathscr{O}_X^\times) &\longrightarrow H^1(X, \mathscr{O}_X^\times) \\ D &\longmapsto [\mathscr{O}_X(D)] \end{aligned}$$

induced by (2) is the coboundary of the long exact sequence in cohomology induced by the short exact sequence

$$0 \to \mathscr{O}_X^\times \to \underline{K(X)}^\times \to \underline{K(X)}^\times / \mathscr{O}_X^\times \to 0.$$

Principal divisors correspond to the image of $K(X)^\times$ in $H^0(X, \underline{K(X)}^\times / \mathscr{O}_X^\times)$.

Example 2.11 An integral hypersurface Y in $\mathbf{P}^n_{\mathbf{k}}$ corresponds to a homogeneous prime ideal of height 1 in $\mathbf{k}[x_0, \ldots, x_n]$, which is therefore (since the ring $\mathbf{k}[x_0, \ldots, x_n]$ is factorial) principal. Hence Y is defined by one (homogeneous) irreducible equation f of degree d (called the *degree of* Y). This defines a surjective morphism

$$\{\text{Cartier divisors on } \mathbf{P}^n_{\mathbf{k}}\} \to \mathbf{Z}.$$

Since f/x_0^d is a rational function on $\mathbf{P}^n_{\mathbf{k}}$ with divisor $Y - dH_0$ (where H_0 is the hyperplane defined by $x_0 = 0$), Y is linearly equivalent to dH_0. Conversely, the divisor of any rational function on $\mathbf{P}^n_{\mathbf{k}}$ has degree 0 (because it is the quotient of two

[3] Let i be the inclusion of D in X. Since this is an exact sequence of sheaves *on* X, the sheaf on the right should be $i_*\mathscr{O}_D$ (a sheaf on X with support on D). However, it is customary to drop i_*. Note that as far as cohomology calculations are concerned, this does not make any difference [8, Lemma III.2.10].

homogeneous polynomials of the same degree), hence we obtain an isomorphism

$$\mathrm{Pic}(\mathbf{P}^n_{\mathbf{k}}) \simeq \mathbf{Z}.$$

We denote by $\mathscr{O}_{\mathbf{P}^n_{\mathbf{k}}}(d)$ the invertible sheaf corresponding to an integer d (it is $\mathscr{O}_{\mathbf{P}^n_{\mathbf{k}}}(D)$ for any divisor D of degree d). One checks that the space of global sections of $\mathscr{O}_{\mathbf{P}^n_{\mathbf{k}}}(d)$ is 0 for $d < 0$ and isomorphic to the vector space of homogeneous polynomials of degree d in $\mathbf{k}[x_0, \dots, x_n]$ for $d \geq 0$. More intrinsically, for any finite dimensional $\mathbf{k}$-vector space W, one has

$$H^0(\mathbf{P}(W), \mathscr{O}_{\mathbf{P}(W)}(d)) = \begin{cases} \mathrm{Sym}^d W^\vee & \text{if } d \geq 0, \\ 0 & \text{if } d < 0. \end{cases}$$

Exercise 2.12 Let X be a normal variety. Prove

$$\mathrm{Pic}(X \times \mathbf{P}^n_{\mathbf{k}}) \simeq \mathrm{Pic}(X) \times \mathbf{Z}$$

(*Hint:* proceed as in [8, Proposition II.6.6 and Example II.6.6.1]). In particular,

$$\mathrm{Pic}(\mathbf{P}^m_{\mathbf{k}} \times \mathbf{P}^n_{\mathbf{k}}) \simeq \mathbf{Z} \times \mathbf{Z}.$$

This can be seen directly as in Example 2.11 by proving first that any hypersurface in $\mathbf{P}^m_{\mathbf{k}} \times \mathbf{P}^n_{\mathbf{k}}$ is defined by a *bihomogeneous* polynomial in $((x_0, \dots, x_m), (y_0, \dots, y_n))$.

2.13. Pullback Let $\pi\colon Y \to X$ be a morphism between integral schemes and let D be a Cartier divisor on X. The pullback $\pi^*\mathscr{O}_X(D)$ is an invertible subsheaf of $\underline{K(Y)}$ hence defines a linear equivalence class of divisors on Y (improperly) denoted by π^*D. Only the linear equivalence class of π^*D is well-defined in general; however, when D is a divisor (U_i, f_i) whose support does not contain the image $\pi(Y)$, the collection $(\pi^{-1}(U_i), f_i \circ \pi)$ defines a divisor π^*D in that class. In particular, it makes sense to restrict a Cartier divisor to a subvariety not contained in its support, and to restrict a Cartier divisor *class* to any subvariety.

2.3 Intersection of Curves and Divisors

2.14. Curves A curve is a projective variety of dimension 1. On a smooth curve C, a (Cartier) divisor D is just a finite formal linear combination of closed points $\sum_{p \in C} n_p p$. We define its degree to be the integer $\sum n_p [k(p) : \mathbf{k}]$. If D is effective ($n_p \geq 0$ for all p), we can view it as a 0-dimensional subscheme of X with (affine) support the set of points p for which $n_p > 0$, where it is defined by the ideal $\mathfrak{m}_{X,p}^{n_p}$

(see 2.4). We have

$$h^0(D, \mathscr{O}_D) = \sum_p \dim_{\mathbf{k}}(\mathscr{O}_{X,p}/\mathfrak{m}_{X,p}^{n_p}) = \sum_p n_p \dim_{\mathbf{k}}(\mathscr{O}_{X,p}/\mathfrak{m}_{X,p}) = \deg(D). \tag{4}$$

This justifies the seemingly strange definition of the degree.

One proves (see [8, Corollary II.6.10]) that the degree of the divisor of a regular function is 0, hence the degree factors through

$$\mathrm{Pic}(C) \simeq \{\text{Cartier divisors on } C\}/\underset{\text{lin}}{\equiv} \to \mathbf{Z}.$$

Let X be a variety. It will be convenient to define a *curve on* X as a morphism $\rho: C \to X$, where C is a smooth (projective) curve. Given any, possibly singular, curve in X, one may consider its normalization as a "curve on X." For any Cartier divisor D on X, we set

$$(D \cdot C) := \deg(\rho^* D). \tag{5}$$

This definition extends to **Q**-Cartier **Q**-divisors, but this intersection number is then only a rational number in general.

An important remark is that when D is effective and $\rho(C)$ is not contained in its support, this number is non-negative (this is because the class $\rho^* D$ can be represented by an effective divisor on C; see 2.13). In general, it is 0 whenever $\rho(C)$ does not meet the support of D.

If $\pi: X \to Y$ is a morphism between varieties, a curve $\rho: C \to X$ can also be considered as a curve on Y via the composition $\pi\rho: C \to Y$. If D is a Cartier divisor on Y, we have the so-called *projection formula*

$$(\pi^* D \cdot C)_X = (D \cdot C)_Y. \tag{6}$$

2.15. Numerically Equivalent Divisors Let X be a variety. We say that two **Q**-Cartier **Q**-divisors D and D' on X are *numerically equivalent* if

$$(D \cdot C) = (D' \cdot C)$$

for all curves $C \to X$. Linearly equivalent Cartier divisors are numerically equivalent. Numerical equivalence classes of Cartier divisors form a torsion-free abelian group for the addition of divisors, denoted by $N^1(X)$; it is a quotient of the Picard group $\mathrm{Pic}(X)$.

One can also define the numerical equivalence class of a **Q**-Cartier **Q**-divisor in the **Q**-vector space

$$N^1(X)_{\mathbf{Q}} := N^1(X) \otimes \mathbf{Q}.$$

Let $\pi: Y \to X$ be a morphism between varieties. By the projection formula (6), pullback of Cartier divisors induces a $\mathbf{Q}$-linear map

$$\pi^*: N^1(Y)_{\mathbf{Q}} \to N^1(X)_{\mathbf{Q}}.$$

Theorem 2.16 *If X is a* proper *variety, the group $N^1(X)$ is free abelian of finite rank, called the* Picard number *of X and denoted by $\rho(X)$.*

This is proved in [10, Proposition 3, p. 334]. Over the complex numbers, we will see in Sect. 3.3 that $N^1(X)_{\mathbf{Q}}$ is a subspace of (the finite-dimensional vector space) $H^2(X, \mathbf{Q})$.

We will prove that many important properties of Cartier divisors are numerical, in the sense that they only depend on their numerical equivalence class.

Example 2.17 (Curves) If X is a curve, any curve $\rho: C \to X$ factors through the normalization $\nu: \widehat{X} \to X$ and, for any Cartier divisor D on X, one has $\deg(\rho^*D) = \deg(\nu^*D)\deg(\rho)$. The numerical equivalence class of a divisor is therefore given by its degree on $\widehat{X}$, hence

$$N^1(X) \xrightarrow{\sim} \mathbf{Z}.$$

Example 2.18 (Blow Up of a Point) One deduces from Example 2.11 isomorphisms

$$\operatorname{Pic}(\mathbf{P}^n_{\mathbf{k}}) \simeq N^1(\mathbf{P}^n_{\mathbf{k}}) \simeq \mathbf{Z}[H].$$

Let O be a point of $\mathbf{P}^n_{\mathbf{k}}$ and let $\varepsilon: \widetilde{\mathbf{P}}^n_{\mathbf{k}} \to \mathbf{P}^n_{\mathbf{k}}$ be its blow up. If H_0 is a hyperplane in $\mathbf{P}^n_{\mathbf{k}}$ which does not contain O, it can be defined as

$$\widetilde{\mathbf{P}}^n_{\mathbf{k}} = \{(x, y) \in \mathbf{P}^n_{\mathbf{k}} \times H_0 \mid x \in \langle Oy \rangle\}$$

and ε is the first projection. The fiber $E := \varepsilon^{-1}(O) \subset \widetilde{\mathbf{P}}^n_{\mathbf{k}} \simeq H_0$ is called the *exceptional divisor* of the blow up and ε induces an isomorphism $\widetilde{\mathbf{P}}^n_{\mathbf{k}} \smallsetminus E \xrightarrow{\sim} \mathbf{P}^n_{\mathbf{k}} \smallsetminus \{O\}$. Assume $n \geq 2$. By [8, Proposition II.6.5], we have isomorphisms $\operatorname{Pic}(\widetilde{\mathbf{P}}^n_{\mathbf{k}} \smallsetminus E) \simeq \operatorname{Pic}(\mathbf{P}^n_k \smallsetminus \{O\}) \simeq \operatorname{Pic}(\mathbf{P}^n_k)$ and an exact sequence

$$\mathbf{Z} \xrightarrow{\alpha} \operatorname{Pic}(\widetilde{\mathbf{P}}^n_{\mathbf{k}}) \to \operatorname{Pic}(\widetilde{\mathbf{P}}^n_{\mathbf{k}} \smallsetminus E) \to 0,$$

where $\alpha(m) = [mE]$.

Let L be a line contained in E and let $H \subset \mathbf{P}^n_{\mathbf{k}}$ be a hyperplane. If H does not contain O, we have $(\varepsilon^*H \cdot L) = 0$. If it does contain O, one checks that ε^*H can be written as $H' + E$, where H' meets E along a hyperplane in E. In particular, $(H' \cdot L) = 1$. This implies

$$0 = (\varepsilon^*H \cdot L) = ((H' + E) \cdot L) = (E \cdot L) + 1,$$

hence $(E \cdot L) = -1$. In particular, the map α is injective and we obtain

$$\mathrm{Pic}(\widetilde{\mathbf{P}^n_{\mathbf{k}}}) \simeq N^1(\widetilde{\mathbf{P}^n_{\mathbf{k}}}) \simeq \mathbf{Z}[\varepsilon^*H] \oplus \mathbf{Z}[E].$$

2.4 Line Bundles

A *line bundle* on a scheme X is a scheme L with a morphism $\pi: L \to X$ which is locally (on the base) "trivial", i.e., isomorphic to $\mathbf{A}^1_U \to U$, in such a way that the changes of trivializations are linear, i.e., given by $(x,t) \mapsto (x, \varphi(x)t)$, for some $\varphi \in \mathscr{O}_X^\times(U)$. A *section* of $\pi: L \to X$ is a morphism $s: X \to L$ such that $\pi \circ s = \mathrm{Id}_X$. One checks that the sheaf of sections of $\pi: L \to X$ is an invertible sheaf on X. Conversely, to any invertible sheaf $\mathscr{L}$ on X, one can associate a line bundle on X: if $\mathscr{L}$ is trivial on an affine cover (U_i), just glue the $\mathbf{A}^1_{U_i}$ together, using the g_{ij} of (1). It is common to use the words "invertible sheaf" and "line bundle" interchangeably.

Assume that X is integral and normal. A non-zero section s of a line bundle $L \to X$ defines an effective Cartier divisor on X (by the equation $s = 0$ on each affine open subset of X over which L is trivial), which we denote by $\mathrm{div}(s)$. With the interpretation (3) of the section s, if D is a Cartier divisor on X and L is the line bundle associated with $\mathscr{O}_X(D)$, we have

$$\mathrm{div}(s) = \mathrm{div}(f) + D.$$

In particular, if D is effective, the function $f = 1$ corresponds to a section of $\mathscr{O}_X(D)$ with divisor D. In general, any non-zero rational function f on X can be seen as a (regular, nowhere vanishing) section of the line bundle $\mathscr{O}_X(-\mathrm{div}(f))$.

Example 2.19 Let $\mathbf{k}$ be a field and let W be a $\mathbf{k}$-vector space. We construct a line bundle $L \to \mathbf{P}(W)$ whose fiber above a point x of $\mathbf{P}(W)$ is the line ℓ_x of W represented by x by setting

$$L := \{(x, v) \in \mathbf{P}(W) \times W \mid v \in \ell_x\}.$$

On the standard open set U_i (defined after choice of a basis for W), L is defined in $U_i \times W$ by the equations $v_j = v_i x_j$, for all $j \neq i$. The trivialization on U_i is given by $(x, v) \mapsto (x, v_i)$, so that $g_{ij}(x) = x_i/x_j$, for $x \in U_i \cap U_j$. One checks that this line bundle corresponds to $\mathscr{O}_{\mathbf{P}W}(-1)$ (see Example 2.11).

2.5 Linear Systems and Morphisms to Projective Spaces

Let X be a normal variety. Let $\mathscr{L}$ be an invertible sheaf on X and let $|\mathscr{L}|$ be the set of (effective) divisors of non-zero global sections of $\mathscr{L}$. It is called the *linear system* associated with $\mathscr{L}$. The quotient of two sections which have the same divisor is a regular function on X which does not vanish. *If X is projective*, this quotient is constant and the map div: $\mathbf{P}(H^0(X,\mathscr{L})) \to |\mathscr{L}|$ is therefore bijective.

Let D be a Cartier divisor on X. We write $|D|$ instead of $|\mathscr{O}_X(D)|$; it is the set of effective divisors on X which are linearly equivalent to D.

2.20. Morphisms to a Projective Space We now come to a very important point: the link between morphisms from X to a projective space and vector spaces of sections of invertible sheaves on the normal projective variety X.

Let W be a $\mathbf{k}$-vector space of finite dimension and let $\psi: X \to \mathbf{P}(W)$ be a regular map. Consider the invertible sheaf $\mathscr{L} = \psi^*\mathscr{O}_{\mathbf{P}(W)}(1)$ and the linear map

$$H^0(\psi): W^\vee \simeq H^0\big(\mathbf{P}(W), \mathscr{O}_{\mathbf{P}(W)}(1)\big) \to H^0(X,\mathscr{L}).$$

A section of $\mathscr{O}_{\mathbf{P}W}(1)$ vanishes on a hyperplane; its image by $H^0(\psi)$ is zero if and only if $\psi(X)$ is contained in this hyperplane. In particular, $H^0(\psi)$ is injective if and only if $\psi(X)$ is not contained in any hyperplane.

If $\psi: X \dashrightarrow \mathbf{P}(W)$ is only a rational map, it is defined on a dense open subset U of X, and we get as above a linear map $W^\vee \to H^0(U,\mathscr{L})$. If X is locally factorial, the invertible sheaf $\mathscr{L}$ is defined on U but extends to X (write $\mathscr{L} = \mathscr{O}_U(D)$ and take the closure of D in X) and, since X is normal, the restriction $H^0(X,\mathscr{L}) \to H^0(U,\mathscr{L})$ is bijective, so we get again a map $W^\vee \to H^0(X,\mathscr{L})$.

Conversely, starting from an invertible sheaf $\mathscr{L}$ on X and a finite-dimensional vector space V of sections of $\mathscr{L}$, we define a rational map

$$\psi_V: X \dashrightarrow \mathbf{P}(V^\vee)$$

(also denoted by $\psi_{\mathscr{L}}$ when $V = H^0(X,\mathscr{L})$) by associating with a point x of X the hyperplane of sections of V that vanish at x. This map is not defined at points where all sections in V vanish (they are called *base-points* of V). If we choose a basis $(s_0,\dots,s_N)$ for V, we have also

$$\psi_V(x) = \big(s_0(x),\dots,s_N(x)\big),$$

where it is understood that the $s_j(x)$ are computed via the same trivialization of $\mathscr{L}$ in a neighborhood of x; the corresponding point of $\mathbf{P}^N_{\mathbf{k}}$ is independent of the choice of this trivialization.

These two constructions are inverse of one another. In particular, regular maps from X to a projective space whose image is not contained in any hyperplane correspond to base-point-free linear systems on X.

Example 2.21 (The Rational Normal Curve) We saw in Example 2.11 that the vector space $H^0(\mathbf{P}^1_{\mathbf{k}}, \mathscr{O}_{\mathbf{P}^1_{\mathbf{k}}}(m))$ has dimension $m+1$. A basis is given by $(s^m, s^{m-1}t,\dots,t^m)$. The corresponding linear system is base-point-free and induces a curve

$$\begin{aligned} \mathbf{P}^1_{\mathbf{k}} &\longrightarrow \mathbf{P}^m_{\mathbf{k}} \\ (s,t) &\longmapsto (s^m, s^{m-1}t, \dots, t^m) \end{aligned}$$

whose image (the *rational normal curve*) can be defined by the vanishing of all 2×2-minors of the matrix

$$\begin{pmatrix} x_0 & \cdots & x_{m-1} \\ x_1 & \cdots & x_m \end{pmatrix}.$$

Example 2.22 (The Veronese Surface) We saw in Example 2.11 that the vector space $H^0(\mathbf{P}^2_{\mathbf{k}}, \mathscr{O}_{\mathbf{P}^2_{\mathbf{k}}}(2))$ has dimension 6. The corresponding linear system is base-point-free and induces a morphism

$$\begin{aligned} \mathbf{P}^2_{\mathbf{k}} &\longrightarrow \mathbf{P}^5_{\mathbf{k}} \\ (s,t,u) &\longmapsto (s^2, st, su, t^2, tu, u^2) \end{aligned}$$

whose image (the *Veronese surface*) can be defined by the vanishing of all 2×2-minors of the symmetric matrix

$$\begin{pmatrix} x_0 & x_1 & x_2 \\ x_1 & x_3 & x_4 \\ x_2 & x_4 & x_5 \end{pmatrix}.$$

Example 2.23 (Cremona Involution) The rational map

$$\begin{aligned} \mathbf{P}^2_{\mathbf{k}} &\dashrightarrow \mathbf{P}^2_{\mathbf{k}} \\ (s,t,u) &\longmapsto (\tfrac{1}{s}, \tfrac{1}{t}, \tfrac{1}{u}) = (tu, su, st) \end{aligned}$$

is defined everywhere except at the three points $(1,0,0)$, $(0,1,0)$, and $(0,0,1)$. It is associated with the subspace $\langle tu, su, st\rangle$ of $H^0(\mathbf{P}^2_{\mathbf{k}}, \mathscr{O}_{\mathbf{P}^2_{\mathbf{k}}}(2))$ (which is the space of all conics passing through these three points).

2.6 Globally Generated Sheaves

Let X be a $\mathbf{k}$-scheme of finite type. A coherent sheaf $\mathscr{F}$ is *generated by its global sections at a point* $x \in X$ (or *globally generated at* x) if the images of the global sections of $\mathscr{F}$ (i.e., elements of $H^0(X, \mathscr{F})$) in the stalk $\mathscr{F}_x$ generate that stalk as a $\mathscr{O}_{X,x}$-module. The set of points at which $\mathscr{F}$ is globally generated is the complement of the support of the cokernel of the *evaluation* map

$$\mathrm{ev}\colon H^0(X, \mathscr{F}) \otimes_{\mathbf{k}} \mathscr{O}_X \to \mathscr{F}.$$

It is therefore open. The sheaf $\mathscr{F}$ is *generated by its global sections* (or *globally generated*) if it is generated by its global sections at each point x of X. This is equivalent to the surjectivity of ev, and to the fact that $\mathscr{F}$ is the quotient of a free sheaf.

Since closed points are dense in X, it is enough to check global generation at every closed point x. This is equivalent, by Nakayama's lemma, to the surjectivity of the $k(x)$-linear map

$$\mathsf{ev}_x : H^0(X, \mathscr{F}) \otimes k(x) \to H^0(X, \mathscr{F} \otimes k(x)).$$

We sometimes say that $\mathscr{F}$ is *generated by finitely many global sections* (at $x \in X$) if there are $s_0, \dots, s_N \in H^0(X, \mathscr{F})$ such that the corresponding evaluation maps, where $H^0(X, \mathscr{F})$ is replaced with the vector subspace generated by $s_0, \dots, s_N$, are surjective.

Any quasi-coherent sheaf on an affine scheme $X = \operatorname{Spec}(A)$ is generated by its global sections (such a sheaf can be written as $\widetilde{M}$, where M is an A-module, and $H^0(X, \widetilde{M}) = M$).

Any quotient of a globally generated sheaf has the same property. Any tensor product of globally generated sheaves has the same property. The restriction of a globally generated sheaf to a subscheme has the same property.

2.24. Globally Generated Invertible Sheaves An invertible sheaf $\mathscr{L}$ on X is generated by its global sections if and only if for each closed point $x \in X$, there exists a global section $s \in H^0(X, \mathscr{L})$ that does not vanish at x (i.e., $s_x \notin \mathfrak{m}_{X,x}\mathscr{L}_x$, or $\mathsf{ev}_x(s) \neq 0$ in $H^0(X, \mathscr{L} \otimes k(x)) \simeq k(x)$).

Another way to phrase this, using the constructions of 2.20, is to say that the invertible sheaf $\mathscr{L}$ is generated by finitely many global sections if and only if there exists a *morphism* $\psi : X \to \mathbf{P}^N_{\mathbf{k}}$ such that $\psi^* \mathscr{O}_{\mathbf{P}^N_{\mathbf{k}}}(1) \simeq \mathscr{L}$.[4]

If D is a Cartier divisor on X, the invertible sheaf $\mathscr{O}_X(D)$ is generated by its global sections (for brevity, we will sometimes say that D is generated by its global sections, or globally generated) if for any $x \in X$, there is a Cartier divisor on X, linearly equivalent to D, whose support does not contain x (use (3)).

Example 2.25 We saw in Example 2.11 that any invertible sheaf on the projective space $\mathbf{P}^n_{\mathbf{k}}$ is of the type $\mathscr{O}_{\mathbf{P}^n_{\mathbf{k}}}(d)$ for some integer d. This sheaf is not generated by its global sections for $d < 0$ because any global section is identically 0. However, when $d > 0$, the vector space $H^0(\mathbf{P}^n_{\mathbf{k}}, \mathscr{O}_{\mathbf{P}^n_{\mathbf{k}}}(d))$ is isomorphic to the space of homogeneous polynomials of degree d in the homogeneous coordinates $x_0, \dots, x_n$ on $\mathbf{P}^n_{\mathbf{k}}$. At each point of $\mathbf{P}^n_{\mathbf{k}}$, one of these coordinates, say x_i, does not vanish, hence the section x_i^d does not vanish either. It follows that $\mathscr{O}_{\mathbf{P}^n_{\mathbf{k}}}(d)$ is generated by its global sections if and only if $d \geq 0$.

---

[4]If $s \in H^0(X, \mathscr{L})$, the subset $X_s = \{x \in X \mid \mathsf{ev}_x(s) \neq 0\}$ is open. A family $(s_i)_{i \in I}$ of sections generates $\mathscr{L}$ if and only if $X = \bigcup_{i \in I} X_{s_i}$. If X is noetherian and $\mathscr{L}$ is globally generated, it is generated by finitely many global sections.

2.7 Ample Divisors

The following definition, although technical, is extremely important.

Definition 2.26 A Cartier divisor D on a scheme X of finite type over a field is *ample* if, for every coherent sheaf $\mathscr{F}$ on X, the sheaf $\mathscr{F}(mD)$ is generated by its global sections for all m large enough.

Any sufficiently high multiple of an ample divisor is therefore globally generated, but an ample divisor may not be globally generated (it may have no non-zero global sections).

Proposition 2.27 *Let D be a Cartier divisor on a scheme of finite type over a field. The following conditions are equivalent:*

(i) *D is ample;*
(ii) *pD is ample for all $p > 0$;*
(iii) *pD is ample for some $p > 0$.*

Proof Both implications (i) $\Rightarrow$ (ii) and (ii) $\Rightarrow$ (iii) are trivial. Assume that pD is ample. Let $\mathscr{F}$ be a coherent sheaf. For each $i \in \{0, \ldots, p-1\}$, the sheaf $\mathscr{F}(iD)(mpD) = \mathscr{F}((i + mp)D)$ is generated by its global sections for $m \gg 0$. It follows that $\mathscr{F}(mD)$ is generated by its global sections for all $m \gg 0$, hence D is ample. □

This proposition allows us to say that a **Q**-Cartier **Q**-divisor is ample if some (integral) positive multiple is ample (all further positive multiples are then ample by the proposition). The restriction of an ample **Q**-Cartier **Q**-divisor to a closed subscheme is ample. The sum of two ample **Q**-Cartier **Q**-divisors is still ample. The sum of an ample **Q**-Cartier **Q**-divisor and a globally generated Cartier divisor is ample. Any **Q**-Cartier **Q**-divisor on an affine scheme of finite type over a field is ample.

Proposition 2.28 *Let A and E be **Q**-Cartier **Q**-divisors on a scheme of finite type over a field. If A is ample, so is $A + tE$ for all t rational small enough.*

Proof Upon multiplying by a large positive integer, we may assume that D and E are Cartier divisors.

Since A is ample, $mA \pm E$ is globally generated for all $m \gg 0$ and $(m+1)A \pm E$ is then ample. We write $A \pm tE = (1 - t(m+1))A + t((m+1)A \pm E)$. When $0 < t < \frac{1}{m+1}$, this divisor is therefore ample. □

Here is the fundamental result that justifies the definition of ampleness.

Theorem 2.29 (Serre) *The hyperplane divisor on $\mathbf{P}^n_{\mathbf{k}}$ is ample.*

More precisely, for any coherent sheaf $\mathscr{F}$ on $\mathbf{P}^n_{\mathbf{k}}$, the sheaf $\mathscr{F}(m)$[5] *is generated by* finitely many *global sections for all $m \gg 0$.*

[5]This is the traditional notation for the tensor product $\mathscr{F} \otimes \mathscr{O}_{\mathbf{P}^n_{\mathbf{k}}}(m)$, which is also the same as $\mathscr{F}(mH)$.

Proof The restriction of $\mathscr{F}$ to each standard affine open subset U_i is generated by finitely many sections $s_{ik} \in H^0(U_i, \mathscr{F})$. We want to show that each $s_{ik}x_i^m \in H^0(U_i, \mathscr{F}(m))$ extends for $m \gg 0$ to a section t_{ik} of $\mathscr{F}(m)$ on $\mathbf{P}^n_{\mathbf{k}}$.

Let $s \in H^0(U_i, \mathscr{F})$. It follows from [8, Lemma II.5.3.(b)] that for each j, the section

$$x_i^p s|_{U_i \cap U_j} \in H^0(U_i \cap U_j, \mathscr{F}(p))$$

extends to a section $t_j \in H^0(U_j, \mathscr{F}(p))$ for $p \gg 0$ (in other words, t_j restricts to $x_i^p s$ on $U_i \cap U_j$). We then have

$$t_j|_{U_i \cap U_j \cap U_k} = t_k|_{U_i \cap U_j \cap U_k}$$

for all j and k hence, upon multiplying again by a power of x_i,

$$x_i^q t_j|_{U_j \cap U_k} = x_i^q t_k|_{U_j \cap U_k}.$$

for $q \gg 0$ [8, Lemma II.5.3.(a)]. This means that the $x_i^q t_j$ glue to a section t of $\mathscr{F}(p+q)$ on $\mathbf{P}^n_{\mathbf{k}}$ which extends $x_i^{p+q}s$.

We thus obtain finitely many global sections t_{ik} of $\mathscr{F}(m)$ which generate $\mathscr{F}(m)$ on each U_i hence on $\mathbf{P}^n_{\mathbf{k}}$. □

An important consequence of Serre's theorem is that a projective scheme over $\mathbf{k}$ (defined as a closed subscheme of some $\mathbf{P}^n_{\mathbf{k}}$) carries an effective ample divisor. We also have more.

Corollary 2.30 *A Cartier divisor on a projective variety is linearly equivalent to the difference of two effective Cartier divisors.*

Proof Let D be a Cartier divisor on the projective variety X and let A be an effective ample divisor on X. For $m \gg 0$, the invertible sheaf $\mathscr{O}_X(D + mA)$ is generated by its global sections. In particular, it has a non-zero section; let E be its (effective) divisor. We have $D \underset{\text{lin}}{\equiv} E - mA$, which proves the proposition. □

Corollary 2.31 (Serre) *Let X be a projective* $\mathbf{k}$*-scheme and let $\mathscr{F}$ be a coherent sheaf on X. For all integers q,*

a) *the* $\mathbf{k}$*-vector space $H^q(X, \mathscr{F})$ has finite dimension;*
b) *the* $\mathbf{k}$*-vector spaces $H^q(X, \mathscr{F}(m))$ all vanish for $m \gg 0$.*

Proof Assume $X \subset \mathbf{P}^n_{\mathbf{k}}$. Since any coherent sheaf on X can be considered as a coherent sheaf on $\mathbf{P}^n_{\mathbf{k}}$ (with the same cohomology), we may assume $X = \mathbf{P}^n_{\mathbf{k}}$. For $q > n$, we have $H^q(X, \mathscr{F}) = 0$ and we proceed by descending induction on q.

By Theorem 2.29, there exist integers r and p and an exact sequence

$$0 \longrightarrow \mathscr{G} \longrightarrow \mathscr{O}_{\mathbf{P}^n_{\mathbf{k}}}(-p)^r \longrightarrow \mathscr{F} \longrightarrow 0$$

of coherent sheaves on $\mathbf{P}^n_{\mathbf{k}}$. The vector spaces $H^q(\mathbf{P}^n_{\mathbf{k}}, \mathscr{O}_{\mathbf{P}^n_{\mathbf{k}}}(-p))$ can be computed by hand and are all finite-dimensional. The exact sequence

$$H^q(\mathbf{P}^n_{\mathbf{k}}, \mathscr{O}_X(-p))^r \longrightarrow H^q(\mathbf{P}^n_{\mathbf{k}}, \mathscr{F}) \longrightarrow H^{q+1}(\mathbf{P}^n_{\mathbf{k}}, \mathscr{G})$$

yields a).

Again, direct calculations show that $H^q(\mathbf{P}^n, \mathscr{O}_{\mathbf{P}^n_{\mathbf{k}}}(m-p))$ vanishes for all $m > p$ and all $q > 0$. The exact sequence

$$H^q(\mathbf{P}^n_{\mathbf{k}}, \mathscr{O}_X(m-p))^r \longrightarrow H^q(\mathbf{P}^n_{\mathbf{k}}, \mathscr{F}(m)) \longrightarrow H^{q+1}(\mathbf{P}^n_{\mathbf{k}}, \mathscr{G}(m))$$

yields b). □

2.32. A Cohomological Characterization of Ample Divisors A further consequence of Serre's theorem is an important characterization of ample divisors by the vanishing of higher cohomology groups.

Theorem 2.33 *Let X be a projective* **k**-*scheme and let D be a Cartier divisor on X. The following properties are equivalent:*

(i) *D is ample;*
(ii) *for each coherent sheaf $\mathscr{F}$ on X, we have $H^q(X, \mathscr{F}(mD)) = 0$ for all $m \gg 0$ and all $q > 0$;*
(iii) *for each coherent sheaf $\mathscr{F}$ on X, we have $H^1(X, \mathscr{F}(mD)) = 0$ for all $m \gg 0$.*

Proof Assume D ample. For $m \gg 0$, the divisor mD is globally generated, by a finite number of sections (see footnote 4). It defines a morphism $\psi : X \to \mathbf{P}^N_{\mathbf{k}}$ such that $\psi^* H \underset{\text{lin}}{\equiv} mD$. This morphism has finite fibers: if it contracts a curve $C \subset X$ to a point, one has $mD|_C \underset{\text{lin}}{\equiv} 0$, which contradicts the fact $D|_C$ is ample. Since X is projective, ψ is finite.[6]

Let $\mathscr{F}$ be a coherent sheaf on X. The sheaf $\psi_*\mathscr{F}$ is then coherent [8, Corollary II.5.20]. Since ψ is finite, if $\mathscr{U}$ is a covering of $\mathbf{P}^N_{\mathbf{k}}$ by affine open subsets, $\psi^{-1}(\mathscr{U})$ is a covering of X by affine open subsets [8, Exercise II.5.17.(b)] and, by definition of $\psi_*\mathscr{F}$, the associated cochain complexes are isomorphic. This implies

$$H^q(X, \mathscr{F}) \simeq H^q(\mathbf{P}^N_{\mathbf{k}}, \psi_*\mathscr{F})$$

for all integers q. By Corollary 2.31.b) and the projection formula [8, Exercise II.5.1.(d)], we have, for all $q > 0$ and $s \gg 0$,

$$0 = H^q(\mathbf{P}^N_{\mathbf{k}}, (\psi_*\mathscr{F})(sH)) \simeq H^q(X, \mathscr{F}(s\psi^* H)) = H^q(X, \mathscr{F}(smD)).$$

[6]The very important fact that a projective morphism with finite fibers is finite is deduced in [8] from the difficult Main Theorem of Zariski. In our case, it can also be proved in an elementary fashion.

Applying this to each of the sheaves $\mathscr{F}, \mathscr{F}(D), \dots, \mathscr{F}((m-1)D)$, we see that (ii) holds.

Condition (ii) trivially implies (iii).

Assume that (iii) holds. Let $\mathscr{F}$ be a coherent sheaf on X, let x be a closed point of X, and let $\mathscr{G}$ be the kernel of the surjection

$$\mathscr{F} \to \mathscr{F} \otimes k(x)$$

of $\mathscr{O}_X$-modules. Since (iii) holds, there exists an integer m_0 such that

$$H^1(X, \mathscr{G}(mD)) = 0$$

for all $m \geq m_0$ (note that the integer m_0 may depend on $\mathscr{F}$ and x). Since the sequence

$$0 \to \mathscr{G}(mD) \to \mathscr{F}(mD) \to \mathscr{F}(mD) \otimes k(x) \to 0$$

is exact, the evaluation

$$H^0(X, \mathscr{F}(mD)) \to H^0(X, \mathscr{F}(mD) \otimes k(x))$$

is surjective. This means that its global sections generate $\mathscr{F}(mD)$ in a neighborhood $U_{\mathscr{F},m}$ of x. In particular, there exists an integer m_1 such that $m_1 D$ is globally generated on $U_{\mathscr{O}_X,m_1}$. For all $m \geq m_0$, the sheaf $\mathscr{F}(mD)$ is globally generated on

$$U_x = U_{\mathscr{O}_X,m_1} \cap U_{\mathscr{F},m_0} \cap U_{\mathscr{F},m_0+1} \cap \dots \cap U_{\mathscr{F},m_0+m_1-1}$$

since it can be written as

$$(\mathscr{F}((m_0+s)D)) \otimes \mathscr{O}_X(r(m_1 D))$$

with $r \geq 0$ and $0 \leq s < m_1$. Cover X with a finite number of open subsets U_x and take the largest corresponding integer m_0. This shows that D is ample and finishes the proof of the theorem. □

Corollary 2.34 *Let X and Y be projective* **k***-schemes and let $\psi: X \to Y$ be a morphism with finite fibers. Let A be an ample* **Q***-Cartier* **Q***-divisor on Y. Then the* **Q***-Cartier* **Q***-divisor ψ^*A is ample.*

Proof We may assume that A is a Cartier divisor. Let $\mathscr{F}$ be a coherent sheaf on X. We use the same tools as in the proof of the theorem: the sheaf $\psi_*\mathscr{F}$ is coherent and since A is ample, $H^q(X, \mathscr{F}(m\psi^*A)) \simeq H^q(Y, (\psi_*\mathscr{F})(mA))$ vanishes for all $q > 0$ and $m \gg 0$. By Theorem 2.33, ψ^*A est ample. □

Exercise 2.35 In the situation of the corollary, if ψ is *not* finite, show that ψ^*A is *not* ample.

Exercise 2.36 Let X be a projective $\mathbf{k}$-scheme, let $\mathscr{F}$ be a coherent sheaf on X, and let $A_1, \dots, A_r$ be ample Cartier divisors on X. Show that for each $i > 0$, the set

$$\{(m_1, \dots, m_r) \in \mathbf{N}^r \mid H^i(X, \mathscr{F}(m_1A_1 + \dots + m_rA_r)) \neq 0\}$$

is finite.

2.8 Ample Divisors on Curves

We define the (arithmetic) genus of a (projective) curve X over a field $\mathbf{k}$ by

$$g(X) := \dim_{\mathbf{k}}(H^1(X, \mathscr{O}_X)) =: h^1(X, \mathscr{O}_X).$$

Example 2.37 The curve $\mathbf{P}^1_{\mathbf{k}}$ has genus 0. This can be obtained by a computation in Čech cohomology: cover X with the two standard affine subsets U_0 and U_1. The Čech complex

$$H^0(U_0, \mathscr{O}_{U_0}) \oplus H^0(U_1, \mathscr{O}_{U_1}) \to H^0(U_{01}, \mathscr{O}_{U_{01}})$$

is $\mathbf{k}[t] \oplus \mathbf{k}[t^{-1}] \to \mathbf{k}[t, t^{-1}]$, hence the result.

Exercise 2.38 Show that the genus of a plane curve of degree d is $(d-1)(d-2)/2$ (*Hint:* assume that $(0, 0, 1)$ is not on the curve, cover it with the affine subsets U_0 and U_1 and compute the Čech cohomology groups as above).

The following theorem is extremely useful.[7]

Theorem 2.39 (Riemann–Roch Theorem) *Let X be a smooth curve. For any divisor D on X, we have*

$$\chi(X, D) = \deg(D) + \chi(X, \mathscr{O}_X) = \deg(D) + 1 - g(X).$$

Proof By Proposition 2.30, we can write $D \underset{\text{lin}}{\equiv} E - F$, where E and F are effective (Cartier) divisors on X. Considering them as (0-dimensional) subschemes of X, we have exact sequences (see Remark 2.9)

$$\begin{array}{l} 0 \to \mathscr{O}_X(E - F) \to \mathscr{O}_X(E) \to \mathscr{O}_F \to 0 \\ 0 \to \qquad \mathscr{O}_X \qquad \to \mathscr{O}_X(E) \to \mathscr{O}_E \to 0 \end{array}$$

(note that the sheaf $\mathscr{O}_F(E)$ is isomorphic to $\mathscr{O}_F$, because $\mathscr{O}_X(E)$ is isomorphic to $\mathscr{O}_X$ in a neighborhood of the (finite) support of F, and similarly, $\mathscr{O}_E(E) \simeq \mathscr{O}_E$). As

[7]This should really be called the Hirzebruch–Riemann–Roch theorem (or a (very) particular case of it). The original Riemann–Roch theorem is our Theorem 2.39 with the dimension of $H^1(X, \mathscr{L})$ replaced with that of its Serre-dual $H^0(X, \omega_X \otimes \mathscr{L}^{-1})$.

remarked in (4), we have

$$\chi(F,\mathcal{O}_F)=h^0(F,\mathcal{O}_F)=\deg(F).$$

Similarly, $\chi(E,\mathcal{O}_E)=\deg(E)$. This implies

$$\begin{aligned}\chi(X,D)&=\chi(X,E)-\chi(F,\mathcal{O}_F)\\&=\chi(X,\mathcal{O}_X)+\chi(E,\mathcal{O}_E)-\deg(F)\\&=\chi(X,\mathcal{O}_X)+\deg(E)-\deg(F)\\&=\chi(X,\mathcal{O}_X)+\deg(D)\end{aligned}$$

and the theorem is proved. □

We can now characterize ample divisors on smooth curves.

Corollary 2.40 *A* **Q**-*divisor* D *on a smooth curve is ample if and only if* $\deg(D)>0$.

Proof We may assume that D is a divisor. Let p be a closed point of the smooth curve X. If D is ample, $mD-p$ is linearly equivalent to an effective divisor for some $m\gg 0$, in which case

$$0\le\deg(mD-p)=m\deg(D)-\deg(p),$$

hence $\deg(D)>0$.

Conversely, assume $\deg(D)>0$. By Riemann–Roch, we have $H^0(X,mD)\neq 0$ for $m\gg 0$, so, upon replacing D by a positive multiple, we can assume that D is effective. As in the proof of the theorem, we then have an exact sequence

$$0\to\mathcal{O}_X((m-1)D)\to\mathcal{O}_X(mD)\to\mathcal{O}_D\to 0,$$

from which we get a surjection

$$H^1(X,(m-1)D))\to H^1(X,mD)\to 0.$$

Since these spaces are finite-dimensional, it will be a bijection for $m\gg 0$, in which case we get a surjection

$$H^0(X,mD)\to H^0(D,\mathcal{O}_D).$$

In particular, the evaluation map ev_x (see Sect. 2.6) for the sheaf $\mathcal{O}_X(mD)$ is surjective at every point x of the support of D. Since it is trivially surjective for x outside of this support (it has a section with divisor mD), the sheaf $\mathcal{O}_X(mD)$ is globally generated.

Its global sections therefore define a morphism $\psi: X \to \mathbf{P}^N_\mathbf{k}$ such that $\mathscr{O}_X(mD) = \psi^*\mathscr{O}_{\mathbf{P}^N_\mathbf{k}}(1)$. Since $\mathscr{O}_X(mD)$ is non trivial, ψ is not constant, hence finite because X is a curve. But then, $\mathscr{O}_X(mD) = \psi^*\mathscr{O}_{\mathbf{P}^N_\mathbf{k}}(1)$ is ample (Corollary 2.34) hence D is ample. □

Exercise 2.41 Let X be a curve and let p be a closed point. Show that $X \smallsetminus \{p\}$ is affine.

2.9 *Nef Divisors*

Let X be a projective $\mathbf{k}$-scheme and let D be an ample $\mathbf{Q}$-Cartier $\mathbf{Q}$-divisor on X. For any non-constant curve $\rho: C \to X$, the $\mathbf{Q}$-divisor ρ^*D is ample (Corollary 2.34) hence its degree $(D{\cdot}C)$ is positive. The converse is not quite true, although producing a counter-example at this point is a bit tricky. On the other hand, it is easier to control the following important property.

Definition 2.42 A $\mathbf{Q}$-Cartier $\mathbf{Q}$-divisor D on a projective $\mathbf{k}$-scheme X is *nef* [8] if $(D \cdot C) \geq 0$ for every curve C on X.

Being nef is by definition a numerical property for $\mathbf{Q}$-divisors (it only depends on their numerical equivalence classes). One can define nef classes in the finite-dimensional $\mathbf{R}$-vector space $N^1(X)_\mathbf{R}$ and they obviously form a closed convex cone

$$\mathrm{Nef}(X) \subset N^1(X)_\mathbf{R}. \tag{7}$$

An ample divisor is nef. The restriction of a nef $\mathbf{Q}$-divisor to a closed subscheme is again nef. More generally, the pullback by any morphism of a nef $\mathbf{Q}$-divisor is again nef. More precisely, if $\psi: X \to Y$ is a morphism between projective $\mathbf{k}$-schemes and D is a nef $\mathbf{Q}$-Cartier $\mathbf{Q}$-divisor on Y, the pullback ψ^*D is nef and, if ψ is not finite (so that it contracts a curve C which will then satisfy $(\psi^*D \cdot C) = 0$), its class is on the boundary of the cone Nef(X).

Example 2.43 If X is a curve, one has an isomorphism $\deg: N^1(X)_\mathbf{R} \xrightarrow{\sim} \mathbf{R}$ given by the degree on the normalization (Example 2.17) and, tautologically,

$$\mathrm{Nef}(X) = \deg^{-1}(\mathbf{R}^{\geq 0}).$$

Example 2.44 One checks that in $N^1(\mathbf{P}^n_\mathbf{k})_\mathbf{R} \simeq \mathbf{R}[H]$, one has

$$\mathrm{Nef}(\mathbf{P}^n_\mathbf{k}) = \mathbf{R}^{\geq 0}[H].$$

Let $\varepsilon: \widetilde{\mathbf{P}^n_\mathbf{k}} \to \mathbf{P}^n_\mathbf{k}$ be the blow up of a point O, with exceptional divisor E, so that $N^1(\widetilde{\mathbf{P}^n_\mathbf{k}})_\mathbf{R} \simeq \mathbf{R}[\varepsilon^*H] \oplus \mathbf{R}[E]$ (Example 2.18). The plane closed convex cone $\mathrm{Nef}(\widetilde{\mathbf{P}^n_\mathbf{k}})$

[8] This acronym comes from "numerically effective," or "numerically eventually free" (according to [17, D.1.3]).

is bounded by two half-lines. Note that ε^*H is nef, as the pullback of a nef divisor, but for any $t > 0$, one has (if L is a line in E) $((\varepsilon^*H+tE)\cdot L) = -t$, hence ε^*H+tE is not nef and $\mathbf{R}^{\geq 0}[\varepsilon^*H]$ is one of these half-lines.

Alternatively, one could have argued that since ε is not finite, the class ε^*H is on the boundary of $\mathrm{Nef}(\widetilde{\mathbf{P}^n_{\mathbf{k}}})$. To find the other boundary half-line, recall that there is another morphism $\widetilde{\mathbf{P}^n_{\mathbf{k}}} \to \mathbf{P}^{n-1}_{\mathbf{k}}$. One checks that the class of the inverse image of the hyperplane in $\mathbf{P}^{n-1}_{\mathbf{k}}$ is $[\varepsilon^*H - E]$ (its intersection with the strict transform of a line passing through O is 0); it generates the other boundary half-line.

Exercise 2.45 Let $\varepsilon: X \to \mathbf{P}^n_{\mathbf{k}}$ be the blow up of two distinct points, with exceptional divisors E_1 and E_2. If H is a hyperplane on $\mathbf{P}^n_{\mathbf{k}}$, prove

$$N^1(X)_{\mathbf{R}} \simeq \mathbf{R}[\varepsilon^*H] \oplus \mathbf{R}[E_1] \oplus \mathbf{R}[E_2],$$
$$\mathrm{Nef}(X) \simeq \mathbf{R}^{\geq 0}[\varepsilon^*H] \oplus \mathbf{R}^{\geq 0}[\varepsilon^*H - E_1] \oplus \mathbf{R}^{\geq 0}[\varepsilon^*H - E_2].$$

2.10 Cones of Divisors

One can also consider classes of ample Cartier divisors. They span a convex cone

$$\mathrm{Amp}(X) \subset \mathrm{Nef}(X) \subset N^1(X)_{\mathbf{R}}$$

which is open by Proposition 2.28. However, we do not know (yet) whether ampleness is a numerical property: if the class of $\mathbf{Q}$-divisor D is in $\mathrm{Amp}(X)$, is D ample? This is an important but hard question. To answer it (positively), we will skip a whole chunk of the theory and accept without proof the following key result.

Theorem 2.46 *On a projective scheme, the sum of two $\mathbf{Q}$-Cartier $\mathbf{Q}$-divisors, one nef and one ample, is ample.*

Corollary 2.47 *On a projective scheme, ampleness is a numerical property and the ample cone is the interior of the nef cone.*

Proof Let X be a projective scheme and let D be a $\mathbf{Q}$-Cartier $\mathbf{Q}$-divisor whose class is in the interior of the nef cone. We want to prove that D is ample. We may assume that D is a Cartier divisor. Let H be an ample divisor on X. Since $[D]$ is in the interior of the nef cone, $[D] - t[H]$ is still in the nef cone for some $t > 0$ (small enough). Then $D - tH$ is nef, and $D = (D - tH) + tH$ is ample by the theorem. □

We complete our collection of cones of divisors with the (convex) *effective cone* $\mathrm{Eff}(X) \subset N^1(X)_{\mathbf{R}}$ generated by classes of effective Cartier divisors. It contains the ample cone (why?). It may happen that $\mathrm{Eff}(X)$ is not closed and we let $\mathrm{Psef}(X)$, the *pseudo-effective cone*, be its closure. Finally, the *big cone*

$$\mathrm{Big}(X) = \mathrm{Eff}(X) + \mathrm{Amp}(X) \subset \mathrm{Eff}(X) \qquad (8)$$

is the interior of the pseudo-effective cone. All in all, we have

$$\begin{array}{ccc} \mathrm{Nef}(X) & \subset \mathrm{Psef}(X) & \subset N^1(X)_{\mathbf{R}} \\ & \cup & \\ \cup & \mathrm{Eff}(X) & \\ & \cup & \\ \mathrm{Amp}(X) & \subset \mathrm{Big}(X). & \end{array} \tag{9}$$

Example 2.48 With the notation of Example 2.44, we have

$$\begin{aligned} N^1(\widetilde{\mathbf{P}}^n_{\mathbf{k}})_{\mathbf{R}} &\simeq \mathbf{R}[\varepsilon^*H] \oplus \mathbf{R}[E], \\ \mathrm{Big}(\widetilde{\mathbf{P}}^n_{\mathbf{k}}) &\simeq \mathbf{R}^{>0}[\varepsilon^*H - E] \oplus \mathbf{R}^{>0}[E], \\ \mathrm{Eff}(\widetilde{\mathbf{P}}^n_{\mathbf{k}}) = \mathrm{Psef}(\widetilde{\mathbf{P}}^n_{\mathbf{k}}) &\simeq \mathbf{R}^{\geq 0}[\varepsilon^*H - E] \oplus \mathbf{R}^{\geq 0}[E], \\ \mathrm{Amp}(\widetilde{\mathbf{P}}^n_{\mathbf{k}}) &\simeq \mathbf{R}^{>0}[\varepsilon^*H] \oplus \mathbf{R}^{>0}[\varepsilon^*H - E], \\ \mathrm{Nef}(\widetilde{\mathbf{P}}^n_{\mathbf{k}}) &\simeq \mathbf{R}^{\geq 0}[\varepsilon^*H] \oplus \mathbf{R}^{\geq 0}[\varepsilon^*H - E]. \end{aligned}$$

Only the effective cone needs to be explained: its description follows from the fact that if $D \subset \widetilde{\mathbf{P}}^n_{\mathbf{k}}$ is an effective divisor other than E, with class $a[\varepsilon^*H - E] + b[E]$, its intersection with a line contained in E (but not in D) is $a-b$, which must therefore be non-negative, and its intersection with the strict transform of a line passing through O is b, which must also be non-negative, hence $a \geq b \geq 0$. The effective cone is therefore contained in $\mathbf{R}^{\geq 0}[\varepsilon^*H - E] \oplus \mathbf{R}^{\geq 0}[E]$, and the reverse inclusion is obvious.

2.11 The Canonical Divisor

Let X be a smooth variety. We define the *canonical sheaf* ω_X as the determinant of the sheaf of differentials Ω_X. It is an invertible sheaf on X. A canonical divisor K_X is any Cartier divisor on X which defines ω_X.

When X is only normal, with regular locus $j\colon U \hookrightarrow X$, we define ω_X as the (not necessarily invertible) sheaf $j_*\omega_U$. A canonical divisor K_X is then any Weil divisor on X which restricts to a canonical divisor on U.

If $Y \subset X$ is a normal Cartier divisor in X, one has the *adjunction formula*

$$K_Y = (K_X + Y)|_Y. \tag{10}$$

Example 2.49 The canonical sheaf on $\mathbf{P}^n_{\mathbf{k}}$ is $\omega_{\mathbf{P}^n_{\mathbf{k}}} = \mathscr{O}_{\mathbf{P}^n_{\mathbf{k}}}(-n-1)$. If follows from the adjunction formula (10) that for a normal hypersurface X of degree d in $\mathbf{P}^n_{\mathbf{k}}$, one has $\omega_X = \mathscr{O}_X(-n-1+d)$.

When X is projective of dimension n and smooth (or only Cohen–Macaulay), the canonical sheaf ω_X is a dualizing sheaf: for any locally free coherent sheaf $\mathscr{F}$ on X, there are isomorphisms (Serre duality)

$$H^i(X, \mathscr{F}) \simeq H^{n-i}(X, \mathscr{F}^\vee \otimes \omega_X)^\vee.$$

3 Riemann–Roch Theorems

If A is an ample Cartier divisor on a projective scheme X, the vector space $H^0(X, mA)$ contains, for $m \gg 0$, "enough elements" to define a finite morphism $\psi_{mA}: X \to \mathbf{P}^N_{\mathbf{k}}$. Riemann–Roch theorems deal with estimations of the dimensions $h^0(X, mA)$ of these spaces of sections. We will not prove much here, but only state the main results.

When X is a smooth curve, the Riemann–Roch theorem is (Theorem 2.39)

$$\chi(X, mD) := h^0(X, mD) - h^1(X, mD) = m \deg(D) + \chi(X, \mathscr{O}_X) = m \deg(D) + 1 - g(X).$$

This is a polynomial of degree 1 in m. This fact is actually very general and we will use it to define the intersection of n Cartier divisors on a projective scheme of dimension n over a field.

3.1 Intersecting Two Curves on a Surface

On a surface, curves and hypersurfaces are the same thing. Formula (5) therefore defines the intersection number of two curves on a surface. We want to give another, more concrete, interpretation of this number.

Let X be a smooth projective surface defined over an algebraically closed field $\mathbf{k}$ and let C_1 and C_2 be two curves on X with no common component. We would like to define the intersection number of C_1 and C_2 as the number of intersection points "counted with multiplicities."

One way to do that is to define the intersection multiplicity of C_1 and C_2 at a point x of $C_1 \cap C_2$. If f_1 and f_2 be respective generators of the ideals of C_1 and C_2 at x, this is

$$m_x(C_1 \cap C_2) = \dim_{\mathbf{k}} \mathscr{O}_{X,x}/(f_1, f_2).$$

By the Nullstellensatz, the ideal (f_1, f_2) contains a power of the maximal ideal $\mathfrak{m}_{X,x}$, hence the number $m_x(C_1 \cap C_2)$ is finite. It is 1 if and only if f_1 and f_2 generate $\mathfrak{m}_{X,x}$, which means that they form a system of parameters at x, i.e., that C_1 and C_2 meet transversally at x.

We then set

$$(C_1 \cdot C_2) = \sum_{x \in C_1 \cap C_2} m_x(C_1 \cap C_2). \tag{11}$$

Another way to understand this definition is to consider the scheme-theoretic intersection $C_1 \cap C_2$. It is a scheme whose support is finite, and by definition, $\mathscr{O}_{C_1 \cap C_2, x} = \mathscr{O}_{X,x}/(f_1, f_2)$. Hence,

$$(C_1 \cdot C_2) = h^0(X, \mathscr{O}_{C_1 \cap C_2}). \tag{12}$$

There is still another way to interpret this number.

Theorem 3.1 *Under the hypotheses above, we have*

$$(C_1 \cdot C_2) = \chi(X, -C_1 - C_2) - \chi(X, -C_1) - \chi(X, -C_2) + \chi(X, \mathscr{O}_X). \tag{13}$$

Proof Let s_1 be a section of $\mathscr{O}_X(C_1)$ with divisor C_1 and let s_2 be a section of $\mathscr{O}_X(C_2)$ with divisor C_2. One checks that we have an exact sequence

$$0 \to \mathscr{O}_X(-C_1 - C_2) \xrightarrow{(s_2, -s_1)} \mathscr{O}_X(-C_1) \oplus \mathscr{O}_X(-C_2) \xrightarrow{\binom{s_1}{s_2}} \mathscr{O}_X \to \mathscr{O}_{C_1 \cap C_2} \to 0.$$

(Use the fact that the local rings of X are factorial and that local equations of C_1 and C_2 have no common factor.) The theorem follows. □

A big advantage is that the right side of (13) now makes sense for any (even non-effective) Cartier divisors C_1 and C_2. This is the approach we will take in the next section to generalize this definition in all dimensions.

Finally, we check that the definition (11) agrees with our former definition (5).

Lemma 3.2 *For any smooth curve $C \subset X$ and any divisor D on X, we have*

$$(D \cdot C) = \deg(D|_C).$$

Proof We have exact sequences

$$0 \to \mathscr{O}_X(-C) \to \mathscr{O}_X \to \mathscr{O}_C \to 0$$

and

$$0 \to \mathscr{O}_X(-C - D) \to \mathscr{O}_X(-D) \to \mathscr{O}_C(-D|_C) \to 0,$$

which give

$$(D \cdot C) = \chi(C, \mathscr{O}_C) - \chi(C, -D|_C) = \deg(D|_C)$$

by the Riemann-Roch theorem on C. □

Example 3.3 If C_1 and C_2 are curves in $\mathbf{P}^2_{\mathbf{k}}$ of respective degrees d_1 and d_2, we have (this is Bézout's theorem)

$$(C_1 \cdot C_2) = d_1 d_2.$$

Indeed, since $\chi(\mathbf{P}^2_{\mathbf{k}}, \mathscr{O}_{\mathbf{P}^2_{\mathbf{k}}}(-d)) = \binom{d}{2}$ for $d \geq 0$, Theorem 3.1 gives

$$(C_1 \cdot C_2) = \binom{d_1 + d_2}{2} - \binom{d_1}{2} - \binom{d_2}{2} + 1 = d_1 d_2.$$

3.2 General Intersection Numbers

Although we will not present it here, the proof of the following theorem is not particularly hard (it proceeds by induction on n, the case $n = 1$ being the Riemann–Roch Theorem 2.39).

Theorem 3.4 *Let X be a projective* **k***-scheme of dimension n.*

a) *Let D be a Cartier divisor on X. The function $m \longmapsto \chi(X, mD)$ takes the same values on* $\mathbf{Z}$ *as a polynomial $P(T) \in \mathbf{Q}[T]$ of degree $\leq n$. We define (D^n) to be $n!$ times the coefficient of T^n in P.*
b) *More generally, if $D_1, \ldots, D_r$ are Cartier divisors on X, the function*

$$(m_1, \ldots, m_r) \longmapsto \chi(X, m_1 D_1 + \cdots + m_r D_r)$$

takes the same values on $\mathbf{Z}^r$ as a polynomial $P(T_1, \ldots, T_r)$ with rational coefficients of total degree $\leq n$. When $r \geq n$, we define the intersection number

$$(D_1 \cdot \ldots \cdot D_r)$$

to be the coefficient of $T_1 \cdots T_r$ in P (it is 0 when $r > n$).
c) *The map*

$$(D_1, \ldots, D_n) \longmapsto (D_1 \cdot \ldots \cdot D_n)$$

is $\mathbf{Z}$*-multilinear, symmetric, and takes integral values.*

The intersection number only depends on the linear equivalence classes of the divisors D_i, since it is defined from the invertible sheaves $\mathscr{O}_X(D_i)$ but in fact only on the numerical equivalence classes of the D_i. This follows from the fact that for any numerically trivial divisor D and any coherent sheaf $\mathscr{F}$ on X, we have $\chi(X, \mathscr{F}(D)) = \chi(X, \mathscr{F})$ [10, Section 2, Theorem 1].

Example 3.5 If X is a subscheme of $\mathbf{P}^N_{\mathbf{k}}$ of dimension n and if $H|_X$ is a hyperplane section of X, the intersection number $((H|_X)^n)$ is the degree of X as defined in [8, Section I.7]. In particular, $(H^n) = 1$ on $\mathbf{P}^n_{\mathbf{k}}$.

Example 3.6 If $D_1, \dots, D_n$ are effective and meet properly in a finite number of points, and if $\mathbf{k}$ is algebraically closed, the intersection number does have a geometric interpretation as the number of points in $D_1 \cap \dots \cap D_n$, counted with multiplicities. This is the length of the 0-dimensional scheme-theoretic intersection $D_1 \cap \dots \cap D_n$ (see [12, Theorem VI.2.8]; compare with (11) and (12)).

By multilinearity, we may define intersection numbers of $\mathbf{Q}$-Cartier $\mathbf{Q}$-divisors. For example, let X be the cone in $\mathbf{P}^3_{\mathbf{k}}$ with equation $x_0x_1 = x_2^2$ (its vertex is $(0,0,0,1)$) and let L be the line defined by $x_0 = x_2 = 0$ (compare with Example 2.6). Then $2L$ is a hyperplane section ($x_0 = 0$) of X, hence $((2L)^2) = \deg(X) = 2$. So we have $(L^2) = 1/2$.

Intersection numbers are seldom computed directly from the definition. Here are two useful tools.

Proposition 3.7 *Let $\pi\colon Y \to X$ be a surjective morphism between projective varieties and let $D_1, \dots, D_r$ be Cartier divisors on X with $r \geq \dim(Y)$.*

a) **(Restriction formula)** *If D_r is effective,*

$$(D_1 \cdot \ldots \cdot D_r) = (D_1|_{D_r} \cdot \ldots \cdot D_{r-1}|_{D_r}).$$

b) **(Pullback formula)** *We have*[9]

$$(\pi^*D_1 \cdot \ldots \cdot \pi^*D_r) = \deg(\pi)(D_1 \cdot \ldots \cdot D_r).$$

Example 3.8 Let again $\varepsilon\colon \widetilde{\mathbf{P}}^n_{\mathbf{k}} \to \mathbf{P}^n_{\mathbf{k}}$ be the blow up of a point, with exceptional divisor E (Example 2.18). If L is a line contained in $E \simeq \mathbf{P}^{n-1}_{\mathbf{k}}$, we saw in that example that $(E \cdot L) = -1$. Since $\mathrm{Pic}(E) \simeq \mathbf{Z}$, this implies $\mathscr{O}_{\widetilde{\mathbf{P}}^n_{\mathbf{k}}}(E)|_E = \mathscr{O}_E(-1)$. The restriction formula then gives

$$(E^n) = ((E|_E)^{n-1}) = (-1)^{n-1}.$$

On the other hand, the divisor class $[\varepsilon^*H - E]$ is the pullback of a hyperplane class via the second projection $\widetilde{\mathbf{P}}^n_{\mathbf{k}} \to \mathbf{P}^{n-1}_{\mathbf{k}}$ (Example 2.44). The pullback formula therefore gives

$$((\varepsilon^*H - E)^n) = 0. \tag{14}$$

---

[9] The number $\deg(\pi)$ is the degree of the field extension $\pi^*\colon K(X) \hookrightarrow K(Y)$ if this extension is finite, and 0 otherwise.

We may choose H such that ε^*H does not meet E. The restriction formula then gives $(D_1 \cdot \ldots \cdot D_{n-2} \cdot E \cdot \varepsilon^*H) = 0$ for all divisors $D_1, \ldots, D_{n-2}$. Expanding (14), we get

$$((\varepsilon^*H)^n) + (-1)^n(E^n) = 0$$

and again, by the projection formula, $(E^n) = (-1)^{n-1}((\varepsilon^*H)^n) = (-1)^{n-1}(H^n) = (-1)^{n-1}$.

Corollary 3.9 *Let D be a* **Q**-*Cartier* **Q**-*divisor on a projective variety X of dimension n.*

If D is ample, $(D^n) > 0$; if D is nef, $(D^n) \geq 0$.

Proof If D is ample, the sections of mD define, for $m \gg 0$, a finite morphism $\psi: X \to \mathbf{P}^N_{\mathbf{k}}$ such that $\psi^*H \underset{\text{lin}}{\equiv} mD$. Since the image $\psi(X)$ has dimension n, we have (Example 3.5) $((H|_{\psi(X)})^n) = \deg(\psi(X)) > 0$. The projection formula then yields

$$((mD)^n) = ((\psi^*H|_{\psi(X)})^n),$$

hence $(D^n) > 0$.

If D is only nef, choose an ample divisor A on X. For all $t \in \mathbf{Q}^{>0}$, the **Q**-divisor $D + tA$ is ample (Theorem 2.46), and we get $(D^n) \geq 0$ by letting t go to 0 in $((D + tA)^n) > 0$. □

Exercise 3.10 Let $D_1, \ldots, D_n$ be **Q**-Cartier **Q**-divisors on a projective variety. Prove the following:

a) if $D_1, \ldots, D_n$ are ample, $(D_1 \cdot \ldots \cdot D_n) > 0$;
b) if $D_1, \ldots, D_n$ are nef, $(D_1 \cdot \ldots \cdot D_n) \geq 0$.

3.3 Intersection of Divisors Over the Complex Numbers

Let X be a complex projective variety of dimension n. There is a short exact sequence of analytic sheaves

$$0 \to \underline{\mathbf{Z}} \xrightarrow{\cdot 2i\pi} \mathcal{O}_{X,\text{an}} \xrightarrow{\exp} \mathcal{O}^*_{X,\text{an}} \to 0$$

which induces a morphism

$$c_1: H^1(X, \mathcal{O}^*_{X,\text{an}}) \to H^2(X, \mathbf{Z})$$

called the *first Chern class*. So we can in particular define the first Chern class of an algebraic invertible sheaf on X and it induces an injection

$$N^1(X)_{\mathbf{Q}} \hookrightarrow H^2(X, \mathbf{Q}).$$

Given divisors $D_1, \ldots, D_n$ on X, the intersection product $(D_1 \cdot \ldots \cdot D_n)$ defined in Theorem 3.4 is the cup product

$$c_1(\mathcal{O}_X(D_1)) \smile \cdots \smile c_1(\mathcal{O}_X(D_n)) \in H^{2n}(X, \mathbf{Z}) \simeq \mathbf{Z}.$$

In particular, the degree of a divisor D on a curve $C \subset X$ is

$$c_1(\nu^* \mathcal{O}_X(D)) \in H^2(\widetilde{C}, \mathbf{Z}) \simeq \mathbf{Z},$$

where $\nu \colon \widetilde{C} \to C$ is the normalization of C.

Remark 3.11 A theorem of Serre says that the canonical map $H^1(X, \mathcal{O}_X^*) \to H^1(X, \mathcal{O}_{X,\mathrm{an}}^*)$ is bijective. In other words, isomorphism classes of holomorphic and algebraic line bundles on X are the same.

3.4 Asymptotic Numbers of Sections

Let D be a Cartier divisor on a projective $\mathbf{k}$-scheme X of dimension n. By definition, we have

$$\chi(X, mD) = \sum_{i=0}^{n} (-1)^i h^i(X, mD) = m^n \frac{(D^n)}{n!} + O(m^{n-1}).$$

The following proposition (whose proof proceeds again by induction on n) gives more information on each term $h^i(X, mD)$.

Proposition 3.12 *Let D be a Cartier divisor on a projective* $\mathbf{k}$*-scheme X of dimension n.*

a) *We have $h^i(X, mD) = O(m^n)$ for all $i \geq 0$.*
b) *If D is nef, we have $h^i(X, mD) = O(m^{n-1})$ for all $i > 0$, hence*

$$h^0(X, mD) = m^n \frac{(D^n)}{n!} + O(m^{n-1}).$$

In particular, if D is ample, we have by Corollary 3.9

$$\lim_{m \to +\infty} \frac{h^0(X, mD)}{m^n} > 0.$$

We defined in (8) a big class as the sum of an effective and an ample class. Let us say that a $\mathbf{Q}$-Cartier $\mathbf{Q}$-divisor D is big if it can be written as the sum of an effective and an ample divisor. This is a numerical property since ampleness is.

Proposition 3.13 *A Cartier divisor D on a projective $\mathbf{k}$-scheme X is big if and only if*

$$\limsup_{m\to+\infty} \frac{h^0(X,mD)}{m^n} > 0. \tag{15}$$

In particular, if a $\mathbf{Q}$-Cartier $\mathbf{Q}$-divisor D is nef, it is big if and only if $(D^n) > 0$.

When D is big, the limsup in (15) is actually a limit, but this is difficult to prove.

Proof If D is big, we write it as $D = E + A$, with E effective and A ample. We may assume that E and A are divisors. Then $h^0(X,mD) \ge h^0(X,mA)$ for all m and (15) follows.

Assume conversely that (15) holds. Let A be an ample effective divisor on X. The exact sequence

$$0 \to \mathscr{O}_X(mD - A) \to \mathscr{O}_X(mD) \to \mathscr{O}_A(mD|_A) \to 0$$

induces an exact sequence

$$0 \to H^0(X, mD - A) \to H^0(X, mD) \to H^0(A, mD|_A).$$

Since A has dimension $n-1$, Proposition 3.12.a) gives $h^0(A, mD|_A) = O(m^{n-1})$, hence (15) implies $H^0(X, mD - A) \neq 0$ for infinitely many $m > 0$. For those m, we can write $mD - A \underset{\text{lin}}{\equiv} E$, with E effective, hence D is big. □

Ample divisors are nef and big, but not conversely (see Example 2.48). Nef and big divisors share many of the properties of ample divisors: for example, Proposition 3.12 shows that the dimensions of the spaces of sections of their successive multiples grow in the same fashion. They are however much more tractable; for instance, the pullback of a nef and big divisor by a generically finite morphism is still nef and big.

3.5 *Kodaira Dimension and Iitaka Fibrations*

Let D be a Cartier divisor on a projective $\mathbf{k}$-scheme X. If D is ample, the sections of mD define, for $m \gg 0$, a finite morphism $\psi_{mD}: X \to \mathbf{P}^N_{\mathbf{k}}$ (one can even show that ψ_{mD} is a closed embedding for all $m \gg 0$).

One may wonder what happens for a general (non-ample) Cartier divisor. When D is only big, we may write $m_0 D = E + A$, with E effective and A ample divisors, for some positive integer m_0. Among the sections of mm_0D, one then finds the sections of mA times s_E^m, where $\operatorname{div}(s_E) = E$. In other words, the composition of the rational map

$$\psi_{mm_0D}: X \dashrightarrow \mathbf{P}^N_{\mathbf{k}}$$

with a suitable linear projection $\mathbf{P}^N_{\mathbf{k}} \dashrightarrow \mathbf{P}^M_{\mathbf{k}}$ is the morphism

$$\psi_{mA}: X \longrightarrow \mathbf{P}^M_{\mathbf{k}}.$$

In particular, $\max_{m>0} \dim(\psi_{mD}(X)) = n$.

We make the following important definition.

Definition 3.14 (Kodaira Dimension) Let X be a projective normal variety and let D be a Cartier divisor on X. We define the *Kodaira dimension* of D by

$$\kappa(D) := \max_{m>0} \dim(\psi_{mD}(X)).$$

We make the convention

$$\kappa(D) = -\infty \quad \Longleftrightarrow \quad \forall m > 0 \quad H^0(X, mD) = 0$$

and we also have

$$\kappa(D) = 0 \quad \Longleftrightarrow \quad \max_{m>0} h^0(X, mD) = 1.$$

We just saw that big divisors have maximal Kodaira dimension $n := \dim(X)$. Conversely, if D has maximal Kodaira dimension n, some $m_0 D$ defines a rational map $\psi: X \dashrightarrow \mathbf{P}^N_{\mathbf{k}}$ with image $Y := \psi(X)$ of dimension n. Let $U \subset X$ be the largest smooth open subset on which ψ is defined. Since X is normal, we have $\operatorname{codim}_X(X \smallsetminus U) \geq 2$ and we can write $m_0 D|_U \underset{\text{lin}}{\equiv} \psi|_U^* H + E$, with E effective Cartier divisor on U. Since X is normal, we have for all $m > 0$

$$\begin{aligned} h^0(X, mm_0D) = h^0(U, mm_0D|_U) &\geq h^0(U, m\psi|_U^* H) \\ &\geq h^0(Y, mH|_Y) = \frac{\deg(Y)}{n!} m^n + O(m^{n-1}), \end{aligned}$$

hence D is big.

The Cartier divisors with maximal Kodaira dimension are therefore exactly the big divisors, so this property is numerical. This is not the case in general: although $\kappa(D)$ only depends on the linear equivalence class of the divisor D, since it is defined from the invertible sheaf $\mathscr{O}_X(D)$, it is not in general invariant under numerical equivalence. If D is a divisor of degree 0 on a curve X (so that $D \underset{\text{num}}{\equiv} 0$), one has $\kappa(D) = -\infty$ if $[D]$ is not a torsion element of $\operatorname{Pic}(X)$, and $\kappa(D) = 0$ otherwise.

When X is a smooth projective variety, the Kodaira dimension $\kappa(X)$ of the canonical divisor K_X (see Sect. 2.11) is an important invariant of the variety (called the Kodaira dimension of X). We say that X is *of general type* is K_X is big, i.e., if $\kappa(X) = \dim(X)$.

Examples 3.15

1) The Kodaira dimension of $\mathbf{P}^n_{\mathbf{k}}$ is $-\infty$.
2) If X is a smooth hypersurface of degree d in $\mathbf{P}^n_{\mathbf{k}}$, its Kodaira dimension is (see Example 2.49)

$$\kappa(X) = \begin{cases} -\infty & \text{if } d \le n; \\ 0 & \text{if } d = n+1; \\ \dim(X) = n-1 & \text{if } d > n+1. \end{cases}$$

3) If X is a curve, its Kodaira dimension is

$$\kappa(X) = \begin{cases} -\infty & \text{if } g(X) = 0; \\ 0 & \text{if } g(X) = 1; \\ 1 & \text{if } g(X) \ge 2. \end{cases}$$

There is a general structure theorem. For a Cartier divisor D on a projective variety X, we define the set

$$N(D) := \{m \ge 0 \mid H^0(X, mD) \ne 0\}.$$

It is a semi-group hence, if $N(D) \ne 0$ (i.e., if $\kappa(D) \ne -\infty$), all sufficiently large elements of $N(D)$ are multiples of a single largest positive integer which we denote by $e(D)$. When D is ample, and even only big, one has $e(D) = 1$.

Example 3.16 Let Y be a projective variety, let A be an ample divisor on Y, let E be an elliptic curve, and let B be an element of order $m \ge 2$ in the group $\mathrm{Pic}(E)$. On $X := Y \times E$, the divisor $D := \mathrm{pr}_1^* A + \mathrm{pr}_2^* B$ has Kodaira dimension $\dim(X) - 1$ and $e(D) = m$.

For each $m \in N(D)$, one can construct the rational map

$$\psi_{mD} \colon X \dashrightarrow \mathbf{P}(H^0(X, mD)^\vee).$$

Theorem 3.17 (Iitaka Fibration) *Let X be a projective normal variety and let D be a Cartier divisor on X. For all $m \in N(D)$ sufficiently large, the induced maps*

$$X \dashrightarrow \psi_{mD}(X)$$

are all birationally equivalent to a fixed algebraic fibration

$$X_\infty \longrightarrow Y_\infty$$

of normal projective varieties, with $\dim(Y_\infty) = \kappa(D)$ and such that the restriction of D to a very general fiber has Kodaira dimension 0.

To understand this theorem, we need to know what an algebraic fibration is. Essentially, we require the fibers to be connected, so when D is big (i.e., $\dim(X_\infty) = \dim(Y_\infty)$), the theorem says in particular that for each m sufficiently large, ψ_{mD} is birational onto its image.

The formal definition is the following.

Definition 3.18 An *(algebraic) fibration* is a projective morphism $\pi: X \to Y$ between varieties such that $\pi_* \mathscr{O}_X \simeq \mathscr{O}_Y$.

The composition of two fibrations is a fibration. When Y is normal, any projective birational morphism $\pi: X \to Y$ is a fibration.[10] Conversely, any fibration $\pi: X \to Y$ with $\dim(X) = \dim(Y)$ is birational (this follows for example from Proposition 3.19 below).

Since the closure of the image of a morphism π is defined by the ideal sheaf kernel of the canonical map $\mathscr{O}_Y \to \pi_* \mathscr{O}_X$, a fibration is surjective and, by Zariski's Main Theorem, its fibers are connected [8, Corollary III.11.3] and even geometrically connected [5, III, Corollaire (4.3.12)]. When Y is normal, the converse is true in characteristic 0.[11] More generally, any projective morphism $\rho: X \to Y$ between varieties factors as

$$\rho: X \xrightarrow{\pi} Y' \xrightarrow{u} Y,$$

where π is a fibration and u is finite. This is the Stein factorization and ρ is a fibration if and only if u is an isomorphism.

Proposition 3.19 *A surjective projective morphism $\rho: X \to Y$ between normal varieties is a fibration if and only if the corresponding finitely generated field extension* $\mathbf{k}(Y) \subset \mathbf{k}(X)$ *is algebraically closed.*

This proposition allows us to extend the definition of a fibration to rational maps between normal varieties.

Proof If the extension $\mathbf{k}(Y) \subset \mathbf{k}(X)$ is algebraically closed, we consider the Stein factorization $\rho: X \to Y' \xrightarrow{u} Y$. The extension $\mathbf{k}(Y) \subset \mathbf{k}(Y')$ is finite, hence algebraic, hence trivial since $\mathbf{k}(Y') \subset \mathbf{k}(X)$. The morphism u is then finite and birational, hence an isomorphism (because Y is normal), and ρ is a fibration.

Assume conversely that ρ is a fibration. Any element of $\mathbf{k}(Y)$ algebraic over $\mathbf{k}(X)$ generates a finite extension $\mathbf{k}(X) \subset \mathbf{K}$ contained in $\mathbf{k}(Y)$ which corresponds to a rational factorization $\rho: X \overset{\pi}{\dashrightarrow} Y' \overset{u}{\dashrightarrow} Y$, where $\mathbf{K} = \mathbf{k}(Y')$ and u is generically finite. Replacing X and Y' by suitable modifications, we may assume that π and u are morphisms. Since X is normal, ρ is still a fibration hence, for any affine open

[10]For any affine open subset $U \subset Y$, the ring extension $H^0(U, \mathscr{O}_Y) \subset H^0(U, \pi_* \mathscr{O}_X)$ is finite because π is projective and the quotient fields are the same because $\mathbf{k}(Y) = \mathbf{k}(X)$. Since $H^0(U, \mathscr{O}_Y)$ is integrally closed in $\mathbf{k}(Y)$, these rings are the same.

[11]In general, one needs to require that the generic fiber of π be geometrically integral. In positive characteristic, u might very well be a bijection without being an isomorphism (even if Y is normal: think of the Frobenius morphism).

subset $U \subset Y$, the inclusions

$$H^0(U, \mathscr{O}_Y) \subset H^0(U, u_*\mathscr{O}_{Y'}) \subset H^0(U, \rho_*\mathscr{O}_X) = H^0(U, \mathscr{O}_Y)$$

are equalities, hence u has degree 1 and $\mathbf{K} = \mathbf{k}(X)$. This proves that the extension $\mathbf{k}(Y) \subset \mathbf{k}(X)$ is algebraically closed. □

Remark 3.20 Here are some other properties of a fibration $\pi: X \to Y$.

- If X is normal, so is Y.
- For any Cartier divisor D on Y, one has $H^0(X, \pi^*D) \simeq H^0(Y, D)$ and $\kappa(X, \pi^*D) \simeq \kappa(Y, D)$.
- The induced map $\pi^*: \mathrm{Pic}(Y) \to \mathrm{Pic}(X)$ is injective.

4 Rational Curves on Varieties

Mori proved in 1979 a conjecture of Hartshorne characterizing projective spaces as the only smooth projective varieties with ample tangent bundle [15]. The techniques that Mori introduced to solve this conjecture have turned out to have more far reaching applications than Hartshorne's conjecture itself and we will explain them in this section.

4.1 Parametrizing Curves

Let $\mathbf{k}$ be a field and let C be a smooth (projective) curve over $\mathbf{k}$. Given a quasi-projective $\mathbf{k}$-variety X, we want to "parametrize" all curves $C \to X$. If T is a $\mathbf{k}$-scheme, a *family of curves from C to X parametrized by T* is a morphism $\rho: C \times T \to X$: for each closed point $t \in T$, one has a curve $c \mapsto \rho_t(c) := \rho(c, t)$ (defined over the field $k(t)$).

We want to construct a $\mathbf{k}$-scheme $\mathrm{Mor}(C, X)$ and a "universal family of curves"

$$\mathsf{ev}: C \times \mathrm{Mor}(C, X) \to X, \tag{16}$$

called the *evaluation map,* such that for any $\mathbf{k}$-scheme T, the correspondance between

- morphisms $\varphi: T \to \mathrm{Mor}(C, X)$ and
- families $\rho: C \times T \to X$ of curves parametrized by T

obtained by sending φ to

$$\rho(c, t) = \mathsf{ev}(c, \varphi(t))$$

is one-to-one. In other words, any family of curves parametrized by T is pulled back from the universal family (16) by a uniquely defined morphism $T \to \mathrm{Mor}(C, X)$.

Taking $T = \mathrm{Spec}(\mathbf{k})$, we see that $\mathbf{k}$-points of $\mathrm{Mor}(C,X)$ should be in one-to-one correspondence with curves $C \to X$.

Taking $T = \mathrm{Spec}(\mathbf{k}[\varepsilon]/(\varepsilon^2))$, we see that the Zariski tangent space to $\mathrm{Mor}(C,X)$ at a $\mathbf{k}$-point $[\rho]$ is isomorphic to the space of extensions of ρ to morphisms

$$\rho_\varepsilon \colon C \times \mathrm{Spec}\,\mathbf{k}[\varepsilon]/(\varepsilon^2) \to X$$

which should be thought of as first-order infinitesimal deformations of ρ.

Theorem 4.1 (Grothendieck, Mori) *Let* $\mathbf{k}$ *be a field, let* C *be a smooth* $\mathbf{k}$*-curve, let* X *be a smooth quasi-projective* $\mathbf{k}$*-variety, and let* $\rho\colon C \to X$ *be a curve.*

a) *There exist a* $\mathbf{k}$*-scheme* $\mathrm{Mor}(C,X)$*, locally of finite type, which parametrizes morphisms from* C *to* X*.*
b) *The Zariski tangent space to* $\mathrm{Mor}(C,X)$ *at* $[\rho]$ *is isomorphic to* $H^0(C,\rho^*T_X)$*.*
c) *Locally around* $[\rho]$*, the scheme* $\mathrm{Mor}(C,X)$ *is defined by* $h^1(C,\rho^*T_X)$ *equations in a smooth* $\mathbf{k}$*-variety of dimension* $h^0(C,\rho^*T_X)$*.*[12] *In particular, by Riemann–Roch,*[13]

$$\dim_{[\rho]} \mathrm{Mor}(C,X) \ge \chi(C,\rho^*T_X) = -(K_X \cdot C) + (1 - g(C))\dim(X). \tag{17}$$

We will not reproduce Grothendieck's general construction, since it is very nicely explained in [7], and we will only explain the much easier case $C = \mathbf{P}^1_{\mathbf{k}}$.

Any $\mathbf{k}$-morphism $\rho\colon \mathbf{P}^1_{\mathbf{k}} \to \mathbf{P}^N_{\mathbf{k}}$ can be written as

$$\rho(u,v) = (F_0(u,v),\dots,F_N(u,v)), \tag{18}$$

where $F_0,\dots,F_N$ are homogeneous polynomials in two variables, of the same degree d, with no non-constant common factor in $\mathbf{k}[U,V]$. Morphisms $\mathbf{P}^1_{\mathbf{k}} \to \mathbf{P}^N_{\mathbf{k}}$ of degree d are therefore parametrized by the (open) complement $\mathrm{Mor}_d(\mathbf{P}^1_{\mathbf{k}},\mathbf{P}^N_{\mathbf{k}})$ in $\mathbf{P}(\mathbf{k}[U,V]_d^{N+1})$ of the union, for all $e \in \{1,\dots,d\}$, of the (closed) images of the morphisms

$$\begin{aligned} \mathbf{P}(\mathbf{k}[U,V]_e) \times \mathbf{P}(\mathbf{k}[U,V]_{d-e}^{N+1}) &\longrightarrow \mathbf{P}(\mathbf{k}[U,V]_d^{N+1}) \\ (G,(G_0,\dots,G_N)) &\longmapsto (GG_0,\dots,GG_N). \end{aligned}$$

The evaluation map is

$$\begin{aligned} \mathrm{ev}\colon \mathbf{P}^1_{\mathbf{k}} \times \mathrm{Mor}_d(\mathbf{P}^1_{\mathbf{k}},\mathbf{P}^N_{\mathbf{k}}) &\longrightarrow \mathbf{P}^N_{\mathbf{k}} \\ ((u,v),\rho) &\longmapsto (F_0(u,v),\dots,F_N(u,v)). \end{aligned}$$

[12] In particular, a sufficient (but not necessary!) condition for $\mathrm{Mor}(C,X)$ to be smooth at $[\rho]$ is $H^1(C,\rho^*T_X) = 0$.

[13] We are using here a generalization of Theorem 2.39 to locally free sheaves of any rank.

Example 4.2 In the case $d = 1$, we can write $F_i(u, v) = a_i u + b_i v$, with $(a_0, \dots, a_N, b_0, \dots, b_N)$ in $\mathbf{P}^{2N+1}_{\mathbf{k}}$. The condition that $F_0, \dots, F_N$ have no common zeroes is equivalent to

$$\operatorname{rank}\begin{pmatrix} a_0 \cdots a_N \\ b_0 \cdots b_N \end{pmatrix} = 2.$$

Its (closed) complement Z in $\mathbf{P}^{2N+1}_{\mathbf{k}}$ is defined by the vanishing $\begin{vmatrix} a_i & a_j \\ b_i & b_j \end{vmatrix} = 0$ of all 2×2-minors. The evaluation map is

$$\begin{array}{rccc} \mathrm{ev}\colon & \mathbf{P}^1_{\mathbf{k}} \times (\mathbf{P}^{2N+1}_{\mathbf{k}} \smallsetminus Z) & \longrightarrow & \mathbf{P}^N_{\mathbf{k}} \\ & \big((u, v), (a_0, \dots, a_N, b_0, \dots, b_N)\big) & \longmapsto & \big(a_0 u + b_0 v, \dots, a_N u + b_N v\big). \end{array}$$

Finally, morphisms from $\mathbf{P}^1_{\mathbf{k}}$ to $\mathbf{P}^N_{\mathbf{k}}$ are parametrized by the disjoint union

$$\operatorname{Mor}(\mathbf{P}^1_{\mathbf{k}}, \mathbf{P}^N_{\mathbf{k}}) = \bigsqcup_{d \geq 0} \operatorname{Mor}_d(\mathbf{P}^1_{\mathbf{k}}, \mathbf{P}^N_{\mathbf{k}}) \tag{19}$$

of quasi-projective schemes.

When X is a (closed) subscheme of $\mathbf{P}^N_{\mathbf{k}}$ defined by homogeneous equations $G_1, \dots, G_m$, morphisms $\mathbf{P}^1_{\mathbf{k}} \to X$ of degree d are parametrized by the subscheme $\operatorname{Mor}_d(\mathbf{P}^1_{\mathbf{k}}, X)$ of $\operatorname{Mor}_d(\mathbf{P}^1_{\mathbf{k}}, \mathbf{P}^N_{\mathbf{k}})$ defined by the equations

$$G_j(F_0, \dots, F_N) = 0 \qquad \text{for all } j \in \{1, \dots, m\}.$$

Again, morphisms from $\mathbf{P}^1_{\mathbf{k}}$ to X are parametrized by the disjoint union

$$\operatorname{Mor}(\mathbf{P}^1_{\mathbf{k}}, X) = \bigsqcup_{d \geq 0} \operatorname{Mor}_d(\mathbf{P}^1_{\mathbf{k}}, X)$$

of quasi-projective schemes.[14]

We now make a very important remark. Assume that X can be defined by homogeneous equations $G_1, \dots, G_m$ with coefficients in a subring R of $\mathbf{k}$. If $\mathfrak{m}$ is a maximal ideal of R, one may consider the reduction $X_{\mathfrak{m}}$ of X modulo $\mathfrak{m}$: this is the subscheme of $\mathbf{P}^N_{R/\mathfrak{m}}$ defined by the reductions of the G_j modulo $\mathfrak{m}$.

---

[14]When X is only quasi-projective, embed it into some projective variety $\overline{X}$. There is an evaluation morphism

$$\mathrm{ev}\colon \mathbf{P}^1_{\mathbf{k}} \times \operatorname{Mor}(\mathbf{P}^1_{\mathbf{k}}, \overline{X}) \longrightarrow \overline{X}$$

and $\operatorname{Mor}(\mathbf{P}^1_{\mathbf{k}}, X)$ is the complement in $\operatorname{Mor}(\mathbf{P}^1_{\mathbf{k}}, \overline{X})$ of the image by the (proper) second projection of the closed subscheme $\mathrm{ev}^{-1}(\overline{X} \smallsetminus X)$.

Because the equations defining the complement of $\mathrm{Mor}_d(\mathbf{P}^1_{\mathbf{k}}, \mathbf{P}^N_{\mathbf{k}})$ in $\mathbf{P}(\mathbf{k}[U,V]_d^{N+1})$ have coefficients in $\mathbf{Z}$ and are the same for all fields, the scheme $\mathrm{Mor}_d(\mathbf{P}^1_{\mathbf{k}}, X)$ is defined over R and $\mathrm{Mor}_d(\mathbf{P}^1_{\mathbf{k}}, X_{\mathfrak{m}})$ is its reduction modulo $\mathfrak{m}$. In fancy terms, one may express this as follows: if $\mathscr{X}$ is a scheme over Spec (R), the R-morphisms $\mathbf{P}^1_R \to \mathscr{X}$ are parametrized by the R-points of a locally noetherian scheme

$$\mathrm{Mor}(\mathbf{P}^1_R, \mathscr{X}) \to \mathrm{Spec}\,(R)$$

and the fiber of a closed point $\mathfrak{m}$ is the space $\mathrm{Mor}(\mathbf{P}^1_{\mathbf{k}}, \mathscr{X}_{\mathfrak{m}})$.

4.2 Free Rational Curves and Uniruled Varieties

Let X be a smooth $\mathbf{k}$-variety of dimension n and let $\rho \colon \mathbf{P}^1_{\mathbf{k}} \to X$ be a non-constant morphism (this is called a *rational curve* on X). Any locally free coherent sheaf on $\mathbf{P}^1_{\mathbf{k}}$ is isomorphic to a direct sum of invertible sheaves, hence we can write

$$\rho^* T_X \simeq \mathscr{O}_{\mathbf{P}^1_{\mathbf{k}}}(a_1) \oplus \cdots \oplus \mathscr{O}_{\mathbf{P}^1_{\mathbf{k}}}(a_n), \tag{20}$$

with $a_1 \geq \cdots \geq a_n$ and $a_1 \geq 2$. If $H^1(\mathbf{P}^1_{\mathbf{k}}, \rho^* T_X)$ vanishes (this happens exactly when $a_n \geq -1$), the scheme $\mathrm{Mor}(\mathbf{P}^1_{\mathbf{k}}, X)$ is smooth at $[\rho]$ (Theorem 4.1). We investigate a stronger condition.

Definition 4.3 Let X be a $\mathbf{k}$-variety. A rational curve $\rho \colon \mathbf{P}^1_{\mathbf{k}} \to X$ is *free* if its image is contained in the smooth locus of X and $\rho^* T_X$ is generated by its global sections (this happens exactly when $a_n \geq 0$).

By Theorem 4.1, the scheme $\mathrm{Mor}(\mathbf{P}^1_{\mathbf{k}}, X)$ is smooth at points corresponding to free rational curves. The K_X-degree of a free curve is $-\sum a_i < 0$.

The importance of free curves comes from the following result. We will say that a $\mathbf{k}$-variety X is covered by rational curves, or is *uniruled*, if there is a $\mathbf{k}$-variety M and a dominant morphism

$$\mathbf{P}^1_{\mathbf{k}} \times M \to X$$

which does not contract $\mathbf{P}^1_{\mathbf{k}} \times \{m\}$ for some (hence for all) geometric point $m \in X(\overline{\mathbf{k}})$.[15] By the universal property, we may assume that M is a component of $\mathrm{Mor}(\mathbf{P}^1_{\mathbf{k}}, X)$. Using again this universal property, one checks that given any field extension L of $\mathbf{k}$, the $\mathbf{k}$-variety X is uniruled if and only if the L-variety X_L is uniruled.

[15] This is automatic if $\dim(M) = \dim(X) - 1$ and we can always reduce to that case, but the seemingly more general definition we gave is more flexible.

If X is uniruled and $\mathbf{k}$ is algebraically closed, there is a rational curve through every point of X. The converse holds if $\mathbf{k}$ is uncountable (use the fact that $\mathrm{Mor}(\mathbf{P}^1_{\mathbf{k}}, X)$ has countably many components). It is therefore common to work on an algebraically closed uncountable extension of the base field.

Proposition 4.4 *Let X be a smooth quasi-projective* **k***-variety.*

a) *If the rational curve $\rho\colon \mathbf{P}^1_{\mathbf{k}} \to X$ is free, the evaluation map*

$$\mathsf{ev}\colon \mathbf{P}^1_{\mathbf{k}} \times \mathrm{Mor}(\mathbf{P}^1_{\mathbf{k}}, X) \to X$$

is smooth at all points of $\mathbf{P}^1_{\mathbf{k}} \times \{[\rho]\}$, hence X is uniruled.

b) *Conversely, if X is uniruled and* **k** *is algebraically closed* of characteristic 0, *there is a* free *rational curve on X through a general point of X.*

Proof The tangent map to ev at $(t, [\rho])$ is the map

$$\begin{aligned} T_{\mathbf{P}^1_{\mathbf{k}},t} \oplus H^0(\mathbf{P}^1_{\mathbf{k}}, \rho^*T_X) &\longrightarrow T_{X,\rho(t)} \simeq (\rho^*T_X)_t \\ (u, \sigma) &\longmapsto T_t\rho(u) + \sigma(t). \end{aligned}$$

If ρ is free, this map is surjective because the evaluation map

$$H^0(\mathbf{P}^1_{\mathbf{k}}, \rho^*T_X) \longrightarrow (\rho^*T_X)_t$$

is. Since $\mathrm{Mor}(\mathbf{P}^1_{\mathbf{k}}, X)$ is smooth at $[\rho]$, the morphism ev is smooth at $(t, [\rho])$ and this proves a).

Conversely, if ev is dominant, it is smooth at a general point $(t, [\rho])$ because we are in characteristic 0. This implies that the map

$$H^0(\mathbf{P}^1_{\mathbf{k}}, \rho^*T_X) \to (\rho^*T_X)_t/\mathrm{Im}(T_t\rho) \tag{21}$$

is surjective. There is a commutative diagram

$$\begin{array}{ccc} H^0(\mathbf{P}^1_{\mathbf{k}}, \rho^*T_X) & \xrightarrow{\ a\ } & (\rho^*T_X)_t \\ \uparrow & & \uparrow {\scriptstyle T_t\rho} \\ H^0(\mathbf{P}^1_{\mathbf{k}}, T_{\mathbf{P}^1_{\mathbf{k}}}) & \xrightarrow{\ a'\ } & T_{\mathbf{P}^1_{\mathbf{k}},t}. \end{array}$$

Since a' is surjective, the image of a contains $\mathrm{Im}(T_t\rho)$. Since the map (21) is surjective, a is surjective. Hence ρ^*T_X is generated by global sections at one point. It is therefore generated by global sections and ρ is free. □

Corollary 4.5 *If X is a smooth projective uniruled variety over a field of characteristic 0, the* plurigenera $p_m(X) := h^0(X, \mathscr{O}_X(mK_X))$ *vanish for all $m > 0$, i.e., $\kappa(X) = -\infty$.*

The converse is conjectured to hold and has been proved in dimensions ≤ 3 (for curves, it is obvious since $p_1(X)$ is the genus of X; for surfaces, we have the Castelnuovo criterion: $p_{12}(X) = 0$ if and only if X is birationally isomorphic to a ruled surface).

Proof We may assume that the base field $\mathbf{k}$ is algebraically closed. By Proposition 4.4.b), there is a free rational curve $\rho\colon \mathbf{P}^1_{\mathbf{k}} \to X$ through a general point of X. Since $\rho^* K_X$ has negative degree, any section of $\mathcal{O}_X(mK_X)$ must vanish on $\rho(\mathbf{P}^1_{\mathbf{k}})$, hence on a dense subset of X, hence on X. □

Example 4.6 Let $\rho\colon \mathbf{P}^1_{\mathbf{k}} \to X$ be a rational curve on a surface X and let $C \subset X$ be its image. If ρ is free, C "moves" on X by Proposition 4.4.a), hence $(C^2) \geq 0$. Conversely, if C is smooth and $(C^2) \geq 0$, the *normal exact sequence*

$$0 \to T_C \to T_X|_C \to \mathcal{O}_C(C) \to 0$$

implies that C is free.

Example 4.7 If $\varepsilon\colon \widetilde{\mathbf{P}^2_{\mathbf{k}}} \to \mathbf{P}^2_{\mathbf{k}}$ is the blow up of one point, the exceptional curve E is not free because $(E^2) = -1$. If $C \subset \widetilde{\mathbf{P}^2_{\mathbf{k}}}$ is any other smooth rational curve, we write $C \underset{\text{lin}}{\equiv} d\varepsilon^* H - mE$ (Example 2.18). From $(C \cdot E) \geq 0$ (because $C \neq E$) and $(C \cdot (\varepsilon^* H - E)) \geq 0$ (because $\varepsilon^* H - E$ is nef by Example 2.44), we get $d \geq m \geq 0$, hence $(C^2) = d^2 - m^2 \geq 0$ and C is free by Example 4.6.

Example 4.8 On the blow up X of $\mathbf{P}^2_{\mathbf{C}}$ at nine general points, there are countably many rational curves with self-intersection -1 [8, Exercise V.4.15.(e)] and none of these curves are free by Example 4.6. The inverse image on X of a general line of $\mathbf{P}^2_{\mathbf{C}}$ is free by Example 4.6 again, since its self-intersection is 1.

Exercise 4.9 Let X be a subscheme of $\mathbf{P}^N_{\mathbf{k}}$ defined by equations of degrees $d_1, \ldots, d_s$ over an algebraically closed field. Assume $d_1 + \cdots + d_s < N$. Show that through any point of X, there is a line contained in X (we say that X *is covered by lines* and it is in particular uniruled; in characteristic 0, such a general line is free by Proposition 4.4.b), but in positive characteristic, it may happen that none of these lines are free; see [3, Exercise 2.5.3] for an example).

4.3 *Bend-and-Break Lemmas*

In this section, we explain the techniques that Mori invented to prove the conjecture of Hartshorne mentioned in the introduction to this section. The main idea is that if a curve deforms on a projective variety while passing through a fixed point, it must at some point break up with at least one rational component, hence the name "bend-and-break."

We work over an *algebraically closed field* $\mathbf{k}$.

The first bend-and-break lemma (which we will not prove) can be found in [15, Theorems 5 and 6]. It says that a curve deforming non-trivially while keeping a point fixed must break into several pieces, including a rational curve passing through the fixed point. We introduce one piece of notation: if c is a point of a curve C and x a point of a variety X, we let $\mathrm{Mor}(C,X;c \mapsto x)$ be the closed subscheme of $\mathrm{Mor}(C,X)$ that parametrizes morphisms $\rho: C \to X$ such that $\rho(c) = x$. In terms of the evaluation map (16), it is defined as

$$\mathrm{Mor}(C,X;c \mapsto x) := \mathrm{pr}_1\big((\{c\} \times \mathrm{Mor}(C,X)) \cap \mathrm{ev}^{-1}(x)\big). \tag{22}$$

Proposition 4.10 (Mori) *Let X be a projective variety defined over an algebraically closed field, let $\rho: C \to X$ be a smooth curve, and let c be a point on C. If* $\dim_{[\rho]} \mathrm{Mor}(C,X;c \mapsto \rho(c)) \geq 1$, *there exists a rational curve on X through $\rho(c)$.*

It follows from (22) that

$$\dim_{[\rho]} \mathrm{Mor}(C,X;c \mapsto \rho(c)) \geq \dim_{[\rho]} \mathrm{Mor}(C,X) - \dim(X).$$

According to (17), *when X is smooth along $\rho(C)$*, the hypothesis of the proposition is therefore fulfilled whenever

$$(-K_X \cdot C) \geq g(C)\dim(X) + 1. \tag{23}$$

Once we know there is a rational curve, it may under certain conditions be broken up into several components. This is the second bend-and-break lemma (which we will not prove either).

Proposition 4.11 (Mori) *Let X be a projective variety and let $\rho: \mathbf{P}^1_{\mathbf{k}} \to X$ be a rational curve, birational onto its image $C \subset X$. If*

$$\dim_{[\rho]}(\mathrm{Mor}(\mathbf{P}^1_{\mathbf{k}}, X; 0 \mapsto \rho(0), \infty \mapsto \rho(\infty))) \geq 2,$$

the curve C can be deformed to a connected union of at least two rational curves on X still passing through $\rho(0)$ and $\rho(\infty)$.

As above, *when X is smooth along $f(\mathbf{P}^1_{\mathbf{k}})$*, the hypothesis is fulfilled whenever

$$(-K_X \cdot \mathbf{P}^1_{\mathbf{k}}) \geq \dim(X) + 2. \tag{24}$$

4.4 Rational Curves on Fano Varieties

A Fano variety is a smooth projective variety X such that $-K_X$ is ample.

Example 4.12 The projective space $\mathbf{P}^n_{\mathbf{k}}$ is a Fano variety. More generally, any smooth complete intersection in $\mathbf{P}^n_{\mathbf{k}}$ defined by equations of degrees $d_1, \dots, d_s$ with $d_1 + \cdots + d_s \leq n$ is a Fano variety. A finite product of Fano varieties is a Fano variety.

We will apply the bend-and-break lemmas to show that any Fano variety X is covered by rational curves. Start from any curve $\rho: C \to X$; we want to show, using the estimate (23), that it deforms non-trivially while keeping a point fixed. Since $-K_X$ is ample, the intersection number $(-K_X \cdot C)$ is positive. But we need it to be greater than $g(C)\dim(X)$. Composing ρ with a cover $C' \to C$ of degree m multiplies $(-K_X \cdot C)$ by m, but also (roughly) multiplies $g(C)$ by m, *except in positive characteristic,* where the Frobenius morphism allows us to increase the degree of ρ without changing the genus of C. This gives in that case the required rational curve on X. Using the second bend-and-break lemma, we can bound the degree of this curve by a constant depending only on the dimension of X, and this is essential for the remaining step: reduction of the characteristic zero case to positive characteristic.

We explain this last step in a simple case. Assume for a moment that X and x are defined over $\mathbf{Z} \subset \mathbf{k}$; for almost all prime numbers p, the reduction of X modulo p is a Fano variety of the same dimension hence there is a rational curve (defined over the algebraic closure of the field $\mathbf{F}_p$) through x. This means that the scheme $\mathrm{Mor}(\mathbf{P}^1_{\mathbf{k}}, X; 0 \mapsto x)$, which is defined over $\mathbf{Z}$, has a geometric point modulo almost all primes p. Since we can moreover bound the degree of the curve by a constant independent of p, we are in fact in a quasi-projective subscheme of $\mathrm{Mor}(\mathbf{P}^1_{\mathbf{k}}, X; 0 \mapsto x)$, and this implies that it has a point over $\overline{\mathbf{Q}}$, hence over $\overline{\mathbf{k}}$. In general, X and x are only defined over some finitely generated ring but these ideas still work.

Theorem 4.13 (Mori) *Let X be a Fano variety of dimension $n > 0$ defined over an algebraically closed field. Through any point of X there is a rational curve of $(-K_X)$-degree at most $n + 1$.*

Even over $\mathbf{C}$, there is no known proof of this theorem that uses only transcendental methods. A consequence of the theorem is that Fano varieties are uniruled (see Sect. 4.2). However, some Fano varieties contain no free rational curves ([11]; by Proposition 4.4.b), this may only happen in positive characteristics).

Proof Assume that the field $\mathbf{k}$ has characteristic $p > 0$; choose a smooth curve $\rho\colon C \to X$ through a point x of X and a point c of C such that $f(c) = x$. Consider the ($\mathbf{k}$-linear) Frobenius morphism $C_1 \to C$;[16] it has degree p, but C_1 and C being isomorphic as abstract schemes have the same genus. Iterating the construction, we get a morphism $F_m\colon C_m \to C$ of degree p^m between curves of the same genus which we compose with ρ. Then,

$$(-K_X \cdot C_m) - ng(C_m) = -p^m(K_X \cdot C) - ng(C)$$

[16] If $F\colon \mathbf{k} \to \mathbf{k}$ is the Frobenius morphism, the $\mathbf{k}$-scheme C_1 fits into the Cartesian diagram

$$\begin{array}{ccc} C_1 & \xrightarrow{F} & C \\ \downarrow & \searrow & \downarrow \\ \mathrm{Spec}\,(\mathbf{k}) & \xrightarrow{F} & \mathrm{Spec}\,(\mathbf{k}.) \end{array}$$

In other words, C_1 is the scheme C, but $\mathbf{k}$ acts on $\mathscr{O}_{C_1}$ via pth powers.

is positive for m large enough. By Proposition 4.10 and (23), there exists a (birational) rational curve $\rho'\colon \mathbf{P}^1_{\mathbf{k}} \to X$, with say $\rho'(0) = x$. If

$$(-K_X \cdot \mathbf{P}^1_{\mathbf{k}}) \geq n+2,$$

one can, by Proposition 4.11 and (23), break up the rational curve $\rho'(\mathbf{P}^1_{\mathbf{k}})$ into at least two (rational) pieces. Since $-K_X$ is ample, the component passing through x has smaller $(-K_X)$-degree, and we can repeat the process as long as $(-K_X \cdot \mathbf{P}^1_{\mathbf{k}}) \geq n+2$, until we get to a rational curve of $(-K_X)$-degree no more than $n+1$.

This proves the theorem in positive characteristic. Assume now that $\mathbf{k}$ has characteristic 0. Embed X in some projective space, where it is defined by a finite set of equations, and let R be the (finitely generated) subring of $\mathbf{k}$ generated by the coefficients of these equations and the coordinates of x.[17] There is a projective scheme $\mathscr{X} \to \operatorname{Spec}(R)$ with an R-point x_R, such that X is obtained from its generic fiber by base change from the quotient field $K(R)$ of R to $\mathbf{k}$. The geometric generic fiber is a Fano variety of dimension n, defined over the subfield $\overline{K(R)}$ of $\mathbf{k}$. There is a dense open subset U of $\operatorname{Spec}(R)$ over which $\mathscr{X}$ is smooth of dimension n [6, th. 12.2.4.(iii)]. Since ampleness is an open property [6, cor. 9.6.4], we may even, upon shrinking U, assume that for each maximal ideal $\mathfrak{m}$ of R in U, the geometric fiber $X_{\mathfrak{m}}$ is a Fano variety of dimension n, defined over $\overline{R/\mathfrak{m}}$.

We state and prove two classical properties of the finitely generated domain R.

Lemma 4.14 *Let R be a finitely generated domain. We have the following:*

- *for each maximal ideal $\mathfrak{m}$ of R, the field $R/\mathfrak{m}$ is finite;*
- *maximal ideals (i.e., closed points) are dense in* $\operatorname{Spec}(R)$.

Proof The first item is proved as follows. The field $R/\mathfrak{m}$ is a finitely generated $(\mathbf{Z}/\mathbf{Z} \cap \mathfrak{m})$-algebra, hence is finite over the quotient field of $\mathbf{Z}/\mathbf{Z} \cap \mathfrak{m}$ by the Nullstellensatz (which says that if k is a field and K a finitely generated k-algebra which is a field, K is a finite extension of k; see [13, Theorem 5.2]). If $\mathbf{Z} \cap \mathfrak{m} = 0$, the field $R/\mathfrak{m}$ is therefore a finite dimensional $\mathbf{Q}$-vector space; pick a basis $(e_1, \dots, e_m)$. If $x_1, \dots, x_r$ generate the $\mathbf{Z}$-algebra $R/\mathfrak{m}$, there exists an integer q such that qx_j belongs to $\mathbf{Z}e_1 \oplus \cdots \oplus \mathbf{Z}e_m$ for each j. This implies

$$\mathbf{Q}e_1 \oplus \cdots \oplus \mathbf{Q}e_m = R/\mathfrak{m} \subset \mathbf{Z}[1/q]e_1 \oplus \cdots \oplus \mathbf{Z}[1/q]e_m,$$

which is absurd; therefore, $\mathbf{Z}/\mathbf{Z} \cap \mathfrak{m}$ is finite and so is $R/\mathfrak{m}$.

For the second item, we need to show that the intersection of all maximal ideals of R is $\{0\}$. Let a be a non-zero element of R and let $\mathfrak{n}$ be a maximal ideal of the localization R_a. The field $R_a/\mathfrak{n}$ is finite by the first item hence its subring $R/R \cap \mathfrak{n}$ is a finite domain hence a field. Therefore $R \cap \mathfrak{n}$ is a maximal ideal of R which is in the open subset $\operatorname{Spec}(R_a)$ of $\operatorname{Spec}(R)$ (in other words, $a \notin \mathfrak{n}$). □

[17] This ring was $\mathbf{Z}$ in the brief description of the proof before the statement of the theorem.

We now go back to the proof of Theorem 4.13. As proved in Sect. 4.1, there is a quasi-projective scheme

$$\rho\colon \mathrm{Mor}_{\le n+1}(\mathbf{P}^1_R, \mathscr{X}; 0 \mapsto x_R) \to \mathrm{Spec}(R)$$

which parametrizes rational curves of degree at most $n+1$ on $\mathscr{X}$ through x_R.

Let $\mathfrak{m}$ be a maximal ideal of R. Since the field $R/\mathfrak{m}$ is finite by the lemma, hence of positive characteristic, what we just saw implies that the (geometric) fiber of ρ over a closed point of the dense open subset U of $\mathrm{Spec}(R)$ is non-empty; it follows that the image of ρ, which is a constructible[18] subset of $\mathrm{Spec}(R)$ by Chevalley's theorem [8, Exercise II.3.19], contains all closed points of U, therefore is dense by the second item of the lemma, hence contains the generic point [8, Exercise II.3.18.(b)]. This implies that the generic fiber is non-empty; it has therefore a geometric point, which corresponds to a rational curve on X through x, of degree at most $n+1$, defined over $\overline{K(R)}$, hence over $\mathbf{k}$.[19] □

4.5 *Rational Curves on Varieties Whose Canonical Divisor is not nef*

We proved in Theorem 4.13 that when X is a smooth projective variety (defined over an algebraically closed field) such that $-K_X$ is ample (i.e., when X is a Fano variety), there is a rational curve through any point of X. The theorem we state in this section considerably weakens the hypothesis: assuming only that K_X has negative degree on *one* curve C, it says that there is a rational curve through any point of C.

Note that the proof of Theorem 4.13 goes through in positive characteristic under this weaker hypothesis and does prove the existence of a rational curve through any point of C. However, to pass to the characteristic 0 case, one needs to bound the degree of this rational curve with respect to some ample divisor by some "universal" constant so that we deal only with a quasi-projective part of a morphism space. Apart from these technical difficulties, the ideas are essentially the same as in Theorem 4.13. This theorem is the main result of [14]. We will not prove it.

Theorem 4.15 (Miyaoka-Mori) *Let X be a projective variety defined over an algebraically closed field, let H be an ample divisor on X, and let $\rho\colon C \to X$ be a smooth curve such that X is smooth along $\rho(C)$ and $(K_X \cdot C) < 0$. Given any point*

[18] A constructible subset is a finite union of locally closed subsets.

[19] The "universal" bound on the degree of the rational curve is essential for the proof.

For those who know some elementary logic, the statement that there exists a rational curve of $(-K_X)$-degree at most some constant on a projective Fano variety X is a first-order statement, so the Lefschetz principle tells us that if it is valid on all algebraically closed fields of positive characteristics, it is valid over all algebraically closed fields.

x on $\rho(C)$, there exists a rational curve Γ on X through x with

$$(H\cdot\Gamma)\le 2\dim(X)\,\frac{(H\cdot C)}{(-K_X\cdot C)}.$$

When X is smooth, the rational curve can be broken up, using Proposition 4.11 and (24), into several pieces (of lower H-degrees) keeping any two points fixed [one of which being on $\rho(C)$], until one gets a rational curve Γ which satisfies $(-K_X\cdot\Gamma)\le\dim(X)+1$ in addition to the bound on the H-degree.

It is nevertheless useful to have a more general statement allowing X to be singular. It implies for example that a normal projective variety X with ample ($\mathbf{Q}$-Cartier) anticanonical divisor is covered by rational curves of $(-K_X)$-degree at most $2\dim(X)$.

Finally, a simple corollary of this theorem is that *the canonical divisor of a smooth projective complex variety which contains no rational curves is nef.*

Our next result generalizes Theorem 4.13 and shows that varieties with nef but not numerically trivial anticanonical divisor are also covered by rational curves. This class of varieties is much larger than the class of Fano varieties.

Theorem 4.16 *If X is a smooth projective variety with $-K_X$ nef,*

- *either K_X is numerically trivial,*
- *or X is uniruled.*

Proof We may assume that the base field is algebraically closed and uncountable. Let H be the restriction to X of a hyperplane in some embedding $X\subset\mathbf{P}^N_{\mathbf{k}}$. Assume $(K_X\cdot H^{n-1})=0$, where $n:=\dim(X)$. For any curve $C\subset X$, there exist hypersurfaces $H_1,\dots,H_{n-1}$ in $\mathbf{P}^N_{\mathbf{k}}$, of respective degrees $d_1,\dots,d_{n-1}$, such that the scheme-theoretic intersection $Z:=X\cap H_1\cap\cdots\cap H_{n-1}$ has pure dimension 1 and contains C. Since $-K_X$ is nef, we have

$$0\le(-K_X\cdot C)\le(-K_X\cdot Z)=d_1\cdots d_{n-1}(-K_X\cdot H^{n-1})=0,$$

hence K_X is numerically trivial and we are in the first case.

Assume now $(K_X\cdot H^{n-1})<0$. Let x be a point of X and let C be the normalization of the intersection of $n-1$ general hyperplane sections through x. By Bertini's theorem, C is an irreducible curve and $(K_X\cdot C)=(K_X\cdot H^{n-1})<0$. By Theorem 4.15, there is a rational curve on X which passes through x and we are in the second case. □

An abelian variety has trivial canonical divisor and contains no rational curves.

Exercise 4.17 Let X be a smooth projective variety with $-K_X$ big. Show that X is covered by rational curves.

Exercise 4.18 Let X be a smooth projective variety, let $Y\subset X$ be a smooth hypersurface, and let $C\to X$ be a curve such that $(K_X\cdot C)=0$ and $(Y\cdot C)<0$. Prove that X contains a rational curve.

5 The Cone Theorem

5.1 *Cone of Curves*

We copy the definition of Weil divisors and define a (real) 1-cycle on a projective scheme X as a formal linear combination $\sum_C t_C C$, where the C are (embedded, possibly singular) integral projective curves in X and the t_C are real numbers.[20] Say that two such 1-cycles C and C' on X are numerically equivalent if

$$(D \cdot C) = (D \cdot C')$$

for all Cartier divisors D on X. The quotient of the vector space of 1-cycles by this equivalence relation is a real vector space $N_1(X)_{\mathbf{R}}$ which is canonically dual to $N^1(X)_{\mathbf{R}}$, hence finite-dimensional (Theorem 2.16).

In $N_1(X)_{\mathbf{R}}$, we define the convex cone $\mathrm{NE}(X)$ of classes of effective 1-cycles and its closure $\overline{\mathrm{NE}}(X)$. The nef cone $\mathrm{Nef}(X) \subset N^1(X)_{\mathbf{R}} = N_1(X)_{\mathbf{R}}^{\vee}$ (defined in (7)) is then simply the dual cone to $\mathrm{NE}(X)$ (or to $\overline{\mathrm{NE}}(X)$). Since the ample cone is the interior of the nef cone (Corollary 2.47), we obtain, from an elementary general property of dual cones, a useful characterization of ample classes.

Theorem 5.1 (Kleiman) *Let X be a projective variety. A Cartier divisor D on X is ample if and only if $D \cdot z > 0$ for all non-zero z in $\overline{\mathrm{NE}}(X)$.*

5.2 *Mori's Cone Theorem*

We fix the following notation: if D is a Cartier divisor on X and S a subset of $N_1(X)_{\mathbf{R}}$, we set

$$S_{D \geq 0} = \{z \in S \mid D \cdot z \geq 0\}$$

and similarly for $S_{D \leq 0}$, $S_{D > 0}$, and $S_{D < 0}$.

Roughly speaking, Mori's cone theorem describes, for a smooth projective variety X, the part of the cone $\overline{\mathrm{NE}}(X)$ which has negative intersection with the canonical class, i.e., $\overline{\mathrm{NE}}(X)_{K_X < 0}$.

An *extremal ray* of a closed convex cone $V \subset \mathbf{R}^m$ is a half-line $\mathbf{R}^{\geq 0} x \subset V$ such that

$$\forall v, v' \in V \qquad v + v' \in \mathbf{R}^{\geq 0} x \implies v, v' \in \mathbf{R}^{\geq 0} x.$$

[20] To be consistent with our previous definition of "curve on X", we should perhaps say instead that C is a smooth projective curve with a morphism $C \to X$ which is birational onto its image.

If V contains no lines, it is the convex hull of its extremal rays. By Theorem 5.1, this is the case for the cone $\overline{\mathrm{NE}}(X)$ when X is projective

Theorem 5.2 (Mori's Cone Theorem) *Let X be a smooth projective variety. There exists a countable family $(\Gamma_i)_{i\in I}$ of rational curves on X such that*

$$0 < (-K_X \cdot \Gamma_i) \le \dim(X) + 1$$

and

$$\overline{\mathrm{NE}}(X) = \overline{\mathrm{NE}}(X)_{K_X \ge 0} + \sum_{i\in I} \mathbf{R}^{\ge 0}[\Gamma_i], \tag{25}$$

where the $\mathbf{R}^{\ge 0}[\Gamma_i]$ are all the extremal rays of $\overline{\mathrm{NE}}(X)$ that meet $N_1(X)_{K_X<0}$; these rays are locally discrete in that half-space.

An extremal ray that meets $N_1(X)_{K_X<0}$ is called *K_X-negative.*

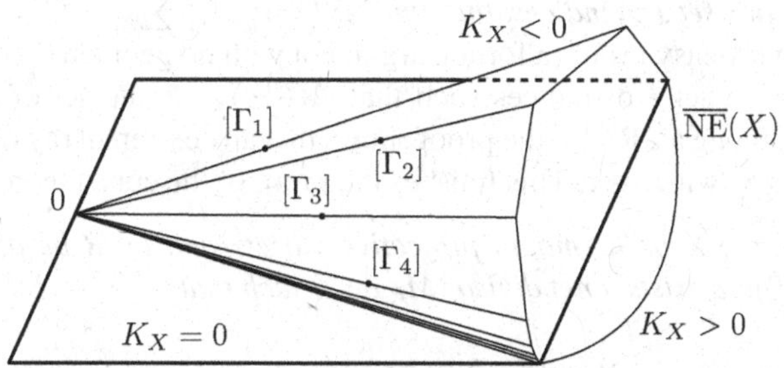

Sketch of Proof. It follows from the description (19) of the scheme $\mathrm{Mor}(\mathbf{P}^1_{\mathbf{k}}, X)$ that there are only countably many classes of rational curves on X. Pick a representative Γ_i for each such class z_i that satisfies $0 < -K_X \cdot z_i \le \dim(X) + 1$.
First step: the rays $\mathbf{R}^{\ge 0} z_i$ are locally discrete in the half-space $N_1(X)_{K_X<0}$.

Let H be an ample divisor on X and let $\varepsilon > 0$. If $z_i \in N_1(X)_{K_X+\varepsilon H<0}$, we have $(H \cdot \Gamma_i) < \frac{1}{\varepsilon}(-K_X \cdot \Gamma_i) \le \frac{1}{\varepsilon}(\dim(X) + 1)$, and there are only finitely many such classes of curves on X (because the corresponding curves lie in a quasi-projective part of $\mathrm{Mor}(\mathbf{P}^1_{\mathbf{k}}, X)$).
Second step: $\overline{\mathrm{NE}}(X)$ is equal to the closure of $V = \overline{\mathrm{NE}}(X)_{K_X\ge 0} + \sum_i \mathbf{R}^{\ge 0} z_i$. If not, there exists an $\mathbf{R}$-divisor M on X which is non-negative on $\overline{\mathrm{NE}}(X)$ (it is in particular nef), positive on $\overline{V} \smallsetminus \{0\}$ and which vanishes at some non-zero point z of $\overline{\mathrm{NE}}(X)$. This point cannot be in V, hence $K_X \cdot z < 0$.

Choose a norm on $N_1(X)_{\mathbf{R}}$ such that $\|[C]\| \ge 1$ for each irreducible curve C (this is possible since the set of classes of irreducible curves is discrete). We may assume, upon replacing M with a multiple, that $M \cdot v \ge 2\|v\|$ for all v in $\overline{V}$. We have

$$2\dim(X)(M \cdot z) = 0 < -K_X \cdot z.$$

Since the class $[M]$ is a limit of classes of ample $\mathbf{Q}$-divisors, and z is a limit of classes of effective rational 1-cycles, there exist an ample $\mathbf{Q}$-divisor H and an effective 1-cycle Z such that

$$2\dim(X)(H \cdot Z) < (-K_X \cdot Z) \qquad \text{and} \qquad H \cdot v \geq \|v\| \tag{26}$$

for all v in $\overline{V}$. We may further assume, by throwing away the other components, that each component C of Z satisfies $(-K_X \cdot C) > 0$.

Since the class of every rational curve Γ on X such that $(-K_X \cdot \Gamma) \leq \dim(X) + 1$ is in $\overline{V}$ (either it is in $\overline{\mathrm{NE}}(X)_{K_X \geq 0}$, or $(-K_X \cdot \Gamma) > 0$ and $[\Gamma]$ is one of the z_i), we have $(H \cdot \Gamma) \geq \|[\Gamma]\| \geq 1$ by (26) and the choice of the norm. Since X is smooth, the bend-and-break Theorem 4.15 implies

$$2\dim(X) \frac{(H \cdot C)}{(-K_X \cdot C)} \geq 1$$

for every component C of Z. This contradicts the first inequality in (26).

Third step: for any set J of indices, the cone $\overline{\mathrm{NE}}(X)_{K_X \geq 0} + \sum_{j \in J} \mathbf{R}^{\geq 0} z_j$ is closed. We skip this relatively easy proof (a formal argument with no geometric content).

If we choose a set I of indices such that $(\mathbf{R}^{\geq 0} z_j)_{j \in I}$ is the set of all (distinct) extremal rays among all $\mathbf{R}^{\geq 0} z_i$, the proof shows that any extremal ray of $\overline{\mathrm{NE}}(X)_{K_X < 0}$ is spanned by a z_i, with $i \in I$. This finishes the proof of the cone theorem. □

Corollary 5.3 *Let X be a smooth projective variety and let R be a K_X-negative extremal ray. There exists a nef divisor M_R on X such that*

$$R = \{z \in \overline{\mathrm{NE}}(X) \mid M_R \cdot z = 0\}.$$

For any such divisor, $mM_R - K_X$ is ample for all $m \gg 0$.

Any such divisor M_R will be called a *supporting divisor* for R.

Proof With the notation of the proof of the cone theorem, there exists a (unique) element i_0 of I such that $R = \mathbf{R}^{\geq 0} z_{i_0}$. By the third step of the proof, the cone

$$V = V_{I \smallsetminus \{i_0\}} = \overline{\mathrm{NE}}(X)_{K_X \geq 0} + \sum_{i \in I,\, i \neq i_0} \mathbf{R}^{\geq 0} z_i$$

is closed and is strictly contained in $\overline{\mathrm{NE}}(X)$ since it does not contain R. This implies that there exists a linear form which is non-negative on $\overline{\mathrm{NE}}(X)$, positive on $V \smallsetminus \{0\}$ and which vanishes at some non-zero point of $\overline{\mathrm{NE}}(X)$, hence on R since $\overline{\mathrm{NE}}(X) = V + R$. The intersection of the interior of the dual cone V^* and the *rational* hyperplane $R^{\perp}$ is therefore non-empty, hence contains an integral point: there exists a divisor M_R on X which is positive on $V \smallsetminus \{0\}$ and vanishes on R. It is in particular nef and the first statement of the corollary is proved.

Choose a norm on $N_1(X)_{\mathbf{R}}$ and let a be the (positive) minimum of M_R on the set of elements of V with norm 1. If b is the maximum of K_X on the same compact, the divisor $mM_R - K_X$ is positive on $V \smallsetminus \{0\}$ for m rational greater than b/a, and positive on $R \smallsetminus \{0\}$ for $m \geq 0$, hence ample for $m > \max(b/a, 0)$ by Kleiman's criterion (Theorem 5.1). This finishes the proof of the corollary. □

Exercise 5.4 Let X be a smooth projective variety and let $A_1, \ldots, A_r$ be ample divisors on X. Show that $K_X + A_1 + \cdots + A_r$ is nef for all $r \geq \dim(X) + 1$.

Exercise 5.5 (A Rationality Result) Let X be a smooth projective variety whose canonical divisor is not nef and let M be a nef divisor on X. Set

$$r := \sup\{t \in \mathbf{R} \mid M + tK_X \text{ nef}\}.$$

a) Show that r is a (finite) non-negative real number.
b) Let $(\Gamma_i)_{i \in I}$ be the (non-empty and countable) set of rational curves on X that appears in the Cone Theorem 5.2. Show

$$r = \inf_{i \in I} \frac{(M \cdot \Gamma_i)}{(-K_X \cdot \Gamma_i)}.$$

c) Deduce that one can write $r = \frac{u}{v}$, with u and v relatively prime integers and $0 < v \leq \dim(X) + 1$, and that there exists a K_X-negative extremal ray R of $\overline{\mathrm{NE}}(X)$ such that

$$((M + rK_X) \cdot R) = 0.$$

5.3 Contractions of K_X-Negative Extremal Rays

Let X be a smooth projective variety and let R be an extremal ray of $\overline{\mathrm{NE}}(X)$. A contraction of R is a fibration $c_R: X \twoheadrightarrow Y$ (see Definition 3.18) which contracts exactly those curves in X whose class is in R. A contraction can only exist when R is generated by the class of a curve; one can show that the contraction is then unique (up to isomorphisms of the bases).

The fact that K_X-negative extremal rays can be contracted is essential to the realization of Mori's minimal model program. This is only known in characteristic 0 (so say over $\mathbf{C}$) in all dimensions (and in any characteristic for surfaces) as a consequence of the following powerful theorem, whose proof is beyond the intended scope (and methods) of these notes.

Theorem 5.6 (Base-Point-Free Theorem (Kawamata)) *Let X be a smooth complex projective variety and let D be a nef divisor on X such that $aD - K_X$ is nef and big for some $a \in \mathbf{Q}^{>0}$. Then mD is generated by its global sections for all $m \gg 0$.*

Corollary 5.7 *Let X be a smooth complex projective variety and let R be a K_X-negative extremal ray of $\overline{\mathrm{NE}}(X)$.*

a) *The contraction $c_R: X \twoheadrightarrow Y$ of R exists. It is given by the Stein factorization of the morphism defined by any sufficiently high multiple of any supporting divisor of R.*
b) *Let C be any curve on X with class in R. There is an exact sequence*

$$\begin{array}{ccccccc} 0 \longrightarrow \mathrm{Pic}(Y) & \xrightarrow{c_R^*} & \mathrm{Pic}(X) & \longrightarrow & \mathbf{Z} \\ & & [D] & \longmapsto & (D \cdot C) \end{array}$$

and $\rho(Y) = \rho(X) - 1$.
c) *The restriction of $-K_X$ to any fiber of c_R is ample.*

Item b) implies that there are isomorphisms

$$N^1(Y)_{\mathbf{R}} \xrightarrow{\sim} R^{\perp} \subset N^1(X)_{\mathbf{R}} \qquad \text{and} \qquad N^1(Y)_{\mathbf{R}} \xrightarrow{\sim} N^1(X)_{\mathbf{R}}/\langle R \rangle .$$

Proof of the corollary. Let M_R be a supporting divisor for R, as in Corollary 5.3. By the same corollary and Theorem 5.6, mM_R is generated by its global sections for $m \gg 0$. The contraction c_R is given by the Stein factorization of the induced morphism $X \to \mathbf{P}^N_{\mathbf{k}}$. This proves a). Note for later use that there exists a Cartier divisor D_m on Y such that $mM_R \underset{\text{lin}}{\equiv} c_R^* D_m$.

For b), we saw in Remark 3.20 that c_R^* is injective. Let now D be a divisor on X such that $(D \cdot C) = 0$. Proceeding as in the proof of Corollary 5.3, we see that the divisor $mM_R + D$ is nef for all $m \gg 0$ and vanishes only on R. It is therefore a supporting divisor for R hence some multiple $m'(mM_R + D)$ also defines its contraction. Since the contraction is unique, it is c_R and there exists a Cartier divisor $E_{m,m'}$ on Y such that $m'(mM_R + D) \underset{\text{lin}}{\equiv} c_R^* E_{m,m'}$. We obtain $D \underset{\text{lin}}{\equiv} c_R^*(E_{m,m'+1} - E_{m,m'} - D_m)$ and this finishes the proof of b).

For c), let $F \subset X$ be a fiber of c_R and let $z \in \overline{\mathrm{NE}}(F)$ be non-zero. Since $z \in \overline{\mathrm{NE}}(X)$ and $mM_R - K_X$ is ample for m sufficiently large (Corollary 5.3), we have by Theorem 5.1 $(mM_R - K_X) \cdot z > 0$. Since $z \in \overline{\mathrm{NE}}(F)$, we have $M_R \cdot z = 0$, hence $(-K_X) \cdot z > 0$. By Theorem 5.1, this proves that $-K_X$ is ample on F. □

5.4 Various Types of Contractions

Let X be a smooth complex projective variety and let R be a K_X-negative extremal ray, with contraction $c_R: X \twoheadrightarrow Y$ (Corollary 5.7). The curves contracted by c_R are exactly those whose class is in R. Their union locus$(R) \subset X$ is called the *locus* of R.

Since c_R is a fibration, either $\dim(Y) < \dim(X)$, or c_R is birational (Proposition 3.19). In the latter case, Zariski's Main Theorem says that locus$(R) = \pi^{-1}(\pi(\text{locus}(R)))$, the fibers of locus$(R) \to c_R(\text{locus}(R))$ are connected and everywhere positive-dimensional (in particular, $c_R(\text{locus}(R))$ has codimension at least 2 in Y), and c_R induces an isomorphism $X \smallsetminus \text{locus}(R) \xrightarrow{\sim} Y \smallsetminus c_R(\text{locus}(R))$.

There are three cases:

- $\text{locus}(R) = X$, so $\dim(c_R(X)) < \dim(X)$ and c_R is a *fiber contraction;*
- $\text{locus}(R)$ is a divisor in X and c_R is a *divisorial contraction;*
- $\text{locus}(R)$ has codimension at least 2 in X and c_R is a *small contraction.*

In the case of a divisorial contraction, the locus is always an irreducible divisor. In the case of a small contraction, the locus may be disconnected.

Proposition 5.8 *Let X be a smooth complex projective variety and let R be a K_X-negative extremal ray of $\overline{\text{NE}}(X)$ with contraction c_R. Then* locus(R) *is covered by* rational *curves contracted by c_R.*

Proof Any point x in $\text{locus}(R)$ is on some irreducible curve C whose class is in R. Let M_R be a (nef) supporting divisor for R (as in Corollary 5.3), let H be an ample divisor on X, and let m be an integer such that

$$m > 2\dim(X)\frac{(H \cdot C)}{(-K_X \cdot C)}.$$

By Proposition 4.15, applied with the ample divisor $mM_R + H$, there exists a rational curve Γ through x such that

$$\begin{aligned} 0 &< ((mM_R + H) \cdot \Gamma) \\ &\le 2\dim(X)\frac{((mM_R + H) \cdot C)}{(-K_X \cdot C)} \\ &= 2\dim(X)\frac{(H \cdot C)}{(-K_X \cdot C)} \\ &< m. \end{aligned}$$

It follows that the integer $(M_R \cdot \Gamma)$ must vanish, and $(H \cdot \Gamma) < m$: the class $[\Gamma]$ is in R hence Γ is contained in $\text{locus}(R)$. This proves the proposition. □

Exercise 5.9 Let $X \to \mathbf{P}^n_{\mathbf{k}}$ be the blow up of two distinct points. Determine the cone $\overline{\text{NE}}(X)$ and its extremal rays, and for each extremal ray, describe its contraction (see Exercise 2.45).

Exercise 5.10 Let X be a smooth projective variety of dimension n over an algebraically closed field of any characteristic. Let $R = \mathbf{R}^{\ge 0}z \subset \overline{\text{NE}}(X)$ be a K_X-negative extremal ray and let M_R be a supporting divisor for R (Corollary 5.3).

a) Let $C \subset X$ be an irreducible curve such that

$$(M_R \cdot C) < \frac{1}{2n}(-K_X \cdot C).$$

Prove that C is contained in locus(R).

b) If $(M_R^n) > 0$, prove that there exists an integral hypersurface $Y \subset X$ such that $Y \cdot z < 0$, hence locus$(R) \neq X$.
c) Show the converse: if locus$(R) \neq X$, then $(M_R^n) > 0$.
d) Explain why, in characteristic 0, b) and c) follow from the existence of a contraction of R (Corollary 5.7).

5.5 *Fiber-Type Contractions*

Let X be a smooth complex projective variety and let R be a K_X-negative extremal ray with contraction $c_R: X \twoheadrightarrow Y$ *of fiber type,* i.e., $\dim(Y) < \dim(X)$. It follows from Proposition 5.8.a) that *X is covered by rational curves* (contained in fibers of c_R). Moreover, a general fiber F of c_R is smooth and $-K_F = (-K_X)|_F$ is ample (Corollary 5.7.c)): F is a Fano variety as defined in Sect. 4.4.

The normal variety Y may be singular, but not too much. Recall that a variety is *locally factorial* if its local rings are unique factorization domains. This is equivalent to saying that all Weil divisors are Cartier divisors.

Proposition 5.11 *Let X be a smooth complex projective variety and let R be a K_X-negative extremal ray of $\overline{\mathrm{NE}}(X)$. If the contraction $c_R: X \twoheadrightarrow Y$ is of fiber type, Y is locally factorial.*

Proof Let C be an irreducible curve whose class generates R (Theorem 5.2). Let D be a prime Weil divisor on Y. Let c_R^0 be the restriction of c_R to $c_R^{-1}(Y_{\mathrm{reg}})$ and let D_X be the closure in X of $(c_R^0)^*(D \cap Y_{\mathrm{reg}})$.

The Cartier divisor D_X is disjoint from a general fiber of c_R hence has intersection 0 with C. By Corollary 5.7.b), there exists a Cartier divisor D_Y on Y such that $D_X \underset{\mathrm{lin}}{\equiv} c_R^* D_Y$. Since $c_{R*}\mathscr{O}_X \simeq \mathscr{O}_Y$, by the projection formula, the Weil divisors D and D_Y are linearly equivalent on Y_{reg} hence on Y [8, Proposition II.6.5.(b)]. This proves that D is a Cartier divisor and Y is locally factorial. □

Example 5.12 (A Projective Bundle is a Fiber Contraction) Let $\mathscr{E}$ be a locally free sheaf of rank $r \geq 2$ over a smooth projective variety Y and let $X = \mathbf{P}(\mathscr{E})$,[21] with projection $\pi: X \to Y$. If ξ is the class of the invertible sheaf $\mathscr{O}_X(1)$, we have

$$K_X = -r\xi + \pi^*(K_Y + \det(\mathscr{E})).$$

If L is a line contained in a fiber of π, we have $(K_X \cdot L) = -r$. The class $[L]$ spans a K_X-negative ray whose contraction is π.

[21] For once, we follow Grothendieck's notation: for a locally free sheaf $\mathscr{E}$, the projectivization $\mathbf{P}(\mathscr{E})$ is the space of *hyperplanes* in the fibers of $\mathscr{E}$.

Example 5.13 (A Fiber Contraction Which is not a Projective Bundle) Let C be a smooth curve of genus g, let d be a positive integer, and let $\mathrm{Pic}^d(C)$ be the Jacobian of C which parametrizes isomorphism classes of invertible sheaves of degree d on C.

Let C_d be the symmetric product of d copies of C; the Abel-Jacobi map $\pi_d \colon C_d \to \mathrm{Pic}^d(C)$ is a $\mathbf{P}^{d-g}$-bundle for $d \geq 2g - 1$ hence is the contraction of a K_{C_d}-negative extremal ray by Example 5.12. In general, the fibers of π_d are still all projective spaces (of varying dimensions). If L_d is a line in a fiber, we have

$$(K_{C_d} \cdot L_d) = g - d - 1.$$

Indeed, the formula holds for $d \geq 2g-1$ by 5.12. Assume it holds for d; use a point of C to get an embedding $\iota \colon C_{d-1} \to C_d$. Then $(\iota^* C_{d-1} \cdot L_d) = 1$ and the adjunction formula yields

$$\begin{aligned}(K_{C_{d-1}} \cdot L_{d-1}) &= (\iota^*(K_{C_d} + C_{d-1}) \cdot L_{d-1}) \\ &= ((K_{C_d} + C_{d-1}) \cdot \iota_* L_{d-1}) \\ &= ((K_{C_d} + C_{d-1}) \cdot L_d), \\ &= (g - d - 1) + 1,\end{aligned}$$

which proves the formula by descending induction on d.

It follows that for $d \geq g$, the (surjective) map π_d is the contraction of the K_{C_d}-negative extremal ray $\mathbf{R}^{\geq 0}[L_d]$. It is a fiber contraction for $d > g$. For $d = g + 1$, the generic fiber is $\mathbf{P}^1_{\mathbf{k}}$, but there are larger-dimensional fibers when $g \geq 3$, so the contraction is not a projective bundle.

5.6 Divisorial Contractions

Let X be a smooth complex projective variety and let R be a K_X-negative extremal ray whose contraction $c_R \colon X \twoheadrightarrow Y$ is *divisorial*. It follows from Proposition 5.8.b) and its proof that the locus of R is an irreducible divisor E such that $E \cdot z < 0$ for all $z \in R \smallsetminus \{0\}$.

Again, Y may be singular (see Example 5.18), but not too much. We say that a scheme is *locally* **Q***-factorial* if any Weil divisor has a non-zero multiple which is a Cartier divisor. One can still intersect any Weil divisor D with a curve C on such a variety: choose a positive integer m such that mD is a Cartier divisor and set

$$(D \cdot C) = \frac{1}{m} \deg \mathcal{O}_C(mD).$$

This number is however only rational in general.

Proposition 5.14 *Let X be a smooth complex projective variety and let R be a* K_X-*negative extremal ray of* $\overline{\mathrm{NE}}(X)$. *If the contraction* $c_R: X \twoheadrightarrow Y$ *is divisorial, Y is locally* **Q**-*factorial.*

Proof Let C be an irreducible curve whose class generates R (Theorem 5.2). Let D be a prime Weil divisor on Y. Let $c_R^0: c_R^{-1}(Y_{\mathrm{reg}}) \to Y_{\mathrm{reg}}$ be the morphism induced by c_R and let D_X be the closure in X of $c_R^{0*}(D \cap Y_{\mathrm{reg}})$.

Let E be the locus of R. Since $(E \cdot C) \neq 0$, there exist integers $a \neq 0$ and b such that $aD_X + bE$ has intersection 0 with C. By Corollary 5.7.b), there exists a Cartier divisor D_Y on Y such that $aD_X + bE \underset{\mathrm{lin}}{\equiv} c_R^* D_Y$.

Lemma 5.15 *Let X and Y be varieties, with Y normal, and let* $\pi: X \to Y$ *be a proper birational morphism. Let F an effective Cartier divisor on X whose support is contained in the exceptional locus of* π. *We have*

$$\pi_* \mathcal{O}_X(F) \simeq \mathcal{O}_Y.$$

Proof Since this is a statement which is local on Y, it is enough to prove $H^0(Y, \mathcal{O}_Y) \simeq H^0(Y, \pi_*\mathcal{O}_X(F))$ when Y is affine. By Zariski's Main Theorem, we have

$$H^0(Y, \mathcal{O}_Y) \simeq H^0(Y, \pi_*\mathcal{O}_X) \simeq H^0(X, \mathcal{O}_X),$$

hence

$$H^0(Y, \mathcal{O}_Y) \simeq H^0(X, \mathcal{O}_X) \subset H^0(X, \mathcal{O}_X(F)) \subset H^0(X \smallsetminus E, \mathcal{O}_X(F))$$

and, letting E be the exceptional locus of π,

$$H^0(X \smallsetminus E, \mathcal{O}_X(F)) \simeq H^0(X \smallsetminus E, \mathcal{O}_X) \simeq H^0(Y \smallsetminus \pi(E), \mathcal{O}_Y) \simeq H^0(Y, \mathcal{O}_Y),$$

where the last isomorphism holds because Y is normal and $\pi(E)$ has codimension at least 2 in Y [8, Exercise III.3.5]. All these spaces are therefore isomorphic, hence the lemma. □

Using the lemma, we get

$$\mathcal{O}_{Y_{\mathrm{reg}}}(D_Y) \simeq c^0_{R*}\mathcal{O}_{c_R^{-1}(Y_{\mathrm{reg}})}(aD_X + bE) \simeq \mathcal{O}_{Y_{\mathrm{reg}}}(aD) \otimes c^0_{R*}\mathcal{O}_{X^0}(bE) \simeq \mathcal{O}_{Y_{\mathrm{reg}}}(aD),$$

hence the Weil divisors aD and D_Y are linearly equivalent on Y. It follows that Y is locally **Q**-factorial. □

Example 5.16 (A Smooth Blow Up is a Divisorial Contraction) Let Y be a smooth projective variety, let Z be a smooth subvariety of Y of codimension c, and let $\pi: X \to Y$ be the blow up of Z, with exceptional divisor E. We have [8, Exercise II.8.5.(b)]

$$K_X = \pi^* K_Y + (c-1)E.$$

Any fiber F of $E \to Z$ is isomorphic to $\mathbf{P}^{c-1}$ and $\mathcal{O}_F(E)$ is isomorphic to $\mathcal{O}_F(-1)$. If L is a line contained in F, we have $(K_X \cdot L) = -(c-1)$; the class $[L]$ therefore spans a K_X-negative ray whose contraction is π.

Example 5.17 (A Divisorial Contraction Which is not a Smooth Blow Up) We keep the notation of Example 5.13. The (surjective) map $\pi_g \colon C_g \twoheadrightarrow \operatorname{Pic}^g(C)$ is the contraction of the K_{C_g}-negative extremal ray $\mathbf{R}^{\geq 0}[L_g]$. Its locus is, by Riemann–Roch, the divisor

$$\{D \in C_g \mid h^0(C, K_C - D) > 0\}$$

and its image in $\operatorname{Pic}^g(C)$ has dimension $g-2$. The general fiber over this image is $\mathbf{P}^1_{\mathbf{k}}$, but there are bigger fibers when $g \geq 6$, because the curve C has a g^1_{g-2}, and the contraction is not a smooth blow up.

Example 5.18 (A Divisorial Contraction with Singular Image) Let Z be a smooth projective threefold and let C be an irreducible curve in Z whose only singularity is a node. The blow up Y of Z along C is normal and its only singularity is an ordinary double point q. This is checked by a local calculation: locally analytically, the ideal of C is generated by xy and z, where x, y, z form a system of parameters. The blow up is

$$\{((x, y, z), [u, v]) \in \mathbf{A}^3_{\mathbf{k}} \times \mathbf{P}^1_{\mathbf{k}} \mid xyv = zu\}.$$

It is smooth except at the point $q = ((0, 0, 0), [0, 1])$. The exceptional divisor is the $\mathbf{P}^1_{\mathbf{k}}$-bundle over C with local equations $xy = z = 0$.

The blow up $\pi \colon X \to Y$ at q is smooth. It contains the proper transform E of the exceptional divisor of Y and an exceptional divisor Q, which is a smooth quadric. The intersection $E \cap Q$ is the union of two lines L_1 and L_2 belonging to the two different rulings of Q. Let $\tilde{E} \to E$ and $\tilde{C} \to C$ be the normalizations; each fiber of $\tilde{E} \to \tilde{C}$ is a smooth rational curve, except over the two preimages p_1 and p_2 of the node of C, where it is the union of two rational curves meeting transversally. Over p_i, one of these curves maps to L_i, the other one to the same rational curve L. *It follows that $L_1 + L$ and $L_2 + L$, hence also L_1 and L_2, are numerically equivalent on X*; they have the same class ℓ.

Any curve contracted by π is contained in Q hence its class is a multiple of ℓ. A local calculation shows that $\mathcal{O}_Q(K_X)$ is of type $(-1, -1)$, hence $K_X \cdot \ell = -1$. The ray $\mathbf{R}^{\geq 0}\ell$ is K_X-negative and its (divisorial) contraction is π (hence $\mathbf{R}^{\geq 0}\ell$ is extremal).[22]

Exercise 5.19 Let X be a smooth complex projective Fano variety with Picard number at least 2. Assume that X has an extremal ray whose contraction $X \to Y$

[22] This situation is very subtle: although the completion of the local ring $\mathcal{O}_{Y,q}$ is not factorial (it is isomorphic to $\mathbf{k}[[x, y, z, u]]/(xy - zu)$, and the equality $xy = zu$ is a decomposition into a product of irreducibles in two different ways) the fact that L_1 is numerically equivalent to L_2 implies that the ring $\mathcal{O}_{Y,q}$ is factorial (see [16, (3.31)]).

maps a hypersurface $E \subset X$ to a point. Show that X also has an extremal contraction whose fibers are all of dimension ≤ 1 (*Hint:* consider a ray R such that $(E \cdot R) > 0$.)

5.7 Small Contractions and Flips

Let X be a smooth complex projective variety and let R be a K_X-negative extremal ray whose contraction $c_R\colon X \twoheadrightarrow Y$ is *small.*

The following proposition shows that Y is very singular: it is not even locally $\mathbf{Q}$-factorial, which means that one cannot intersect Weil divisors and curves on Y.

Proposition 5.20 *Let Y be a normal and locally $\mathbf{Q}$-factorial variety and let $\pi\colon X \to Y$ be a birational proper morphism. Every irreducible component of the exceptional locus of π has codimension 1 in X.*

Proof Let E be the exceptional locus of π, let $x \in E$, and set $y = \pi(x)$; identify the quotient fields $K(Y)$ and $K(X)$ by the isomorphism π^*, so that $\mathscr{O}_{Y,y} \subsetneq \mathscr{O}_{X,x} \subset K(X)$. Since $\mathscr{O}_{Y,y}/\mathfrak{m}_{Y,y} \simeq \mathscr{O}_{X,x}/\mathfrak{m}_{X,x} \simeq \mathbf{k}$ (because $\mathbf{k}$ is algebraically closed and x and y are closed points), there exists $t \in \mathfrak{m}_{X,x} \smallsetminus \mathscr{O}_{Y,y}$. Since $t \in K(Y)$, we may write its divisor on Y as the difference of two effective (Weil) divisors D' and D'' without common components.

Since Y is locally $\mathbf{Q}$-factorial, there exists a positive integer m such that mD' and mD'' are Cartier divisors, hence define elements u and v of $\mathscr{O}_{Y,y}$ such that $t^m = \frac{u}{v}$. Both are actually in $\mathfrak{m}_{Y,y}$: v because t^m is not in $\mathscr{O}_{Y,y}$ (otherwise, t would be since $\mathscr{O}_{Y,y}$ is integrally closed), and $u = t^m v$ because it is in $\mathfrak{m}_{X,x} \cap \mathscr{O}_{Y,y} = \mathfrak{m}_{Y,y}$. But $u = v = 0$ defines a subscheme Z of Y containing y, of codimension 2 in some neighborhood of y (it is the intersection of the codimension 1 subschemes mD' and mD''), whereas $\pi^{-1}(Z)$ is defined by $t^m v = v = 0$ hence by the sole equation $v = 0$: it has codimension 1 in X, hence is contained in E. It follows that there is a codimension 1 component of E through every point of E, which proves the proposition. □

Since it is impossible to do anything useful with Y, Mori's idea is that there should exist instead another (mildly singular) projective variety X^+ with a small contraction $c^+\colon X^+ \to Y$ such that K_{X^+} has positive degree on curves contracted by c^+. The map c^+ (or sometimes the resulting rational map $(c^+)^{-1} \circ c\colon X \dashrightarrow X^+$) is called a *flip.*

Definition 5.21 Let $c\colon X \twoheadrightarrow Y$ be a small contraction between normal projective varieties. Assume that K_X is $\mathbf{Q}$-Cartier and $-K_X$ is ample on all fibers of c. A *flip* of c is a small contraction $c^+\colon X^+ \to Y$ such that

- X^+ is a projective normal variety;
- K_{X^+} is $\mathbf{Q}$-Cartier and ample on all fibers of c^+.

The *existence* of a flip of the small contraction of a negative extremal ray has only been shown recently in all dimensions in characteristic 0 ([2]; see also [4, cor. 2.5], and [1] in dimension 3 and characteristics > 5).

Proposition 5.22 *Let X be a locally $\mathbf{Q}$-factorial complex projective variety and let $c: X \twoheadrightarrow Y$ be a* small *contraction of a K_X-negative extremal ray R. If the flip $X^+ \twoheadrightarrow Y$ exists, the variety X^+ is locally $\mathbf{Q}$-factorial with Picard number $\rho(X)$.*

Proof The composition $\varphi = c^{-1} \circ c^+ : X^+ \dashrightarrow X$ is an isomorphism in codimension 1, hence induces an isomorphism between the Weil divisor class groups of the normal varieties X and X^+ [8, Proposition II.6.5.(b)]. Let D^+ be a Weil divisor on X^+ and let D be the corresponding Weil divisor on X. Let C be an irreducible curve whose class generates R, let r be a rational number such that $((D + rK_X) \cdot C) = 0$, and let m be an integer such that mD, mrK_X, and mrK_{X^+} are Cartier divisors (the fact that K_{X^+} is $\mathbf{Q}$-Cartier is part of the definition of a flip!). By Corollary 5.7.b), there exists a Cartier divisor D_Y on Y such that $m(D + rK_X) \underset{\text{lin}}{\equiv} c^* D_Y$, and

$$mD^+ = \varphi^*(mD) \underset{\text{lin}}{\equiv} (c^+)^* D_Y - \varphi^*(mrK_X) \underset{\text{lin}}{\equiv} (c^+)^* D_Y - mrK_{X^+}$$

is a Cartier divisor. This proves that X^+ is locally $\mathbf{Q}$-factorial. Moreover, φ^* induces an isomorphism between $N^1(X)_{\mathbf{R}}$ and $N^1(X^+)_{\mathbf{R}}$, hence the Picard numbers are the same. □

Exercise 5.23 Let V be a $\mathbf{k}$-vector space of dimension n and let $r \in \{1, \ldots, n-1\}$. Let $\mathsf{Gr}(r, V)$ be the *Grassmannian* that parametrizes vector subspaces of V of codimension r and set

$$X := \{(W, [u]) \in \mathsf{Gr}(r, V) \times \mathbf{P}(\mathrm{End}(V)) \mid u(W) = 0\}.$$

a) Show that X is smooth irreducible of dimension $r(2n - r) - 1$, that $\mathrm{Pic}(X) \simeq \mathbf{Z}^2$, and that the projection $\mathsf{pr}_1 : X \to \mathsf{Gr}(r, V)$ is a K_X-negative extremal contraction.
b) Show that

$$Y := \mathsf{pr}_2(X) = \{[u] \in \mathbf{P}(\mathrm{End}(V)) \mid \mathrm{rank}(u) \le r\}$$

is irreducible of dimension $r(2n - r) - 1$. It can be proved that Y is normal. If $r \ge 2$, show that Y is not locally $\mathbf{Q}$-factorial and that $\mathrm{Pic}(Y) \simeq \mathbf{Z}[\mathscr{O}_Y(1)]$. What happens when $r = 1$?

5.8 The Minimal Model Program

Given a projective variety X defined over an algebraically closed field $\mathbf{k}$, one may try to find another projective variety birationally isomorphic to X and which is "as simple as possible." More formally, we define, on the set $\mathscr{C}_X$ of all (isomorphism classes of) smooth projective varieties birationally isomorphic to X, a relation as follows: if X_0 and X_1 are in $\mathscr{C}_X$, we write $X_0 \succeq X_1$ if there is a birational *morphism*

$X_0 \to X_1$. This defines an ordering on $\mathscr{C}_X$ and we look for *minimal elements* in $\mathscr{C}_X$ or even, if we are optimisitic, for the *smallest element* of $\mathscr{C}_X$.

When X is a smooth surface, it has a smooth minimal model obtained by contracting all exceptional curves on X. If X is not uniruled, this minimal model has nef canonical divisor and is the smallest element in $\mathscr{C}_X$. When X is uniruled, this minimal model is not unique, and is either a ruled surface or $\mathbf{P}^2_{\mathbf{k}}$.

The next proposition (which we will not prove) shows that smooth projective varieties with nef canonical bundles are minimal in the above sense. They are called *minimal models*.

Proposition 5.24 *Let X and Y be smooth projective varieties and let $\pi : X \to Y$ be a birational morphism which is not an isomorphism. There exists a rational curve C on X contracted by π such that $(K_X \cdot C) < 0$.*

In particular, if K_X is nef, X is a minimal element in $\mathscr{C}_X$.

A few warnings about minimal models:

- uniruled varieties do not have minimal models, since they carry free curves, on which the canonical class has negative degree;
- there exist smooth projective varieties which are not uniruled but which are not birational to any *smooth* projective variety with nef canonical bundle;[23]
- in dimension at least 3, minimal models may not be unique, although any two are isomorphic in codimension 1 [3, 7.18].

Starting from X, Mori's idea is to try to simplify X by contracting K_X-negative extremal rays, hoping to end up with a variety X_0 which either has a fiber contraction [in which case X_0, hence also X, is covered by rational curves (see Sect. 5.5)] or has nef canonical divisor (hence no K_{X_0}-negative extremal rays). However, three main problems arise.

- The end-product of a contraction is usually singular. This means that to continue Mori's program, *we must allow singularities.* This is very bad from our point of view, since most of our methods do not work on singular varieties. Completely different methods are required.
- One must determine what kind of singularities must be allowed. Whichever choices we make, the singularities of the target of a small contraction are too severe and one needs to perform a flip. So we have the problem of *existence of flips.*
- One needs to know that the process terminates. The Picard number decreases for a fiber or divisorial contraction, but not for a flip! So we have the additional problem of *termination of flips:* do there exist infinite sequences of flips?

In characteristic 0, the first two problems have been overcome: the first one by the introduction of cohomological methods to prove the cone theorem on (mildly) singular varieties, the second one more recently in [2] (see [4, cor. 2.5]). As for the

[23]This is the case for any desingularization of the quotient X of an abelian variety of dimension 3 by the involution $x \mapsto -x$ [18, 16.17]; a minimal model here is X itself, but it is singular.

third point, flips are only known to terminate in dimension ≤ 4 ([4, cor. 2.8], [9, Theorem 5-1-15]).

Many additional problems arise over fields of positive characteristic: resolution of singularities is not known (in dimensions ≥ 4), which makes even the definition of the kind of singularities that one wants to allow difficult, and most vanishing theorems fail, which makes cohomological methods difficult to apply. For a nice overview of the main differences with the characteristic 0 case and the latest developments, we refer the reader to [1].

Acknowledgements These are notes for the five 1-hour lectures I gave for the CIMPA-CIMAT-ICTP School "Moduli of Curves," which took place at the Centro de Investigación en Matemáticas (CIMAT) in Guanajuato, México, 22 February–4 March 2016. I would like to thank the organizers, and especially Leticia Brambila-Paz, for making this event possible in this very warm place and for attracting a great number of students from all over the world.

References

1. C. Birkar, Existence of flips and minimal models for 3-folds in char *p*. Ann. Sci. Éc. Norm. Supér. **49**, 169–212 (2016)
2. C. Birkar, P. Cascini, C.D. Hacon, J.M. McKernan, Existence of minimal models for varieties of log general type. J. Am. Math. Soc. **23**, 405–468 (2010)
3. O. Debarre, *Higher-Dimensional Algebraic Geometry*. Universitext (Springer, New York, 2001)
4. S. Druel, Existence de modèles minimaux pour les variétés de type général (d'après Birkar, Cascini, Hacon et McKernan), Séminaire Bourbaki, Exp. 982, 2007/2008. Astérisque **326**, 1–38 (2009)
5. A. Grothendieck, Eléments de géométrie algébrique III, 1. Inst. Hautes Études Sci. Publ. Math. **11** (1966)
6. A. Grothendieck, Eléments de géométrie algébrique IV, 3. Inst. Hautes Études Sci. Publ. Math. **28** (1966)
7. A. Grothendieck, Techniques de construction et théorèmes d'existence en géométrie algébrique IV : les schémas de Hilbert, Séminaire Bourbaki. Exp. 221 (1960/61). Astérisque hors série 6, Soc. Math. Fr. (1997)
8. R. Hartshorne, *Algebraic Geometry*. Graduate Texts in Mathematics vol. 52 (Springer, New York, 1977)
9. Y. Kawamata, K. Matsuda, K. Matsuki, Introduction to the minimal model problem, in *Algebraic Geometry, Sendai, 1985*, ed. by T. Oda. Advanced Studies in Pure Mathematics, vol. 10 (North-Holland, Amsterdam, 1987), pp. 283–360
10. S. Kleiman, Towards a numerical theory of ampleness. Ann. Math. **84**, 293–344 (1966)
11. J. Kollár, Nonrational hypersurfaces. J. Am. Math. Soc. **8**, 241–249 (1995)
12. J. Kollár, *Rational Curves on Algebraic Varieties*. Ergebnisse der Mathematik und ihrer Grenzgebiete, vol. 32 (Springer, Berlin, 1996)
13. H. Matsumura, *Commutative Ring Theory*, Cambridge Studies in Advanced Mathematics, 2nd edn., vol. 8 (Cambridge University Press, Cambridge/New York, 1989)
14. Y. Miyaoka, S. Mori, A numerical criterion for uniruledness. Ann. Math. **124**, 65–69 (1986)
15. S. Mori, Projective manifolds with ample tangent bundles. Ann. Math. **110**, 593–606 (1979)
16. S. Mori, Threefolds whose canonical bundles are not numerically effective. Ann. Math. **116**, 133–176 (1982)

17. M. Reid, Chapters on algebraic surfaces, in *Complex Algebraic Geometry (Park City, UT, 1993)*, IAS/Park City Mathematics Series, vol. 3 (American Mathematical Society, Providence, RI, 1997), pp. 3–159
18. K. Ueno, *Classification Theory of Algebraic Varieties and Compact Complex Spaces*. Springer Lecture Notes, vol. 439 (Springer, New York, 1975)

Progress on Syzygies of Algebraic Curves

Gavril Farkas

1 Introduction

The terms *syzygy* was originally used in astronomy and refers to three celestial bodies (for instance Earth-Sun-Moon) lying on a straight line. In mathematics, the term was introduced in 1850 by Sylvester, one of the greatest mathematical innovators and neologists[1] of all times. For Sylvester a syzygy is a linear relation between certain functions with arbitrary functional coefficients, which he called syzygetic multipliers. We quote from Sylvester's original paper in the Cambridge and Dublin Journal of Mathematics **5** (1850), page 276: *The members of any group of functions, more than two in number, whose nullity is implied in the relation of double contact, whether such group form a complete system or not, must be in syzygy.*

Most of the original applications, due to Cayley, Sylvester and others involved Classical Invariant Theory, where the syzygies in question were algebraic relations between the invariants and covariants of binary forms. An illustrative example in this sense is represented by the case of a general binary cubic form $f(x, y) = ax^3 + 3bx^2y + 3cxy^2 + dy^3$. One can show that there exists a covariant T of degree 3 and

[1]Sylvester is responsible for a remarkable number of standard mathematical terms like matrix, discriminant, minor, Jacobian, Hessian, invariant, covariant and many others. Some of his other terms have not stuck, for instance his *derogatory matrices*, that is, matrices whose characteristic polynomial differs from the minimal polynomial are all but forgotten today.

G. Farkas (✉)
Humboldt-Universität zu Berlin, Institut für Mathematik, Unter den Linden 6, 10099 Berlin, Germany
e-mail: farkas@math.hu-berlin.de

L. Brambila Paz et al. (eds.), *Moduli of Curves*, Lecture Notes of the Unione Matematica Italiana 21, DOI 10.1007/978-3-319-59486-6_4

order 3 (that is, of bidegree $(3, 3)$ in the coefficients of f and in the variables x and y respectively), having the rather unforgiving form

$$T := (a^2d - 3abc + 2b^3)x^3 + 3(abd + b^2c - 2ac^2)x^2y - 3(acd - 2b^2d + bc^2)xy^2 \\ -(ad^2 - 3bcd + 2c^3)y^3.$$

Denoting by D and by H the discriminant and the Hessian of the form f respectively, these three covariants are related by the following *syzygy* of order 6 and degree 6:

$$4H^3 = Df^2 - T^2.$$

The mathematical literature of the nineteenth century is full of results and methods of finding syzygies between invariants of binary forms, and sophisticated algorithms, often based on experience and intuition rather than on solid proofs, have been devised.

Sylvester pursued an ill-fated attempt to unite mathematics and poetry, which he deemed both to be guided by comparable concerns with formal relations between quantities. In his treatise *Laws of verse*, he even introduced the concept of *phonetic syzygy* as a repetition of syllables in certain rhymes and wrote poems to illustrate the principle of poetic syzygy. Not surprisingly, his ideas and terminology in this direction have not become widespread.

Returning to mathematics, it was Hilbert's landmark paper [27] from 1890 that not only put an end to Classical Invariant Theory in its constructive form propagated by the German and the British schools, but also introduced syzygies as objects of pure algebra. *Hilbert's Syzygy Theorem* led to a new world of free resolutions, higher syzygies and homological algebra. Although Hilbert's original motivation was Invariant Theory, his ideas had immediate and widespread impact, influencing the entire development of commutative algebra and algebraic geometry.

In algebraic geometry, the first forays into syzygies of algebraic varieties came from two different directions. In Germany, Brill and M. Noether pursued a long-standing program of bringing algebra into the realm of Riemann surfaces and thus making Riemann's work rigorous. Although the curves Brill and Noether were primarily concerned with were plane curves with ordinary singularities, they had a profound understanding of the importance of the canonical linear system of a curve and raised for the first time the question of describing canonical curves by algebraic equations. In Italy, in 1893 Castelnuovo's [8] using purely geometric methods which he also employed in the proof of his bound on the degree of a curve in projective space, showed that for a curve $C \subseteq \mathbf{P}^3$ of degree d, hypersurfaces of degree $n \geq d-2$ cut out the complete linear system $|\mathcal{O}_C(n)|$. At this point, we mention the work of Petri [31] (a student of Lindemann, who became a teacher but remained under the strong influence of Max and Emmy Noether). It revisits a topic that had already been considered by Enriques in 1919 and it gives a complete proof on the presentation of the generators of the ideal of a canonical curve using the algebraic methods of Brill, Noether (and Hilbert). In Petri's work, whose importance would only be recognized much later with the advent of modern Brill-Noether theory, the structure of the equations of canonical curves come to the very center of investigation.

Once the foundations of algebraic geometry had been rigorously laid out, Serre's sheaf cohomology had been developed and people could return to the central problems of the subject, the idea of using homological algebra in order to study systematically the geometry of projective varieties can be traced back at least to Grothendieck and Mumford. In 1966 Mumford introduced a fundamental homological invariant, the *Castelunovo-Mumford regularity*, in order to describe qualitatively the equations of an algebraic variety. He gave a fundamental bound for this invariant in terms of the degree and recovered in this way Castelnuovo's classical result [8] for curves. Grothendieck's construction [22] of Hilbert schemes parametrizing all subschemes $X \subseteq \mathbf{P}^n$ having a fixed Hilbert polynomial relies on the possibility of effectively bounding the degree of all equations of a variety with fixed Hilbert polynomial. Syzygies per se however became mainstream in algebraic geometry only after Green [23] introduced Koszul cohomology and repackaged in ways appealing to algebraic geometers all the information contained in the minimal free resolution of the coordinate ring of an algebraic variety. Striking new relationships between free resolutions on one side and moduli spaces on the other have been found. For instance, matrix factorizations discovered in algebraic context by Eisenbud turned out to fit into the framework of A_{∞}-algebras, and as such had recent important applications to mirror symmetry and enumerative geometry.

2 Syzygies of Graded Modules Over Polynomial Algebras

We fix the polynomial ring $S := \mathbb{C}[x_0, \ldots, x_r]$ in $r+1$ variables and let $M = \oplus_{d \geq 0} M_d$ be a finitely generated graded S-module. Choose a minimal set of (homogeneous) generators $(m_1, \ldots, m_t)$ of M, where m_i is an element of M of degree a_i for $i = 1, \ldots, t$. We denote by K_1 the module of relations between the elements m_i, that is, defined by the exact sequence

$$0 \longleftarrow M \stackrel{(m_i)}{\longleftarrow} \bigoplus_{i=1}^{t} S(-a_i) \longleftarrow K_1 \longleftarrow 0.$$

The module K_1 is a submodule of a finitely generated module, thus by the Hilbert Basis Theorem, it is a finitely generated graded S-module itself. Its degree d piece consists of t-tuples of homogeneous polynomials $(f_1, \ldots, f_t)$, where $\deg(f_i) = d - a_i$ such that

$$\sum_{i=1}^{t} f_i m_i = 0 \in M_d.$$

Elements of K_1 are called first order syzygies of M. Let us choose a minimal set of generators of K_1 consisting of relations $(R_1, \ldots, R_s)$, where

$$R_j := (f_1^{(j)}, \ldots, f_t^{(j)}), \text{ with } f_1^{(j)} m_1 + \cdots + f_t^{(j)} m_r = 0,$$

for $j = 1, \ldots, s$. Here $f_i^{(j)}$ is a homogeneous polynomial of degree $b_j - a_i$ for $i = 1, \ldots, t$ and $j = 1, \ldots, s$. Thus we have an induced map of *free* S-modules

$$F_0 := \bigoplus_{i=1}^{t} S(-a_i) \stackrel{(f_i^{(j)})}{\longleftarrow} \bigoplus_{j=1}^{s} S(-b_j) =: F_1.$$

Then we move on and resolve K_1, to find a minimal set of relations among the relations between the generators of M, that is, we consider the finitely generated S-module K_2 defined by the following exact sequence:

$$0 \longleftarrow K_1 \stackrel{(R_j)}{\longleftarrow} \bigoplus_{j=1}^{s} S(-b_j) \longleftarrow K_2 \longleftarrow 0.$$

Elements of K_2 are relations between the relations of the generators of M and as such, they are called second order syzygies of M. One can now continue and resolve the finitely generated S-module K_2. The fact that this process terminates after at most $r + 1$ steps is the content of *Hilbert's Syzygy Theorem*, see for instance [14] Theorem 1.1:

Theorem 2.1 *Every finitely generated graded S-module M admits a minimal free graded S-resolution*

$$\mathbb{F}_\bullet : 0 \longleftarrow M \longleftarrow F_0 \longleftarrow F_1 \longleftarrow \cdots \longleftarrow F_r \longleftarrow F_{r+1} \longleftarrow 0.$$

The minimal free resolution $\mathbb{F}_\bullet$ is uniquely determined up to an isomorphism of complexes of free S-modules. In particular, each two resolutions have the same length. Every individual piece F_p of the resolution is uniquely determined, as a graded module, by its numbers of generators and their degrees and can be written as

$$F_p := \bigoplus_{q>0} S(-p-q)^{\oplus b_{pq}(M)}.$$

The quantities $b_{p,q}(M)$ have an intrinsic meaning and depend only on M. In fact, by the very definition of the Tor functor, one has

$$b_{p,q}(M) = \dim_{\mathbb{C}} \mathrm{Tor}^p(M, \mathbb{C})_{p+q},$$

see also [14] Proposition 1.7. There is a convenient way of packaging together the numerical information contained in the resolution $\mathbb{F}_\bullet$, which due to the computer algebra software system *Macaulay*, has become widespread:

Definition 2.2 The graded Betti diagram of the S-module M is obtained by placing in column p and row q the Betti number $b_{p,q}(M)$.

Thus the column p of the Betti diagram encoded the number of p-th syzygies of M of various weights. Since it is customary to write the columns of a table from left to right, it is for this reason that the rows in the resolution $\mathbb{F}_\bullet$ go from right to left, which requires some getting used to.

A much coarser invariant of the S-module M than the Betti diagram is its *Hilbert function* $h_M : \mathbb{Z} \to \mathbb{Z}$, given by $h_M(d) := \dim_{\mathbb{C}}(M_d)$. The Betti diagram determines the Hilbert function of M via the following formula:

$$h_M(d) = \sum_{p \geq 0} (-1)^p \dim_{\mathbb{C}} F_p(d) = \sum_{p \geq 0, q > 0} (-1)^p b_{p,q}(M) \binom{d+r-p-q}{r}.$$

Conversely, the Hilbert function of M determines the alternating sum of Betti numbers on each diagonal of the Betti diagram. For fixed $k \geq 0$, we denote by $B_k := \sum_p (-1)^p b_{p,k-p}(M)$ the corresponding alternating sum of Betti numbers in one of the diagonals of the Betti diagram of M. The quantities B_k can then be determined inductively from the Hilbert function, using the formula:

$$B_k = h_M(k) - \sum_{\ell < k} B_\ell \binom{r+k-\ell}{r}.$$

In algebro-geometric applications, the alternating sum of Betti numbers on diagonals correspond to the geometric constraints of the problem at hand. The central question in syzygy theory is thus to determine the possible Betti diagrams corresponding to a given Hilbert function.

In order to explicitly compute the Betti numbers $b_{p,q}(M)$ it is useful to remember that the Tor functor is symmetric in its two variables. In particular, there exists a canonical isomorphisms $\mathrm{Tor}^p(M, \mathbb{C}) \cong \mathrm{Tor}^p(\mathbb{C}, M)$. To compute the last Tor group one is thus led to take an explicit resolution of the S-module $\mathbb{C}$ by free graded S-modules. This is given by the *Koszul complex*. We denote by $V := S_1 = \mathbb{C}[x_0, \ldots, x_r]_1$ the vector space of linear polynomials in $r+1$ variables.

Theorem 2.3 *The minimal free S-resolution of the module $\mathbb{C}$ is computed by the Koszul complex in $r+1$ variables:*

$$0 \to \bigwedge^{r+1} V \otimes S(-r-1) \to \cdots \to \bigwedge^{p} V \otimes S(-p) \to \bigwedge^{p-1} V \otimes S(-p+1)$$
$$\to \cdots \to V \otimes S(-1) \to \mathbb{C} \to 0.$$

The p-th differential in degree $p+q$ of this complex, denoted by $d_{p,q} : \bigwedge^p V \otimes S_q \to \bigwedge^{p-1} V \otimes S_{q+1}$, is given by the following formula

$$d_{p,q}(f_1 \wedge \ldots \wedge f_p \otimes u) = \sum_{\ell=1}^{p} (-1)^{\ell} f_1 \wedge \ldots \wedge \hat{f_\ell} \wedge \ldots \wedge f_p \otimes (uf_\ell),$$

where $f_1, \ldots, f_p \in V$ and $u \in S_q$.

In order to compute the Betti numbers of M, we tensor the Koszul complex with the S-module M and take cohomology. One is thus naturally led to the definition of *Koszul cohomology* of M, due to Green [23]. Even though Green's repackaging of the higher Tor functors amounted to little new information, the importance of [23] cannot be overstated, for it brought syzygies in the realm of mainstream algebraic geometry. For integers p and q, one defines the Koszul cohomology group $K_{p,q}(M, V)$ to be the cohomology of the complex

$$\bigwedge^{p+1} V \otimes M_{q-1} \stackrel{d_{p+1,q-1}}{\longrightarrow} \bigwedge^{p} V \otimes M_q \stackrel{d_{p,q}}{\longrightarrow} \bigwedge^{p-1} V \otimes M_{q+1}.$$

As already pointed out, one has

$$b_{p,q}(M) = \dim_{\mathbb{C}} K_{p,q}(M, V).$$

From the definition it follows that Koszul cohomology is functorial. If $f : A \to B$ is a morphism of graded S-modules, one has an induced morphism

$$f_* : K_{p,q}(A, V) \to K_{p,q}(B, V)$$

of Koszul cohomology groups. More generally, if

$$0 \longrightarrow A \longrightarrow B \longrightarrow C \longrightarrow 0$$

is a short exact sequence of graded S-modules, one has an associated long exact sequence in Koszul cohomology:

$$\cdots \to K_{p,q}(A, V) \to K_{p,q}(B, V) \to K_{p,q}(C, V) \to K_{p-1,q+1}(A, V) \to K_{p-1,q+1}(B, V) \to \cdots \quad (1)$$

3 Syzygies in Algebraic Geometry

In algebraic geometry, one is primarily interested in resolving (twisted) coordinate rings of projective algebraic varieties. A very good general reference for Koszul cohomology in algebraic geometry is the book of Aprodu and Nagel [4], or the survey paper [3].

We begin by setting notation. Let X be a projective variety, L a globally generated line bundle on X and $\mathcal{F}$ a sheaf on X. Set $r = r(L) = h^0(X, L) - 1$ and denote by $\varphi_L : X \to \mathbf{P}^r$ the morphism induced by the linear system $|L|$. Although strictly speaking this is not necessary, let us assume that L is very ample, therefore φ_L s an embedding. We set $S := \operatorname{Sym} H^0(X, L) \cong \mathbb{C}[x_0, \ldots, x_r]$ and form the twisted coordinate S-module

$$\Gamma_X(\mathcal{F}, L) := \bigoplus_q H^0(X, \mathcal{F} \otimes L^{\otimes q}).$$

Following notation of Green's [23], one introduces the Koszul cohomology groups

$$K_{p,q}(X, \mathcal{F}, L) := K_{p,q}\Big(\Gamma_X(\mathcal{F}, L), H^0(X, L)\Big)$$

and accordingly, one defines the Betti numbers

$$b_{p,q}(X, \mathcal{F}, L) := b_{p,q}\big(\Gamma_X(\mathcal{F}, L)\big).$$

In most geometric applications, one has $\mathcal{F} = \mathcal{O}_X$, in which case $\Gamma_X(L) := \Gamma_X(\mathcal{O}_X, L)$ is the coordinate ring of the variety X under the map φ_L. One writes $b_{p,q}(X, L) := b_{p,q}(X, \mathcal{O}_X, L)$. It turns out that the calculation of Koszul cohomology groups of line bundles can be reduced to usual cohomology of the exterior powers of a certain vector bundle on the variety X.

Definition 3.1 For a globally generated line bundle L on a variety X, we define the *Lazarsfeld vector bundle* M_L via the exact sequence

$$0 \longrightarrow M_L \longrightarrow H^0(X, L) \otimes \mathcal{O}_X \longrightarrow L \longrightarrow 0,$$

where the above map is given by evaluation of the global sections of L.
One also denotes by $Q_L := M_L^\vee$ the dual of the Lazarsfeld bundle. Note that we have a canonical injection $H^0(X, L)^\vee \hookrightarrow H^0(X, Q_L)$ obtained by dualizing the defining sequence for M_L. To make the role of Lazarsfeld bundles more transparent, we recall the description of the tangent bundle of the projective space $\mathbf{P}^r$ provided by the *Euler sequence* (see [25] Example 8.20.1):

$$0 \longrightarrow \mathcal{O}_{\mathbf{P}^r} \longrightarrow H^0(\mathbf{P}^r, \mathcal{O}_{\mathbf{P}^r}(1))^\vee \otimes \mathcal{O}_{\mathbf{P}^r}(1) \longrightarrow T_{\mathbf{P}^r} \longrightarrow 0,$$

By pulling-back the Euler sequence via the map φ_L and dualizing, we observe that the Lazarsfeld bundle is a twist of the restricted cotangent bundle:

$$M_L \cong \Omega_{\mathbf{P}^r}(1) \otimes \mathcal{O}_X.$$

We now take exterior powers in the exact sequence defining M_L, to obtain the following exact sequences for each $p \geq 1$:

$$0 \longrightarrow \bigwedge^{p} M_L \longrightarrow \bigwedge^{p} H^0(X,L) \otimes \mathcal{O}_X \longrightarrow \bigwedge^{p-1} M_L \otimes L \longrightarrow 0.$$

After tensoring and taking cohomology, we link these sequences to the Koszul complex computing $K_{p,q}(X,\mathcal{F},L)$ and obtain the following description of Koszul cohomology groups in terms of ordinary cohomology of powers of twisted Lazarsfeld bundles. For a complete proof we refer to [4] Proposition 2.5:

Proposition 3.2 *One has the following canonical isomorphisms:*

$$K_{p,q}(X,\mathcal{F},L) \cong Coker\Big\{\bigwedge^{p+1} H^0(X,L) \otimes H^0(X,\mathcal{F}\otimes L^{q-1}) \to H^0\big(X, \bigwedge^{p} M_L \otimes \mathcal{F} \otimes L^q\big)\Big\}$$

$$\cong Ker\Big\{H^1\big(X, \bigwedge^{p+1} M_L \otimes \mathcal{F} \otimes L^{q-1}\big) \to \bigwedge^{p+1} H^0(X,L) \otimes H^1(X,\mathcal{F}\otimes L^{q-1})\Big\}.$$

The description of Koszul cohomology given in Proposition 3.2 brings syzygy theory firmly in the realm of algebraic geometry, for usual vector bundle techniques involving stability and geometric constructions come to the fore in order to compute Koszul cohomology groups. We now give some examples and to keep things intuitive, let us assume $\mathcal{F} = \mathcal{O}_X$. Then

$$K_{1,1}(X,L) \cong H^0(X, M_L \otimes L) / \bigwedge^{2} H^0(X,L) \cong I_2(X,L)$$

is the space of quadrics containing the image of the map φ_L. Similarly, the group

$$K_{0,2}(X,L) \cong H^0(X,L^2)/\mathrm{Sym}^2 H^0(X,L)$$

measures the failure of X to be quadratically normal. More generally, if

$$I_X := \bigoplus_{q \geq 2} I_X(q) \subseteq S$$

is the graded ideal of X, then one has the following isomorphism, cf. [4] Proposition 2.8:

$$K_{p,1}(X,L) \cong K_{p-1,2}(\mathbf{P}^r, I_X, \mathcal{O}_{\mathbf{P}^r}(1)).$$

In particular, $K_{2,1}(X,L) \cong \mathrm{Ker}\{I_X(2) \otimes H^0(X,\mathcal{O}_X(1)) \to I_X(3)\}$.

Koszul cohomology shares many of the features of a cohomology theory. We single out one aspect which will play a role later in these lectures:

Theorem 3.3 *Koszul cohomology satisfies the Lefschetz hyperplane principle. If X is a projective variety and $L \in \mathrm{Pic}(X)$, assuming $H^1(X, L^{\otimes q}) = 0$ for all q, then for any divisor $D \in |L|$ one has an isomorphism $K_{p,q}(X, L) \cong K_{p,q}(D, L_{|D})$.*

In general, it is not easy to determine the syzygies of any variety by direct methods, using just the definition. One of the few instances when this is possible is given by the twisted cubic curve $R \subseteq \mathbf{P}^3$. Denoting the coordinates in $\mathbf{P}^3$ by x_0, x_1, x_2 and x_3, the ideal of R is given by the 2×2-minors of the following matrix:

$$\begin{pmatrix} x_0 & x_1 & x_2 \\ x_1 & x_2 & x_3 \end{pmatrix}$$

Therefore the ideal of R is generated by three quadratic equations

$$q_1 := x_0x_2 - x_1^2, \; q_2 := x_0x_3 - x_1x_2 \; \text{ and } \; q_3 := x_1x_3 - x_2^2,$$

among which there exists two linear syzygies, that is, syzygies with linear coefficients:

$$R_1 := x_0q_3 - x_1q_2 + x_2q_1 \; \text{ and } \; R_2 := x_1q_3 - x_2q_2 + x_3q_1.$$

The resolution of the twisted cubic curve is therefore the following:

$$0 \longleftarrow \Gamma_R\big(\mathcal{O}_R(1)\big) \longleftarrow S \longleftarrow S(-2)^{\oplus 3} \longleftarrow S(-3)^{\oplus 2} \longleftarrow 0,$$

and the corresponding Betti diagram is the following (the entries left open being zero):

1		
	3	2

In order to distinguish the Betti diagrams having the simplest possible resolution for a number of steps, we recall the definition due to Green and Lazarsfeld [24]:

Definition 3.4 One says that a polarized variety (X, L) satisfies property (N_p) if it is projectively normal and $b_{j,q}(X, L) = 0$ for $j \leq p$ and $q \geq 2$.

In other words, a variety $\varphi_L : X \hookrightarrow \mathbf{P}^r$ has property (N_1) if it is projectively normal and its ideal $I_{X/\mathbf{P}^r}$ is generated by quadrics $q_1, \ldots, q_s$. The number of these quadrics is equal to $s := b_{1,1}(X, L) = \binom{r+2}{2} - h^0(X, L^2)$ and is thus determined by the numerical characters of X. We say that (X, L) has property (N_2) if all of the above hold and, in addition, all the syzygies between these quadrics are generated by linear syzygies of the type $\ell_1 q_1 + \cdots + \ell_s q_s = 0$, where the ℓ_i are *linear forms*.

Castelnuovo, using what came to be referred to as the *Base Point Free Pencil Trick*, has proven that if L is a line bundle of degree $\deg(L) \geq 2g+2$ on a smooth curve C of genus g, then the curve embedded by the complete linear system $|L|$ is projectively normal and its ideal is generated by quadrics. In other words, using modern terminology, it verifies property (N_1). This fact has been generalized by Green [23] to include the case of higher syzygies as well. This result illustrates a general philosophy that at least for curves, the more positive a line bundle is, the simpler its syzygies are up to an order that is linear in the degree of the line bundle.

Theorem 3.5 *Let L be a line bundle of degree $d \geq 2g+p+1$ on a smooth curve C of genus g. Then C verifies property (N_p).*

Proof Using the description of Koszul cohomology given in Proposition 3.2, we have the equivalence

$$K_{p,2}(C,L)=0 \Longleftrightarrow H^1\Bigl(C, \bigwedge^{p+1} M_L \otimes L\Bigr)=0.$$

Denoting by $Q_L := M_L^{\vee}$, by Serre duality this amounts to the vanishing

$$H^1\Bigl(C, \bigwedge^{p+1} Q_L \otimes \omega_C \otimes L^{\vee}\Bigr)=0.$$

To establish this vanishing, we use a filtration on the vector bundle M_L used several times by Lazarsfeld, for instance in [30] Lemma 1.4.1. We choose general points $p_1,\ldots,p_{r-1} \in C$, where $r := r(L) = d-g$. Then by induction on r one has the following exact sequence on C:

$$0 \longrightarrow M_{L(-p_1-\cdots-p_{r-1})} \longrightarrow M_L \longrightarrow \bigoplus_{i=1}^{r-1} \mathcal{O}_C(-p_i) \longrightarrow 0.$$

Noting that $L(-p_1-\cdots-p_{r-1})$ is a pencil, by the Base Point Free Pencil Trick, one has the identification

$$M_{L(-p_1-\cdots-p_{r-1})} \cong L^{\vee}(p_1+\cdots+p_{r-1}).$$

Thus by dualizing, the exact sequence above becomes:

$$0 \longrightarrow \bigoplus_{i=1}^{r-1} \mathcal{O}_C(p_i) \longrightarrow Q_L \longrightarrow L(-p_1-\cdots-p_{r-1}) \longrightarrow 0.$$

Taking $(p+1)$-st exterior powers in this sequence and tensoring with $\omega_C \otimes L^{\vee}$, we obtain that the vanishing $H^0\bigl(C, \bigwedge^{p+1} Q_L \otimes \omega_C \otimes L^{\vee}\bigr)=0$ holds, once we establish that for each subdivisor D_{p+1} of degree $p+1$ of the divisor $p_1+\ldots+p_{r-1}$, one has

$$H^0(C, \omega_C \otimes L^{\vee}(D_{p+1}))=0,$$

Table 1 The Betti table of a non-special curve of genus g

1	2	...	$p-1$	p	$p+1$	...	r
$b_{1,1}$	$b_{2,1}$	...	$b_{p-1,1}$	$b_{p,1}$	$b_{p+1,1}$	...	$b_{r,1}$
$b_{1,2}$	$b_{2,2}$	...	$b_{p-1,2}$	$b_{p,2}$	$b_{p+1,2}$	...	$b_{r,2}$

and for each subdivisor D_{r-1-p} of $p_1 + \cdots + p_{r-1}$ one has

$$H^0(C, \omega_C(-D_{r-1-p})) = 0.$$

The first vanishing follows immediately for degree reasons, the second is implied by the inequality $r - 1 - p = d - g - p - 1 \geq p$, which is precisely our hypothesis. □

The same filtration argument on the vector bundle M_L shows that in the case of curves, the Betti diagram consists of only two rows, namely that of linear syzygies and that of quadratic syzygies respectively.

Proposition 3.6 *Let L be a globally generated non-special line bundle L on a smooth curve C. Then $K_{p,q}(C, L) = 0$ for all $q \geq 3$ and all p.*

Thus making abstraction of the 0-th row in which the only non-zero entry is $b_{0,0} = 1$, the resolution of each non-special curve $C \subseteq \mathbf{P}^r$ has the shape shown in Table 1.

As already pointed out, the Hilbert function $h_C(t) = dt + 1 - g$ of the curve C determines the difference of Betti numbers of each diagonal in the Betti diagram.

Theorem 3.7 *The difference of Betti numbers of each diagonal diagonal of the Betti diagram of a non-special line bundle L on a curve C is an Euler characteristic of a vector bundle on C. Precisely,*

$$b_{p+1,1}(C, L) - b_{p,2}(C, L) = (p + 1) \cdot \binom{d-g}{p+1} \Big(\frac{d+1-g}{p+2} - \frac{d}{d-g} \Big). \qquad (2)$$

Green's Theorem ensures that $b_{p,2}(C, L) = 0$ as long $p \leq d - 2g - 1$. On the other hand formula (2) indicates that when p is relatively high, then $b_{p,2} > b_{p+1,1}(C, L) \geq 0$, that is, there will certainly appear p-th order non-linear syzygies. One can now distinguish two main goals concerning syzygies of curves:

1. Given a line bundle L of degree d on a genus g curve, determine which Betti numbers are zero.
2. For those Betti numbers $b_{p,1}(C, L)$ and $b_{p,2}(C, L)$ which are non-zero, determine their exact value.

As we shall explain, there is a satisfactory answer to the first question in many important situations, both for the linear and for the quadratic row of syzygies. The second question remains to date largely unanswered.

A crucial aspect of syzygy theory is to identify the geometric sources of syzygies and obtain in this ways guidance as to which Koszul cohomology groups vanish.

The first, and in some sense most important instance of such a phenomenon, when non-trivial geometry implies non-trivial syzygies is given by the *Green-Lazarsfeld Non-Vanishing Theorem* [23].

Theorem 3.8 *Let L be a globally generated line bundle on a variety X and suppose one can write $L = L_1 \otimes L_2$, where $r_i := r(L_i) \geq 1$. Then*

$$K_{r_1+r_2-1,1}(X, L_1 \otimes L_2) \neq 0.$$

Proof We follow an argument due to Voisin [36]. We choose two sections $\sigma \in H^0(X, L_1)$ and $\tau \in H^0(X, L_2)$ and introduce the vector space

$$W := H^0(X,L)/\big(\sigma \cdot H^0(X,L_2) + \tau \cdot H^0(X,L_1)\big).$$

We then have the following short exact sequence:

$$0 \longrightarrow M_{L_1} \oplus M_{L_2} \longrightarrow M_L \longrightarrow W \otimes \mathcal{O}_X \longrightarrow 0.$$

Take $(r_1 + r_2 - 1)$-st exterior powers in this short exact sequence and obtain an injection

$$\bigwedge^{r_1+r_2-1} (M_{L_1} \oplus M_{L_2}) \hookrightarrow \bigwedge^{r_1+r_2-1} M_L. \tag{3}$$

Since $\bigwedge^{r_1-1} M_{L_1} \cong Q_{L_1} \otimes L_1^\vee$ and $\bigwedge^{r_2-1} M_{L_2} \cong Q_{L_2} \otimes L_2^\vee$, whereas clearly $\bigwedge^{r_1} M_{L_1} \cong L_1^\vee$ and $\bigwedge^{r_2} M_{L_2} \cong L_2^\vee$, by tensoring the injection (3) with $L \cong L_1 \otimes L_2$, we obtain the following injection $Q_{L_1} \oplus Q_{L_2} \hookrightarrow \bigwedge^{r_1+r_2-1} M_L \otimes L$, leading to an injection at the level of global sections

$$H^0(X,L_1)^\vee \oplus H^0(X,L_2)^\vee \hookrightarrow H^0\big(X, \bigwedge^{r_1+r_2-1} M_L \otimes L\big). \tag{4}$$

We recall the description of the Koszul cohomology group

$$K_{r_1+r_2-1,1}(X,L) \cong H^0\big(X, \bigwedge^{r_1+r_2-1} M_L \otimes L\big)/ \bigwedge^{r_1+r_2} H^0(X,L).$$

The proof will be complete once one shows that at least one element of $H^0\big(X, \bigwedge^{r_1+r_2-1} M_L \otimes L\big)$ produced via the injection (4), does not lie in the image of $\bigwedge^{r_1+r_2} H^0(X,L)$.

In order to achieve this, let us choose a basis $(\sigma_0 = \sigma, \sigma_1, \ldots, \sigma_{r_1})$ of $H^0(X,L_1)$ and a basis $(\tau_0 = \tau, \tau_1, \ldots, \tau_{r_2})$ of $H^0(X,L_2)$ respectively. Then one syzygy obtained by the inclusion (4) which gives rise to a non-zero element in

$K_{r_1+r_2-1}(X,L)$ is given by the following explicit formula, see also [36] Eq. (1.11.1):

$$\sum_{i=0}^{r_1}\sum_{j=1}^{r_2}(-1)^{i+j}(\tau_0\sigma_0)\wedge\ldots\wedge\widehat{(\tau_0\sigma_i)}\wedge\ldots\wedge(\tau_0\sigma_{r_1})$$

$$\wedge(\sigma_0\tau_1)\wedge\ldots\wedge\widehat{(\sigma_0\tau_j)}\wedge\ldots\wedge(\sigma_0\tau_{r_2})\otimes(\sigma_i\tau_j).$$

Under the isomorphisms $H^0(X,L_1)^\vee \cong \bigwedge^{r_1} H^0(X,L_1)$ and $H^0(X,L_2)^\vee \cong \bigwedge^{r_2} H^0(X,L_2)$, under the injection (4), this last syzygy corresponds to the element $(0,\tau_1\wedge\ldots\wedge\tau_{r_2})$. □

Remark 3.9 Particularly instructive is the case when $r_1 = r_2 = 1$, that is, both L_1 and L_2 are pencils. Keeping the notation above, the non-zero syzygy provided by the Green-Lazarsfeld Non-Vanishing Theorem has the following form:

$$\gamma := -(\tau_0\sigma_1)\otimes(\sigma_0\tau_1)+(\tau_0\sigma_0)\otimes(\sigma_1\tau_1)\in H^0(X,L)\otimes H^0(X,L),$$

giving rise to a non-zero element in $K_{1,1}(X,L)$. The geometric interpretation of this syzygy is transparent. The map φ_L factor through the map $(\varphi_{L_1},\varphi_{L_2}) : X \dashrightarrow \mathbf{P}^1\times\mathbf{P}^1$ induced by the two pencils L_1 and L_2. The quadric γ is then the pull-back of the rank 4 quadric $\mathbf{P}^1\times\mathbf{P}^1 \hookrightarrow \mathbf{P}^3$ under the projection $\mathbf{P}^r \dashrightarrow \mathbf{P}^3$ induced by the space W.

4 Green's Conjecture

Mark Green's Conjecture formulated in 1984 in [23] is an elegant and deceptively simple statement concerning the syzygies of a canonically embedded curve C of genus g. Despite the fact that it has generated a lot of attention and that many important results have been established, Green's Conjecture, in its maximal generality, remains open. Especially in the 1980s and 1990s, progress on Green's Conjecture guided much of both the research on syzygies of algebraic varieties, as well as the development of the computer algebra system *Macaulay*.

Suppose $C \subseteq \mathbf{P}^{g-1}$ is a non-hyperelliptic canonically embedded curve of genus g. Our main goal is to determine the Betti numbers of C, in particular understand in what way the geometry of C influences the shape of the Betti diagram of its canonical embedding. As already pointed out, the Betti diagram of a curve embedded by a non-special line bundle has only two non-trivial rows, corresponding to linear and quadratic syzygies respectively. For canonical curves the situation is even simpler, for the two rows contain the same information.

Proposition 4.1 *One has the following duality for the Koszul cohomology groups of a canonical curve:*

$$K_{p,q}(C,\omega_C)\cong K_{g-p-2,3-q}(C,\omega_C)^\vee.$$

Proof Follows via Serre duality for vector bundles, using the description of Koszul cohomology in terms of Lazarsfeld bundles provided in Proposition 3.2. □

Setting $b_{p,q} := b_{p,q}(C, \omega_C)$, we observe that, in particular, $b_{g-2,3} = b_{0,0} = 1$. Unlike in the case of non-special curves, the Betti diagram of a canonical curve has a unique non-trivial entry in the third row. Furthermore, $b_{p,1} = b_{g-2-p,2}$, that is, the row of quadratic syzygies is a reflection of the linear strand.

We now apply the Green-Lazarsfeld Non-Vanishing Theorem to the case of canonical curves. We write

$$K_C = L + (K_C - L),$$

where we assume that $r := r(L) = h^0(C, L) - 1 \geq 1$ and $r(K_C - L) \geq 1$. By Riemann-Roch, $r(K_C - L) = g + r - d - 1$, therefore Theorem 3.8 implies that the following equivalent non-vanishing statements hold:

$$K_{g+2r-d-2,1}(C, \omega_C) \neq 0 \Longleftrightarrow K_{d-2r,2}(C, \omega_C) \neq 0.$$

Thus existence of linear series of type $\mathfrak{g}^r_d$ always leads to a non-linear syzygy of order $d - 2r$. This leads us to the definition of the *Clifford index*, as a way of measuring the complexity of a curve of fixed genus.

Definition 4.2 Let C be a smooth curve of genus g. We define the *Clifford index* of C as the following quantity:

$$\mathrm{Cliff}(C) := \min\Big\{d - 2r : L \in \mathrm{Pic}^d(C),\ d = \deg(L) \leq g - 1,\ r := r(L) \geq 1\Big\}.$$

It follows from the classical Clifford Theorem that $\mathrm{Cliff}(C) \geq 0$ and that $\mathrm{Cliff}(C) = 0$ if and only if C is hyperelliptic. The Clifford index is lower semicontinuous in families and offers a stratification of the moduli space $\mathcal{M}_g$ of smooth curves of genus g. The Clifford index is closely related to another important invariant, the *gonality* of C, defined as the minimal degree of a finite map $C \to \mathbf{P}^1$. By definition $\mathrm{Cliff}(C) \leq \mathrm{gon}(C) - 2$. It can be proved [11] that always

$$\mathrm{gon}(C) - 3 \leq \mathrm{Cliff}(C) \leq \mathrm{gon}(C) - 2.$$

For a general curve of fixed gonality, one has $\mathrm{Cliff}(C) = \mathrm{gon}(C) - 2$, see [11] Corollary 2.3.2.

It follows from general Brill-Noether theory [6] that for a general curve C of genus g, one has

$$\mathrm{Cliff}(C) = \Big\lfloor \frac{g-1}{2} \Big\rfloor \text{ and } \mathrm{gon}(C) = \Big\lfloor \frac{g+3}{2} \Big\rfloor. \tag{5}$$

We can summarize the discussion above as follows:

Proposition 4.3 *If C is a smooth curve of genus g then $K_{\mathrm{Cliff}(C),2}(C,\omega_C) \neq 0$.*

Green's Conjecture amounts to a converse of Proposition 4.3. By the geometric version of the Riemann-Roch Theorem, the existence of an effective divisor D of degree $d \leq g-1$ with $h^0(C,\mathcal{O}_C(D)) = r+1$, amounts to the statement that under the canonical embedding, we have $\langle D\rangle \cong \mathbf{P}^{d-r-1} \subseteq \mathbf{P}^{g-1}$. Thus tautologically, special linear systems on C amount to special secant planes to its canonical embedding. Under this equivalence, Green's Conjecture is saying that the universal source of non-linear syzygies for the canonical curve is given by secants.

Conjecture 4.4 For every smooth curve of genus g, one has the following equivalence:

$$K_{p,2}(C,\omega_C) = 0 \Longleftrightarrow p < \mathrm{Cliff}(C).$$

In light of Proposition 4.3, the non-trivial part of Green's Conjecture is the establishing of the following vanishing:

$$K_{p,2}(C,\omega_C) = 0, \quad \text{for all } p < \mathrm{Cliff}(C).$$

The appeal of Green's Conjecture lies in the fact that it allows one to read off the complexity of the curve (in the guise of its Clifford index), from the equations of its canonical embedding. The Clifford index is simply the order of the first non-linear syzygy of the canonical curve $C \subseteq \mathbf{P}^{g-1}$.

Spelling out the conclusions of Conjecture 4.4, it follows that if one denotes by $c := \mathrm{Cliff}(C)$, then the resolution of every canonical curve of genus g should have the form shown in Table 2.

The content of Conjecture 4.4 is that $b_{j,2} = 0$ for $j < c$. This can be reduced to a single vanishing, namely $b_{c-1,2} = 0$.

It has contributed to the appeal of Green's Conjecture that its first cases specialize to famous classical results in the theory of algebraic curves. When $p = 0$, Conjecture 4.4 predicts that a non-hyperelliptic canonical curve $C \subseteq \mathbf{P}^{g-1}$ is

Table 2 The Betti table of a canonical curve of genus g and Clifford index c

0	1	...	$c-1$	c	...	$g-c-2$	$g-c-1$	...	$g-3$	$g-2$
1	0	...	0	0	...	0	0	...	0	0
0	$b_{1,1}$	...	$b_{c-1,1}$	$b_{c,1}$	...	$b_{g-c-2,1}$	0	...	0	0
0	0	...	0	$b_{c,2}$	...	$b_{g-c-2,2}$	$b_{g-c-1,2}$	...	$b_{g-2,2}$	0
0	0	...	0	0	...	0	0	...	0	1

projectively normal, that is, all multiplication maps

$$\mathrm{Sym}^k H^0(C, \omega_C) \to H^0(C, \omega_C^{\otimes k})$$

are surjective. This is precisely the content of Max Noether's Theorem, see [6] page 117.

When $p = 1$, Conjecture 4.4 predicts that as long as $\mathrm{Cliff}(C) > 1$ (that is, C is neither trigonal nor a smooth plane quintic, when necessarily $g = 6$), one has $K_{1,2}(C, \omega_C) = 0$, that is, the ideal of the canonical curve is generated by quadrics. This is the content of the Enriques-Babbage Theorem, see [6] page 124.

The first non-classical case of Green's Conjecture is $p = 2$, when one has to show that if C is not tetragonal, then $K_{2,2}(C, \omega_C) = 0$. This has been established almost simultaneously, using vector bundle methods by Voisin [35] and using Gröbner basis techniques by Schreyer [33]. Even before that, the conjecture has been proved for all $g \leq 8$ in [33], in a paper in which many techniques that would ultimately play a major role in the study of syzygies have been introduced.

4.1 The Resolution of a General Canonical Curve

Assume now that C is a general canonical curve of odd genus $g = 2i + 3$. Then $\mathrm{Cliff}(C) = i + 1$ and in fact C has a one dimensional family $W^1_{i+3}(C)$ of pencils of degree $i+3$ computing the Clifford index. The predicted resolution of the canonical image of C has the shape shown in Table 3, where this time we retain only the linear and quadratic strands.

Furthermore, since in each diagonal of the Betti table precisely one entry is non-zero, we can explicitly compute all the Betti numbers and we find:

$$b_{p,1}(C, \omega_C) = \frac{(2i+2-p)(2i-2p+2)}{p+1}\binom{2i+2}{p-1}, \quad \text{for } p \leq i,$$

and

$$b_{p,2}(C, \omega_C) = \frac{(2i+1-p)(2p-2i)}{p+2}\binom{2i+2}{p}, \quad \text{for } 2i \geq p \geq i+1.$$

Table 3 The Betti table of a general canonical curve of genus $g = 2i + 3$

1	2	...	$i-1$	i	$i+1$	$i+2$	...	$2i$
$b_{1,1}$	$b_{2,1}$	...	$b_{i-1,1}$	$b_{i,1}$	0	0	...	0
0	0	...	0	0	$b_{i+1,2}$	$b_{i+2,2}$	...	$b_{2i,2}$

Table 4 The Betti table of a general canonical curve of genus $g = 2i + 2$

1	2	...	$i-1$	i	$i+1$	$i+2$	...	$2i-1$
$b_{1,1}$	$b_{2,1}$	...	$b_{i-1,1}$	$b_{i,1}$	0	0	...	0
0	0	...	0	$b_{i,2}$	$b_{i+1,2}$	$b_{i+2,2}$	...	$b_{2i-1,2}$

Quite remarkably, in this case Green's Conjecture predicts not only which Betti numbers vanish, but also their precise value.

Let us now move to a general curve C of even genus $g = 2i + 2$. In this case $\text{Cliff}(C) = i$. This Clifford index is computed by one of the finitely many pencils of minimal degree $i + 1$. The predicted resolution of the canonical model of C has the shape shown in Table 4.

Note that also in this case, in each diagonal of the Betti table precisely one entry is non-zero, which allows us to determine the entire resolution. We find:

$$b_{p,1}(C, \omega_C) = \frac{(2i - p + 1)(2i - 2p + 1)}{p + 1}\binom{2i + 1}{p - 1}, \quad \text{for } p \le i,$$

and

$$b_{p,2}(C, \omega_C) = \frac{(2i - p)(2p - 2i + 1)}{p + 2}\binom{2i + 1}{p}, \quad \text{for } 2i - 1 \ge p \ge i.$$

There is a qualitative difference between the resolution of a general canonical curve of odd or even genus respectively. In the former case the resolution is *pure*, that is, at each step there are only syzygies of one given degree. In the latter case, the resolution is not pure in the middle, for there exist both linear and quadratic syzygies of order i.

4.2 The Resolution of Canonical Curves of Small Genus

To familiarize ourselves with the content of Green's Conjecture, we concentrate on small genus and we begin with the case $g = 5$. In this case we distinguish two cases, depending on whether the curve C is trigonal or not. Note that the trigonal locus $\mathcal{M}^1_{5,3}$ is an effective divisor on $\mathcal{M}_5$ (Table 5).

From the Enriques-Babbage Theorem, if C is not trigonal, then it is a complete intersection of three quadrics $Q_1, Q_2, Q_3 \subseteq \mathbf{P}^4$. The resolution of C has the form:

$$0 \longleftarrow \Gamma_C(\omega_C) \longleftarrow S \longleftarrow S(-2)^{\oplus 3} \longleftarrow S(-4)^{\oplus 3} \longleftarrow S(-5) \longleftarrow 0.$$

Table 5 The Betti table of a non-trigonal curve of genus $g = 5$

1			
	3		
		3	
			1

Table 6 The Betti table of a trigonal curve of genus $g = 5$

1			
	3	2	
	2	3	
			1

The syzygies between the three quadrics are quadratic and of the trivial type $Q_i \cdot Q_j - Q_j \cdot Q_i$, for $i \neq j$.

Assume now that C is trigonal. Then it turns out that the resolution has the shape shown in Table 6.

The interesting feature of this table is that $b_{1,2}(C, \omega_C) = b_{2,1}(C, \omega_C) = 2$. To give a geometric explanation of this fact, we recall that the linear system on C residual to the degree 3 pencil, realizes C as a plane quintic with one node. Concretely, let $X := \mathrm{Bl}_q(\mathbf{P}^2)$ be the Hirzebruch surface $\mathbb{F}_1$ and denote by $h \in \mathrm{Pic}(X)$ the pull-back of the line class and by $E \in \mathrm{Pic}(X)$ the exceptional divisor respectively. The image of X under the linear system $|2h - E|$ realizes $X \subseteq \mathbf{P}^4$ as a cubic scroll (Observe that $(2h - E)^2 = 3$). As we already mentioned $C \equiv 5h - 2E \in \mathrm{Pic}(X)$. From the adjunction formula, $K_C \equiv \mathcal{O}_C(1)$, that is, we have the following inclusions for the canonical curve of genus 5:

$$C \subseteq X \subseteq \mathbf{P}^4.$$

From the Enriques-Babbage Theorem it follows that the intersection of the three quadrics containing C is precisely the cubic scroll X, that is $I_C(2) \cong I_X(2)$. It follows that one has an inclusion $K_{2,1}(X, \mathcal{O}_X(1)) \subseteq K_{2,1}(C, \omega_C)$. Note that

$$K_{2,1}(X, \mathcal{O}_X(1)) = \mathrm{Ker}\Big\{I_X(2) \otimes H^0(X, \mathcal{O}_X(1)) \to I_X(3)\Big\}.$$

By direct calculation we find that $\dim I_X(3) = 13$. Since $\dim I_X(2) = 3$, it follows that $b_{2,1}(X, \mathcal{O}_X(1)) \geq 2$, therefore $b_{2,1}(C, \omega_C) \geq 2$. It can now be easily showed that one actually has equality, that is, $b_{2,1}(C, \omega_C) = 2$.

To summarize, we have a *set-theoretic* equality of divisors on $\mathcal{M}_5$:

$$\Big\{[C] \in \mathcal{M}_5 : K_{2,1}(C, \omega_C) \neq 0\Big\} = \mathcal{M}^1_{5,3}.$$

Table 7 The Betti table of a non-tetragonal curve of genus $g = 7$

1					
	10	16			
			16	10	
					1

Table 8 The Betti table of a general tetragonal curve of genus $g = 7$

1					
	10	16	3		
		3	16	10	
					1

Let us now move on to the case of curves of genus 7. In some sense this is the first non-trivial case, for 7 is the first genus when a general canonical curve is no longer a complete intersection. If C is not 4-gonal, Green's Conjecture predicts that $b_{2,2}(C, \omega_C) = b_{3,1}(C, \omega_C) = 0$, and C admits the resolution shown in Table 7.

Here $b_{1,1}(C, \omega_C) = \dim I_C(2) = 10$ and since $\dim I_C(3) = \binom{9}{3} - 5(g-1) = 54$, we find that

$$b_{2,1}(C, \omega_C) = \dim \ \mathrm{Ker}\Big\{I_C(2) \otimes H^0(C, \omega_C) \to I_C(3)\Big\} = 16.$$

Assume now that C is a general tetragonal curve of genus 7, in particular $W^2_6(C) = \emptyset$. The fibres of the degree 4 pencil on C span a 3-dimensional scroll $X \subseteq \mathbf{P}^6$. The resolution of C has the form shown in Table 8. The novelty compared to the previous case is that $b_{3,1}(C, \omega_C) = b_{2,2}(C, \omega_C) = 3$. All these syzygies are induced from the 3-dimensional scroll X. Using the Eagon-Northcott complex [32] Section 1, one can easily show that $b_{3,1}(X, \mathcal{O}_X(1)) = 3$.

The last case we treat is that when $W^2_6(C) \neq 0$, that is, C admits a degree 6 plane model. The subvariety

$$\mathcal{M}^2_{7,6} := \Big\{[C] \in \mathcal{M}_7 : W^2_6(C) \neq \emptyset\Big\}$$

is an irreducible codimension 2 subvariety of $\mathcal{M}_7$. A general element $[C] \in \mathcal{M}^2_{7,6}$ corresponds to a plane sextic curve with three nodes. The lines through any of these nodes induces a pencil of degree 4 on C, that is, such a curve C has three pencils of degree 4. In fact, these are the only pencils of minimal degree on C. The resolution of the canonical curve $C \subseteq \mathbf{P}^6$ has the shape shown in Table 9.

Intuitively, one explains the value $b_{3,1}(C, \omega_C) = 9$, by referring to the previous case, of curves having a single $\mathfrak{g}^1_4$. Each such pencil gives a contribution of three syzygies to the Koszul cohomology group $K_{3,1}(C, \omega_C)$ and it turns out that the three pencils on a general curve $[C] \in \mathcal{M}^2_{7,6}$ contribute independently, leading to nine independent syzygies.

Table 9 The Betti table of a general plane sextic of genus $g = 7$

1					
	10	16	9		
		9	16	10	
					1

The previous two Betti tables illustrate vividly the limitations of Green's Conjecture, for they have the same shape since they correspond to curves of the same Clifford index, yet they have different values, depending on the number of minimal pencils of the curve in question.

4.3 Voisin's Proof of the Generic Green Conjecture

Without a doubt, the most important work on Green's Conjecture is Voisin's proof [37, 38] of Green's Conjecture for generic curves of even and odd genus respectively. Her key idea is to specialize to curves lying on a $K3$ surface and interpret the Koszul cohomology groups in question as cohomology groups of certain tautological vector bundles on the Hilbert scheme classifying 0-dimensional subschemes of the ambient $K3$ surface. This novel approach to Koszul cohomology also bore fruit later (though replacing the $K3$ surface with the symmetric product of the curve) in the sensationally simple proof of the *Gonality Conjecture* by Ein and Lazarsfeld [13], which however is not a subject of these lectures.

We start with the case of curves of even genus $g = 2i + 2$. Therefore a generic curve $[C] \in \mathcal{M}_g$ has $\mathrm{Cliff}(C) = i$. Due to the semicontinuity of dimensions of Koszul cohomology groups in flat families, in order to prove the generic Green Conjecture it suffices to exhibit a single smooth curve C of *maximal* Clifford index which satisfies the two equivalent vanishing statements

$$K_{i-1,2}(C, \omega_C) = 0 \Longleftrightarrow K_{i+1,1}(C, \omega_C) = 0.$$

We now quote the main result from [37]. Explaining the proof would take us too far afield and we refer to [4] Section 6.3 for a good summary of the main points in Voisin's proof.

Theorem 4.5 *Let (S, H) be a polarized $K3$ surface with $H^2 = 4i+2$ and* $\mathrm{Pic}(S) = \mathbb{Z} \cdot H$. *Then*

$$K_{i-1,2}(S, H) = 0.$$

Theorem 4.5 immediately leads to a proof of Green's Conjecture for generic curves of even genus. Indeed, first we notice that a general $C \in |H|$ is

Brill-Noether-Petri general due to Lazarsfeld's result [29]. In particular, C has the maximal Clifford index, that is, $\mathrm{Cliff}(C) = i$. On the other hand, by the already mentioned Lefschetz Hyperplane Principle, one has

$$K_{i-1,2}(C, \omega_C) \cong K_{i-1,2}(S, H),$$

thus Theorem 4.5 provides the vanishing required by Green's Conjecture.

The case of generic curves of odd genus $g = 2i+3$ is treated in [38]. The strategy is to specialize again to a curve C lying on a $K3$ surface S, but this case is harder because of an extra difficulty. A general curve of genus $2i+3$ has a one dimensional family of minimal pencils of minimal degree $i+3$. This is in contrast with the even genus case. A general curve $C \in |H|$ as in Theorem 4.5, where $g(C) = 2i+2$, has a finite number of minimal pencils $A \in W^1_{i+2}(C)$. Each such pencil induces a rank 2 *Lazarsfeld-Mukai* vector bundle F_A on S, defined by an elementary transformation on the surface S:

$$0 \longrightarrow F_A \longrightarrow H^0(C, A) \otimes \mathcal{O}_S \longrightarrow A \longrightarrow 0.$$

The geometry of the bundle F_A is essential in the proof of Theorem 4.5 and, crucially, F_A does *not* depend on the choice of $A \in W^1_{i+2}(C)$, so that it is an object one can canonically attach to the curve $C \subseteq S$. This feature no longer holds true in odd genus, and in order to circumvent this difficulty, Voisin uses a more special $K3$ surface instead:

Theorem 4.6 *Let S be a smooth $K3$ surface such that* $\mathrm{Pic}(S) \cong \mathbb{Z}\cdot C \oplus \mathbb{Z}\cdot \Delta$, *where Δ is a smooth rational curve with $\Delta^2 = -2$ and C is a smooth curve of genus $g = 2i+3$, such that $C \cdot \Delta = 2$. Then the following hold:*

$$K_{i,2}(S, C+\Delta) = 0 \ \ and \ \ K_{i,2}(S, C) = 0.$$

Note that $C + \Delta$ can be regarded as a semi-stable curve of genus $2i+4$. The vanishing $K_{i,2}(S, C+\Delta) = 0$ is what one would expect from Theorem 4.5 (although that result has been established only for $K3$ surfaces of Picard number one, an extension of the argument shows that the statement holds in this case as well). The most difficult part of [38] is showing how one can pass from this vanishing to the statement $K_{i,2}(S, C) = 0$, which shows that $C \subseteq S$ verifies Green's Conjecture.

4.4 The Result of Hirschowitz and Ramanan

We are now going to discuss the beautiful paper [28], which predates Voisin's papers [37, 38]. Although at the time the results of Hirschowitz and Ramanan were conditional, they provided substantial evidence for Green's Conjecture. Once the Generic Green Conjecture became a theorem, the results in [28] could be used effectively to extend the range of validity for Green's Conjecture for various classes of non-generic curves.

We fix an odd genus $g = 2i + 3$ and observe that the Hurwitz locus

$$\mathcal{M}^1_{g,i+2} := \Big\{[C] \in \mathcal{M}_g : \exists C \stackrel{i+2:1}{\longrightarrow} \mathbf{P}^1\Big\}$$

is an irreducible divisor on $\mathcal{M}_g$. This divisor has been studied in detail by Harris and Mumford [26] in their course of proving that $\overline{\mathcal{M}}_g$ is of general type for large odd genus. In particular, they determined the class of the closure $\overline{\mathcal{M}}^1_{g,i+2}$ in $\overline{\mathcal{M}}_g$ of the Hurwitz divisor in terms of the standard generators of $\mathrm{Pic}(\overline{\mathcal{M}}_g)$.

On the other hand, one can consider the Koszul divisor

$$\mathfrak{Kosz}_g := \Big\{[C] \in \mathcal{M}_g : K_{\frac{g-3}{2},2}(C, \omega_C) \neq 0\Big\}.$$

We have already explained that in genus 3, the Koszul divisor is the degeneracy locus of the map of vector bundles over $\mathcal{M}_3$, globalizing the morphisms

$$\mathrm{Sym}^2 H^0(C, \omega_C) \longrightarrow H^0(C, \omega_C^2).$$

Max Noether's Theorem, that is, Green's Conjecture in genus 3, implies the already discussed set-theoretic equality $\mathfrak{Kosz}_3 = \mathcal{M}^1_{3,2}$. We have also explained how in genus 5, we have the following description of the Koszul divisor

$$\mathfrak{Kosz}_5 := \Big\{[C] \in \mathcal{M}_5 : I_C(2) \otimes H^0(C, \omega_C) \stackrel{\not\cong}{\longrightarrow} I_C(3)\Big\}.$$

It is relatively standard to show that for any odd genus $\mathfrak{Kosz}_g$ is a virtual divisor, that is, the degeneracy locus of a map between vector bundles of the same rank over $\mathcal{M}_g$. Precisely, one can show that $[C] \in \mathfrak{Kosz}_g$ if and only if the restriction map

$$H^0\Big(\mathbf{P}^{g-1}, \bigwedge^i M_{\mathbf{P}^{g-1}}(2)\Big) \longrightarrow H^0\Big(C, \bigwedge^i M_{\omega_C} \otimes \omega_C^2\Big)$$

is not an isomorphism. Both vector spaces appearing above have the same dimension, independent of C. Therefore either $\mathfrak{Kosz}_g$ is a genuine divisor on $\mathcal{M}_g$, or else, $\mathfrak{Kosz}_g = \mathcal{M}_g$. Voisin's result [38] rules out the second possibility, for $K_{i,2}(C, \omega_C) = 0$, for a general curve $[C] \in \mathcal{M}_g$.

Hirschowitz and Ramanan generalized to arbitrary odd genus the equalities of cycles in moduli already discussed in small genus. Putting together their work with that of Voisin [38], we obtain the following result, present in a slightly revisionist fashion, for as we mentioned, [28] predates the papers [37] and [38]:

Theorem 4.7 *For odd genus $g = 2i + 3$, one has the following equality of effective divisors:*

$$[\mathfrak{Kosz}_g] = (i+1)[\mathcal{M}^1_{g,i+2}] \in \text{Pic}(\mathcal{M}_{2i+3}).$$

It follows that Green's Conjecture holds for every smooth curve of genus g and maximal Clifford index i + 1. Equivalently, the following equivalence holds for any smooth curve C of genus g:

$$K_{i,2}(C, \omega_C) \neq 0 \Longleftrightarrow \text{Cliff}(C) \leq i.$$

This idea bearing fruit in [28], of treating Green's Conjecture variationally as a moduli question, is highly innovative and has been put to use in other non-trivial contexts, for instance in [15, 16] or [17].

4.5 Aprodu's Work and Other Applications

Theorem 4.7 singles out an explicit class of curves of odd genus for which Green's Conjecture is known to hold. It is also rather clear that Theorem 4.7 can be extended to certain classes of stable curves, for instance to all *irreducible* stable curves of genus $g = 2i + 3$, for which one still has the equivalence

$$K_{i,2}(C, \omega_C) \neq 0 \Longleftrightarrow [C] \in \overline{\mathcal{M}}^1_{g,i+2}.$$

Indeed, the definition of $\mathfrak{Kosz}_g$ makes sense for irreducible nodal canonical curves, whereas Harris and Mumford [26] constructed a compactification of Hurwitz spaces, and thus in particular of $\mathcal{M}^1_{g,i+2}$, by means of *admissible covers*. Using such a degenerate form of Theorem 4.7, Aprodu [1] provided a sufficient Brill-Noether type condition for curves of *any* gonality which implies Green's Conjecture.

Theorem 4.8 *Let C be a smooth k-gonal curve of genus $g \geq 2k - 2$. Assume that*

$$\dim W^1_{g-k+2}(C) = g - 2k + 2. \tag{6}$$

Then C verifies Green's Conjecture.

Since C is k-gonal, by adding $g - 2k + 2$ arbitrary base points to a pencil A of minimal degree on C, we observe that $\{A\} + C_{g-2k+2} \subseteq W^1_{g-k+2}(C)$. In particular, $\dim W^1_{g-k+2}(C) \geq g - 2k + 2$. Thus condition (6) requires that there be no more pencils of degree $g - k + 2$ on C than one would normally expect. The main use of Theorem 4.8 is that it reduces Green's Conjecture, which is undoubtedly a very difficult question, to Brill-Noether theory, which is by comparison easier. Indeed, using Kodaira-Spencer theory it was showed in [5] that condition (6) holds for general k-gonal curves of any genus. This implies the following result:

Theorem 4.9 *Let C be a general k-gonal curve of genus g, where $2 \leq k \leq \frac{g+2}{2}$. Then C verifies Green's Conjecture.*

Theorem 4.9 has been first proven without using the bound given in Theorem 4.8 by Teixidor [34] in the range $k \leq \frac{g}{3}$ and by Voisin [37] in the range $h \geq \frac{g}{3}$. Note that in each gonality stratum $\mathcal{M}^1_{g,k} \subseteq \mathcal{M}_g$, Green's Conjecture amounts to a *different* vanishing statement, that is, one does not have a uniform statement of Green's Conjecture over $\mathcal{M}_g$. Theorem 4.9 has thus to be treated one gonality stratum at a time.

The last application we mention involves curves lying on $K3$ surfaces and discusses results from the paper [2]:

Definition 4.10 A polarized $K3$ surface of genus g consists of a pair (S, H), where S is a smooth $K3$ surface and $H \in \text{Pic}(S)$ is an ample class with $H^2 = 2g - 2$. We denote by $\mathcal{F}_g$ the irreducible 19-dimensional moduli space of polarized $K3$ surfaces of genus g.

The highly interesting geometry of $\mathcal{F}_g$ is not a subject of these lectures. We refer instead to [12] for a general reference.

As already discussed, if $[S, H] \in \mathcal{F}_g$ is a general polarized $K3$ surface and $C \in |H|$ is any smooth hyperplane section of S, Voisin proved that C verifies Green's Conjecture. Making decisive use of Theorem 4.8, one can extend this result to arbitrary polarized $K3$ surfaces. We quote from [2]:

Theorem 4.11 *Green's Conjecture holds for a smooth curve C lying on any $K3$ surface.*

Observe a significant difference between this result and Theorems 4.5 and 4.6. Whereas the lattice condition on $\text{Pic}(S)$ in the latter cases forces that $C \in |H|$ has maximal Clifford index $\text{Cliff}(C) = \lfloor \frac{g-1}{2} \rfloor$, Theorem 4.11 applies to curves in every gonality stratum in $\mathcal{M}_g$.

5 The Prym-Green Conjecture

In this section we would like to discuss a relatively new conjecture concerning the resolution of a general paracanonical curve and we shall begin with a general definition.

Definition 5.1 Let C be a smooth curve of genus g and $L \in \text{Pic}^d(C)$ a very ample line bundle inducing an embedding $\varphi_L : C \hookrightarrow \mathbf{P}^r$. We say that the the pair (C, L) has a *natural resolution*, if for every p one has

$$b_{p,2}(C, L) \cdot b_{p+1,1}(C, L) = 0.$$

The naturality of the resolution of $C \subseteq \mathbf{P}^r$ implies that the lower bounds on the number of syzygies of C given by the Hilbert function of C are attained, that is, the minimal resolution of C is as simple as the degree and genus of C allow it. Recall the statement of Theorem 3.7: When $h^1(C,L) = 0$ and thus $r = d-g$, the difference in Betti numbers on each diagonal of the Betti table does not vary with C and L and is given by the following formula:

$$b_{p+1,1}(C,L) - b_{p,2}(C,L) = (p+1)\binom{d-g}{p+1}\Big(\frac{d+1-g}{p+2} - \frac{d}{d-g}\Big). \tag{7}$$

Thus if one knows that $b_{p+1,1}(C,L) \cdot b_{p,2}(C,L) = 0$, then for any given p, depending on the sign of the formula appearing in the right hand side of (7), one can determine which Betti number has to vanish, as well as the exact value of the remaining number on the same diagonal of the Betti table.

Using the concept of natural resolution, one obtains a very elegant and compact reformulation of Green's Conjecture for generic curves. By inspecting again Tables 3 and 4, Voisin's Theorems 4.5 and 4.6 can be summarized in one single sentence:

Theorem 5.2 *The minimal resolution of a generical canonical curve of genus g is natural.*

Note that this is the only case when Green's Conjecture is equivalent to the resolution being natural. For a curve C of non-maximal Clifford index, that is, $\mathrm{Cliff}(C) \leq \lfloor \frac{g-3}{2} \rfloor$, the resolution of the canonical curve is not natural, irrespective of whether Green's Conjecture is valid for C or not. Indeed, for integers p such that

$$\mathrm{Cliff}(C) \leq p < g - \mathrm{Cliff}(C) - 2,$$

the Green-Lazarsfeld Non-Vanishing Theorem implies $b_{p,2}(C,\omega_C) \neq 0$ and $b_{p+1,1}(C,\omega_C) \neq 0$. This observation suggests that, more generally, naturality might be suitable to capture the resolution of a *general* point of a moduli space of curves, rather than that of arbitrary objects with certain numerical properties. A concept very similar to naturality appears in the formulation [21] of the *Minimal Resolution Conjecture* for general sets of points on projective varieties.

Definition 5.3 A *paracanonical* curve of genus g is a smooth genus g curve embedded by a linear system

$$\varphi_{\omega_C \otimes \eta} : C \hookrightarrow \mathbf{P}^{g-2},$$

where $\eta \in \mathrm{Pic}^0(C)$ is a non-trivial line bundle. When η is an ℓ-torsion point in $\mathrm{Pic}^0(C)$ for some $\ell \geq 2$, we refer to a *level ℓ paracanonical* curve.

The case studied by far the most is that of *Prym canonical* curves, when $\ell = 2$. Due to work of Mumford, Beauville, Clemens, Tyurin and others, it has been known for a long time that properties of theta divisors on Prym varieties can be reformulated in terms of the projective geometry of the corresponding Prym canonical curve. We refer to [6] Section 6 for an introduction to this beautiful circle of ideas.

Pairs $[C, \eta]$, where C is a smooth curve of genus g and $\eta \in \mathrm{Pic}^0(C)[\ell]$ is an ℓ-torsion point form an irreducible moduli space $\mathcal{R}_{g,\ell}$, which is a finite cover of $\mathcal{M}_g$. The morphism $\mathcal{R}_{g,\ell} \to \mathcal{M}_g$ is given by forgetting the level ℓ structure η. The geometry of the moduli space $\mathcal{R}_{g,\ell}$ is discussed in detail in [9].

We now fix a general point $[C, \eta] \in \mathcal{R}_{g,\ell}$, where $g \geq 5$. The paracanonical linear system $\omega_C \otimes \eta$ induces an embedding

$$\varphi_{\omega_C \otimes \eta} : C \hookrightarrow \mathbf{P}^{g-2}.$$

The Prym-Green Conjecture, originally formulated for $\ell = 2$ in [20] and in the general case in [9], concerns the resolution of the paracanonical ring $\Gamma_C(\omega_C \otimes \eta)$.

Conjecture 5.4 The resolution of a general level ℓ paracanonical curve of genus g is natural.

There is an obvious weakening of the Prym-Green Conjecture, by allowing η to be a general line bundle of degree zero on C, rather than an ℓ-torsion bundle.

Conjecture 5.5 The resolution of a general paracanonical curve $C \subseteq \mathbf{P}^{g-2}$ of genus g and degree $2g - 2$ is natural.

Since one has an embedding of $\mathcal{R}_{g,\ell}$ in the universal Jacobian variety of degree $2g - 2$ over $\mathcal{M}_g$, clearly Conjecture 5.5 is a weaker statement than the Prym-Green Conjecture. The numerology in Conjectures 5.4 and 5.5 is the same, one conjecture is just a refinement of the other.

We would like to spell-out some of he implications of the Prym-Green Conjecture. For odd g, the Prym-Green Conjecture amounts to the following independent vanishing statements:

$$K_{\frac{g-3}{2},1}\big(C, \omega_C \otimes \eta\big) = 0 \ \text{ and } \ K_{\frac{g-7}{2},2}\big(C, \omega_C \otimes \eta\big) = 0. \tag{8}$$

Assuming naturality of the resolution, one is able to determine explicitly the entire resolution. Setting $g := 2i + 5$, one obtains the resolution shown in Table 10

Table 10 The Betti table of a general paracanonical curve of genus $g = 2i + 5$

1	2	...	$i-1$	i	$i+1$	$i+2$	...	$2i+2$
$b_{1,1}$	$b_{2,1}$	...	$b_{i-1,1}$	$b_{i,1}$	0	0	...	0
0	0	...	0	$b_{i,2}$	$b_{i+1,2}$	$b_{i+2,2}$	...	$b_{2i+2,2}$

where,

$$b_{p,1} = \frac{p(2i-2p+1)}{2i+3}\binom{2i+4}{p+1} \quad \text{for } p \leq i, \text{ and}$$

$$b_{p,2} = \frac{(p+1)(2p-2i+1)}{2i+3}\binom{2i+4}{p+2} \quad \text{for } p \geq i.$$

The resolution is *natural*, but fails to be *pure* in column i, for both Koszul cohomology groups $K_{i,1}(C, \omega_C \otimes \eta)$ and $K_{i,2}(C, \omega_C \otimes \eta)$ are non-zero. Note also the striking resemblance of the minimal resolution of the general level ℓ paracanonical curve of *odd* genus and the resolution of the general canonical curve of *even* genus.

For even genus, the Prym-Green Conjecture reduces to a single vanishing statement, namely:

$$K_{\frac{g-4}{2},1}(C, \omega_C \otimes \eta) = K_{\frac{g-6}{2},2}(C, \omega_C \otimes \eta) = 0. \tag{9}$$

Indeed, by applying (7), one always has $b_{\frac{g-4}{2},1}(C, \omega_C \otimes \eta) = b_{\frac{g-6}{2},2}(C, \omega_C \otimes \eta)$, so the Prym-Green Conjecture in even genus amounts to one single vanishing statement. Like in the previous case, naturality determines the resolution completely.

We write $g := 2i + 6$ and using (3.7), obtain (Table 11).

Here

$$b_{p,1} = \frac{p(i+1-p)}{i+2}\binom{2i+5}{p+1} \text{ for } p \leq i \text{ and } b_{p,2} = \frac{(p+1)(p-i)}{i+2}\binom{2i+5}{p+2} \text{ for } p > i.$$

In this case the resolution is both natural and pure. Therefore, the expected resolution of a general paracanonical curve of *even* genus has the same features as the resolution of a general canonical curve of *odd* genus.

The first non-trivial case of the Prym-Green Conjecture is $g = 6$, when one has to show that the multiplication map

$$\mathrm{Sym}^2 H^0(C, \omega_C \otimes \eta) \longrightarrow H^0(C, \omega_C^2 \otimes \eta^2)$$

is an isomorphism for a generic choice of $[C, \eta] \in \mathcal{R}_{6,\ell}$. Observe that $h^0(C, \omega \otimes \eta) = 5$ whereas $h^0(C, \omega_C^2 \otimes \eta^2) = 15$, therefore both vector spaces appearing in the previous map have he same dimension 15.

Table 11 The Betti table of a general paracanonical curve of genus $g = 2i + 6$

1	2	...	$i-1$	i	$i+1$	$i+2$	...	$2i+3$
$b_{1,1}$	$b_{2,1}$	...	$b_{i-1,1}$	$b_{i,1}$	0	0	...	0
0	0	...	0	0	$b_{i+1,2}$	$b_{i+2,2}$	...	$b_{2i+3,2}$

A systematic study of the Prym-Green Conjecture has been undertaken in [9]. It has been proved with the use of *Macaulay* that the conjecture holds for all $g \leq 18$ and $\ell \leq 5$ with two possible exceptions for $\ell = 2$, when $g = 8, 16$. In those cases, it has been showed that in the case the underlying curve C is a rational g-nodal curve, the level 2 paracanonical curve $C \subseteq \mathbf{P}^{g-2}$ has one unexpected syzygy. This finding strongly suggests that the Prym-Green Conjecture might fail for level $\ell = 2$ and for genera which have strong divisibility properties by 2. Testing with Macaulay the next relevant case $g = 24$ is at the moment out of reach. The weaker Conjecture 5.5 is expected to hold for every g without exceptions.

5.1 The Prym-Green Conjecture for Curves of Odd Genus

We shall now discuss the main ideas of the proof presented in [17] and [18] of the Prym-Green Conjecure for paracanonical curves of odd genus. We set $g := 2i + 5$. One aims to exhibit a smooth level ℓ curve $[C, \eta] \in \mathcal{R}_{g,\ell}$ whose Koszul cohomology satisfies the following two vanishing properties:

$$K_{i+1,1}(C, \omega_C \otimes \eta) = 0 \quad \text{and} \quad K_{i-1,2}(C, \omega_C \otimes \eta) = 0. \tag{10}$$

To construct such a curve, one can resort to $K3$ surfaces, but there is an extra difficulty in comparison to the usual Green's Conjecture, because one should produce both a general curve of genus g, as well as a distinguished ℓ-torsion point in its Jacobian. Since this point has to be sufficiently explicit to be able to compute the Koszul cohomology of the corresponding paracanonical line bundle, it is natural to attempt to realize it as the restriction of a line bundle defined on the ambient $K3$ surface. This program has been carried out using special $K3$ surfaces, which are of *Nikulin* type when $\ell = 2$, or of *Barth–Verra* type for high ℓ. The final solution to the Prym-Green Conjecture in odd genus is presented in the paper [19], using certain elliptic ruled surface which are limits of polarized $K3$ surfaces.

We let S be a smooth $K3$ surface having the following Picard lattice

$$\mathrm{Pic}(S) = \mathbb{Z} \cdot L \oplus \mathbb{Z} \cdot H,$$

where $L^2 = L \cdot H = 2g - 2$ and $H^2 = 2g - 6$. For each smooth curve $C \in |L| \cong \mathbf{P}^g$, the restriction $\mathcal{O}_C(H)$ is a line bundle of degree $2g-2$, hence $\mathcal{O}_C(H-C) \in \mathrm{Pic}^0(C)$. Since the Jacobian of C has the same dimension g as the linear system $|L|$, it is natural to expect that there will exist finitely many curves $C \in |L|$ such that

$$\eta_C := \mathcal{O}_C(H - C) = \mathcal{O}_C(H) \otimes \omega_C^\vee \in \mathrm{Pic}^0(C)$$

is an ℓ-torsion point. A priori, one is not sure that these curves are smooth or even nodal. We name such surfaces after *Barth–Verra*, for they were first introduced in the beautiful paper [7]. In [18], we show that for the general Barth–Verra surface, the expectation outlined above can be realized.

Theorem 5.6 *For a general Barth–Verra surface S of genus $g \geq 3$ and an integer ℓ, there exist precisely*

$$\binom{2\ell^2 - 2}{g}$$

curves $C \in |L|$ such that $\eta_C^{\otimes \ell} \cong \mathcal{O}_C$. All such curves C are smooth and irreducible. The number of curves C such that η_C has order exactly ℓ is strictly positive.

With this very effective tool at hand, we can now state the last result we wish to discuss in these lectures:

Theorem 5.7 *The Prym-Green Conjecture holds for paracanonical level ℓ curves of odd genus.*

This result is proved for $\ell = 2$ in [17] and for $\ell \geq \sqrt{\frac{g+2}{2}}$ in [18]. It is this last proof that we shall discuss in what follows. For the remaining levels, the proof of the Prym-Green Conjecture is completed in [19] using elliptic surfaces.

Assume we are in the situation of Theorem 5.6 and we take a Barth–Verra $K3$ surface S, together with smooth curves $C \in |L|$ having genus $2i + 5$ and $D \in |H|$ having genus $2i+3$ respectively. The ℓ-torsion point on $\eta_C = \mathcal{O}_C(H-C)$ is obtained as a restriction from the surface.

To get a grip on the Koszul cohomology groups which according to (10) should vanish, we first use the functoriality of Koszul cohomology and write-down the following exact sequence:

$$\cdots \longrightarrow K_{p,q}(S, H) \longrightarrow K_{p,q}(C, H_C) \longrightarrow K_{p-1,q+1}(S, -C, H) \longrightarrow \cdots, \tag{11}$$

where, we recall that the mixed Koszul cohomology group $K_{p-1,q+1}(S, -C, H)$ is computed by the following part of the Koszul complex:

$$\cdots \longrightarrow \bigwedge^{p} H^0(S, H) \otimes H^0(S, qH-C) \xrightarrow{d_{p,q}} \bigwedge^{p-1} H^0(S, H) \otimes H^0(S, (q+1)H-C) \xrightarrow{d_{p-1,q+1}}$$

$$\xrightarrow{d_{p-1,q+1}} \bigwedge^{p-2} H^0(S, H) \otimes H^0(X, (q+2)H - C) \longrightarrow \cdots.$$

Using this sequence for $(p, q) = (i + 1, 1)$ and $(p, q) = (i - 1, 2)$ respectively, in order to prove the Prym-Green Conjecture for $g = 2i + 5$, it suffices to show

$$K_{i+1,1}(S, H) = 0 \text{ and } K_{i-1,2}(S, H) = 0, \tag{12}$$

respectively

$$K_{i,2}(S,-C,H)=0 \text{ and } K_{i-2,3}(S,-C,H)=0. \tag{13}$$

Via the Lefschetz Hyperplane Principle for Koszul cohomology, vanishing (12) lies in the regime governed by the classical Green's Conjecture, which has been proved in [2] for (curves lying on) *all* smooth $K3$ surfaces, in particular for Barth-Verra surfaces as well. Since D is a curve of genus $g-2=2i+3$, if one shows that the Clifford index of D is maximal, that is, $\mathrm{Cliff}(D)=i+1$, then by Theorem 4.7 it follows

$$K_{i-1,2}(D,\omega_D)\cong K_{i-1,2}(S,H)=0 \text{ and } K_{i+1,1}(S,H)\cong K_{i+1,1}(D,\omega_D)=0.$$

The fact that $\mathrm{Cliff}(D)=i+1$ amounts to a simple lattice-theoretic calculation in $\mathrm{Pic}(S)$. Once this is carried out, we conclude that (12) holds.

In order to prove (13), one restricts the Koszul cohomology on the surface to a general curve $D\in|H|$ to obtain isomorphisms:

$$K_{i,2}(S,-C,H)\cong K_{i,2}(D,-C_D,\omega_D) \text{ and } K_{i-2,3}(S,-C,H)\cong K_{i-2,3}(D,-C_D,\omega_D).$$

Via the description of twisted Koszul cohomology in terms of Lazarsfeld bundles given in Proposition 3.2, we obtain the following isomorphisms:

$$K_{i,2}(D,-C_D,\omega_D)\cong H^0\Big(D,\bigwedge^{i}M_{\omega_D}\otimes(2K_D-C_D)\Big) \quad\text{and}$$

$$K_{i-2,3}(D,-C_D,\omega_D)\cong H^1\Big(D,\bigwedge^{i-1}M_{\omega_D}\otimes(2K_D-C_D)\Big)$$

$$\cong H^0\Big(D,\bigwedge^{i-1}M^{\vee}_{\omega_D}\otimes(C_D-K_D)\Big)^{\vee}.$$

Although at first sight, these new cohomology groups look opaque, showing that they vanish is easier to prove than the original Green's Conjecture. One specializes the Barth-Verra surface further until both C and D become hyperelliptic curves. Precisely, we specialize S to a $K3$ surface having the following lattice, with respect to an ordered basis (L,η,E), where $\eta=H-L$.

$$\begin{pmatrix} 4i+8 & 0 & 2\\ 0 & -4 & 0\\ 2 & 0 & 0\end{pmatrix}.$$

Note that E is an elliptic pencil cutting out a divisor of degree 2 both on a curve $C\in|L|$ and on a curve $D\in|H|$. The curve D being hyperelliptic, the Lazarsfeld bundle M_{ω_D} splits as a direct sum of degree 2 line bundles. The vanishing (13) becomes the statement that a certain line bundle of degree $\le g(D)-1$ on D has no sections, which can be easily verified. Full details can be found in [17] and [18].

5.2 The Failure of the Prym-Green Conjecture in Genus 8

The paper [10] is dedicated to understanding the failure of the Prym-Green Conjecture in genus 8. We choose a general Prym-canonical curve of genus 8

$$\varphi_{\omega_C \otimes \eta} : C \hookrightarrow \mathbf{P}^6,$$

set $L := \omega_C \otimes \eta$ and denote $I_{C,L}(k) := \mathrm{Ker}\{\mathrm{Sym}^k H^0(C,L) \to H^0(C,L^k)\}$ for all $k \geq 2$. Observe that $\dim I_{C,L}(2) = 7$ and $\dim I_{C,L}(3) = 49$, therefore the map

$$\mu_{C,L} : I_{C,L}(2) \otimes H^0(C,L) \to I_{C,L}(3)$$

globalizes to a morphism of vector bundles of the *same rank* over the stack $\mathcal{R}_{8,2}$. In [10] we present *three* proofs that $\mu_{C,L}$ is never an isomorphism (equivalently $K_{2,1}(C,L) \neq 0$). We believe that one of these approaches will generalize to higher genus and offer hints into the exceptions to the Prym-Green Conjecture in even genus. One of the approaches, links the non-vanishing of $K_{2,1}(C, \omega_C \otimes \eta)$ to the existence of quartic hypersurfaces in $\mathbf{P}^6$ vanishing doubly along the Prym-canonical curve C, but which are not in the image of the map

$$\mathrm{Sym}^2 I_{C,L}(2) \longrightarrow I_{C,L}(4).$$

References

1. M. Aprodu, Remarks on syzygies of d-gonal curves. Math. Res. Lett. **12**, 387–400 (2005)
2. M. Aprodu, G. Farkas, The Green Conjecture for smooth curves lying on arbitrary $K3$ surfaces. Compos. Math. **147**, 839–851 (2011)
3. M. Aprodu, G. Farkas, Koszul cohomology and applications to moduli, in *Grassmanians, Moduli Spaces and Vector Bundles, Clay Mathematics Proceedings*, vol. 14, ed. by E. Previato (2011), pp. 25–50
4. M. Aprodu, J. Nagel, *Koszul Cohomology and Algebraic Geometry*. University Lecture Series, vol. 52 (American Mathematical Society, Providence, RI, 2010)
5. E. Arbarello, M. Cornalba, Footnotes to a paper of Beniamino Segre. Math. Ann. **256**, 341–362 (1981)
6. E. Arbarello, M. Cornalba, P.A. Griffiths, J. Harris, *Geometry of Algebraic Curves*, vol. I. Grundlehren der mathematischen Wissenschaften, vol. 267 (Springer, New York, 1985)
7. W. Barth, A. Verra, Torsions of $K3$ sections, in *Problems in the Theory of Surfaces and Their Classification (Cortona, 1988), Symposia Mathematica*, Vol. XXXII (Academic, London, 1991), pp. 1–24
8. G. Castelnuovo, Sui multipli di una serie lineare di gruppi di punti appartenente ad una curva algebrica. Rendiconti di Circolo Matematico di Palermo **7**, 89–110 (1893)
9. A. Chiodo, D. Eisenbud, G. Farkas, F.-O. Schreyer, Syzygies of torsion bundles and the geometry of the level ℓ modular varieties over $\overline{\mathcal{M}}_g$. Invent. Math. **194**, 73–118 (2013)
10. E. Colombo, G. Farkas, A. Verra, C. Voisin, Syzygies of Prym and paracanonical curves of genus 8. arXiv:1612.01026

11. M. Coppens, G. Martens, Secant spaces and Clifford's theorem. Compos. Math. **78**, 193–212 (1991)
12. I. Dolgachev, Mirror symmetry for lattice polarized $K3$ surfaces. J. Math. Sci. **81**, 2599–2630 (1996)
13. L. Ein, R. Lazarsfeld, The gonality conjecture on syzygies of algebraic curves of large degree. Publ. Math. Inst. Hautes Études Sci. **122**, 301–313 (2015)
14. D. Eisenbud, *The Geometry of Syzygies*. Graduate Texts in Mathematics, vol. 229 (Springer, New York, 2005)
15. G. Farkas, Syzygies of curves and the efective cone of $\overline{\mathcal{M}}_g$. Duke Math. J. **135**, 53–98 (2006)
16. G. Farkas, Koszul divisors on moduli spaces of curves. Am. J. Math. **131**, 819–869 (2009)
17. G. Farkas, M. Kemeny, The generic Green-Lazarsfeld Conjecture. Invent. Math. **203**, 265–301 (2016)
18. G. Farkas, M. Kemeny, The Prym-Green Conjecture for line bundles of high order. Duke Math. J. **166**, 1103–1124 (2017)
19. G. Farkas, M. Kemeny, Minimal resolutions of paracanonical curves of odd genus. Preprint
20. G. Farkas, K. Ludwig, The Kodaira dimension of the moduli space of Prym varieties. J. Eur. Math. Soc. **12**, 755–795 (2010)
21. G. Farkas, M. Mustaţă, M. Popa, Divisors on $\mathcal{M}_{g,g+1}$ and the Minimal Resolution Conjecture for points on canonical curves. Ann. Sci. de L'École Normale Supérieure **36**, 553–581 (2003)
22. A. Grothendieck, Techniques de construction et théorèmes d'existence en géomètrie algébrique IV : les schémas de Hilbert. Séminaire Bourbaki **6**, 249–276 (1960–1961)
23. M. Green, Koszul cohomology and the cohomology of projective varieties. J. Differ. Geom. **19**, 125–171 (1984)
24. M. Green, R. Lazarsfeld, On the projective normality of complete linear series on an algebraic curve. Invent. Math. **83**, 73–90 (1986)
25. R. Hartshorne, *Algebraic Geometry*. Graduate Texts in Mathematics, vol. 52 (Springer, New York, 1977)
26. J. Harris, D. Mumford, On the Kodaira dimension of $\overline{\mathcal{M}}_g$. Invent. Math. **67**, 23–88 (1982)
27. D. Hilbert, Über die Theorie der algebraischen Formen. Math. Ann. **36**, 473–530 (1890)
28. A. Hirschowitz, S. Ramanan, New evidence for Green's Conjecture on syzygies of canonical curves. Annales Scientifiques de l'École Normale Supérieure **31**, 145–152 (1998)
29. R. Lazarsfeld, Brill-Noether-Petri without degenerations. J. Differ. Geom. **23**, 299–307 (1986)
30. R. Lazarsfeld, A sampling of vector bundle techniques in the study of linear series, in *Lectures on Riemann Surfaces*, ed. by M. Cornalba, X. Gomez-Mont, A. Verjovsky (International Centre for Theoretical Physics/World Scientific/ College on Riemann Surfaces/ Trieste/Singapore/Trieste, 1987), pp. 500–559
31. K. Petri, Über die invariante darstellung algebraischer funktionen einer veränderlichen. Math. Ann. **88**, 242–289 (1922)
32. F.-O. Schreyer, Syzygies of canonical curves and special linear series. Math. Ann. **275**, 105–137 (1986)
33. F.-O. Schreyer, A standard basis approach to syzygies of canonical curves. J. Reine Angew. Math. **421**, 83–123 (1991)
34. M. Teixidor, Green's conjecture for the generic r-gonal curve of genus $g \geq 3r-7$. Duke Math. J. **111**, 195–222 (2002)
35. C. Voisin, Courbes tétragonales et cohomologie de Koszul. J. Reine Angew. Math. **387**, 111–121 (1988)
36. C. Voisin, Déformation de syzygies et théorie de Brill-Noether. Proc. Lond. Math. Soc. **67**, 493–515 (1993)
37. C. Voisin, Green's generic syzygy conjecture for curves of even genus lying on a $K3$ surface. J. Eur. Math. Soc. **4**, 363–404 (2002)
38. C. Voisin, Green's canonical syzygy conjecture for generic curves of odd genus. Compos. Math. **141**, 1163–1190 (2005)

Lectures on Bridgeland Stability

Emanuele Macrì and Benjamin Schmidt

1 Introduction

Stability conditions in derived categories provide a framework for the study of moduli spaces of complexes of sheaves. They have been introduced in [30], with inspiration from work in string theory [41]. It turns out that the theory has far further reach. A non exhaustive list of influenced areas is given by counting invariants, geometry of moduli spaces of sheaves, representation theory, homological mirror symmetry, and classical algebraic geometry. This article will focus on the basic theory of Bridgeland stability on smooth projective varieties and give some applications to the geometry of moduli spaces sheaves. We pay particular attention to the case of complex surfaces.

Stability on Curves The theory starts with vector bundles on curves. We give an overview of the classical theory in Sect. 2. Let C be a smooth projective curve. In

E. Macrì (✉)
Department of Mathematics, Northeastern University, 360 Huntington Avenue, Boston, MA 02115-5000, USA
e-mail: e.macri@northeastern.edu

B. Schmidt
Department of Mathematics, The Ohio State University, 231 W 18th Avenue, Columbus, OH 43210-1174, USA

Department of Mathematics, The University of Texas at Austin, 2515 Speedway Stop C1200, Austin, TX 78712-1202, USA
e-mail: schmidt@math.utexas.edu

L. Brambila Paz et al. (eds.), *Moduli of Curves*, Lecture Notes of the Unione Matematica Italiana 21, DOI 10.1007/978-3-319-59486-6_5

order to obtain a well behaved moduli space one has to restrict oneself to so called *semistable* vector bundles. Any vector bundle E has a *slope* defined as $\mu(E) = \frac{d(E)}{r(E)}$, where $d(E)$ is the *degree* and $r(E)$ is the *rank*. It is called semistable if for all subbundles $F \subset E$ the inequality $\mu(F) \leq \mu(E)$ holds.

The key properties of this notion that one wants to generalize to higher dimensions are the following. Let $\mathcal{A} = \mathrm{Coh}(C)$ denote the category of coherent sheaves. One can recast the information of rank and degree as an additive homomorphism

$$Z : K_0(C) \to \mathbb{C}, \ v \mapsto -d(v) + \sqrt{-1}\, r(v),$$

where $K_0(C)$ denotes the Grothendieck group of C, generated by classes of vector bundles. Then:

(1) For any $E \in \mathcal{A}$, we have $\Im Z(E) \geq 0$.
(2) If $\Im Z(E) = 0$ for some non trivial $E \in \mathcal{A}$, then $\Re Z(E) < 0$.
(3) For any $E \in \mathcal{A}$ there is a filtration

$$0 = E_0 \subset E_1 \subset \ldots \subset E_{n-1} \subset E_n = E$$

of objects $E_i \in \mathcal{A}$ such that $A_i = E_i/E_{i-1}$ is semistable for all $i = 1, \ldots, n$ and $\mu(A_1) > \ldots > \mu(A_n)$.

Higher Dimensions The first issue one has to deal with is that if one asks for the same properties to be true for coherent sheaves on a higher dimensional smooth projective variety X, it is not so hard to see that property (2) cannot be achieved (by any possible group homomorphism Z). The key idea is then to change the category in which to define stability. The bounded derived category of coherent sheaves $\mathrm{D^b}(X)$ contains many full abelian subcategories with similar properties as $\mathrm{Coh}(X)$ known as *hearts of bounded t-structures*. A *Bridgeland stability condition* on $\mathrm{D^b}(X)$ is such a heart $\mathcal{A} \subset \mathrm{D^b}(X)$ together with an additive homomorphism $Z : K_0(X) \to \mathbb{C}$ satisfying the three properties above (together with a technical condition, called the *support property*, which will be fundamental for the deformation properties below). The precise definition is in Sect. 5.

Other classical generalizations of stability on curves such as slope stability or Gieseker stability (see Sect. 3) do not directly fit into the framework of Bridgeland stability. However, for most known constructions, their moduli spaces still appear as special cases of Bridgeland stable objects. We will explain this in the case of surfaces.

Bridgeland's Deformation Theorem The main theorem in [30] (see Theorem 5.15) is that the set of stability conditions $\mathrm{Stab}(X)$ can be given the structure of a complex manifold in such a way that the map $(\mathcal{A}, Z) \mapsto Z$ which forgets the heart is a local homeomorphism. For general X is not even known whether $\mathrm{Stab}(X) \neq \emptyset$. However, if $\dim X = 2$, the situation is much more well understood. In Sect. 6 we construct stability conditions $\sigma_{\omega,B}$ for each choice of ample $\mathbb{R}$-divisor class ω and arbitrary $\mathbb{R}$-divisor class B. This construction originated in [31] in the case of K3 surfaces. Arcara and Bertram realized that the construction can be generalized to any surface by using the Bogomolov Inequality in [6]. The proof of the support property is in [15, 18, 19]. As a consequence, these stability conditions vary continuously in ω and B.

Moduli Spaces If we fix a numerical class v, it turns out that semistable objects with class v vary nicely within Stab(X) due to [31]. More precisely, there is a locally finite wall and chamber structure such that the set of semistable objects with class v is constant within each chamber. In the case of surfaces, as mentioned before, there will be a chamber where Bridgeland semistable objects of class v are exactly (twisted) Gieseker semistable sheaves. The precise statement is Corollary 6.25. The first easy applications of the wall and chamber structure are in Sect. 7.

The next question is about moduli spaces. Stability of sheaves on curves (or more generally, Gieseker stability on higher dimensional varieties) is associated to a GIT problem. This guarantees that moduli spaces exist as projective schemes. For Bridgeland stability, there is no natural GIT problem associated to it. Hence, the question of existence and the fact that moduli spaces are indeed well-behaved is not clear in general. Again, in the surface case, for the stability conditions $\sigma_{\omega,B}$, it is now known [109] that moduli spaces exist as Artin stacks of finite-type over $\mathbb{C}$. In some particular surfaces, it is also known coarse moduli spaces parameterizing S-equivalence classes of semistable objects exist, and they are projective varieties. We review this in Sect. 6.6.

Birational Geometry of Moduli Spaces of Sheaves on Surfaces The birational geometry of moduli spaces of sheaves on surfaces has been heavily studied by using wall crossing techniques in Bridgeland's theory. Typical question are what their nef and effective cones are and what the stable base locus decomposition of the effective cone is. The case of $\mathbb{P}^2$ was studied in many articles such as [8, 38, 74, 75, 115]. The study of the abelian surfaces case started in [78] and was completed in [86, 118]. The case of K3 surfaces was handled in [16, 17, 86]. Enriques surfaces were studied in [91]. We will showcase some of the techniques in Sect. 8 by explaining how to compute the nef cone of the Hilbert scheme of n points for surfaces of Picard rank one if $n \gg 0$. In this generality it was done in [25] and then later generalized to moduli of vector bundles with large discriminant in [37].

The above proofs are based on the so called *Positivity Lemma* from [17] (see Theorem 8.1). Roughly, the idea is that to any Bridgeland stability condition σ and to any family $\mathcal{E}$ of σ-semistable objects parameterized by a proper scheme S, there is a nef divisor class $D_{\sigma,\mathcal{E}}$ on S. Moreover, if there exists a curve C in S such that $D_{\sigma,\mathcal{E}}.C = 0$, then all objects $\mathcal{E}_{|X\times\{c\}}$ are S-equivalent, for all $c \in C$. In examples, the divisor class will induce an ample divisor class on the moduli space of stable objects. Hence, we can use the Positivity Lemma in two ways: until we move in a chamber in the space of stability conditions, this gives us a subset of the ample cone of the moduli space. Once we hit a wall, we have control when we hit also a boundary of the nef cone if we can find a curve of σ-stable objects in a chamber which becomes properly semistable on the wall.

Bridgeland Stability for Threefolds As mentioned before, the original motivation for Bridgeland stability comes from string theory. In particular, it requires the construction of stability conditions on Calabi-Yau threefolds. It is an open problem to even find a single example of a stability condition on a simply connected projective Calabi-Yau threefold where skyscraper sheaves are stable (examples where they are semistable are in [19]). Most successful attempts on threefolds trace back to a conjecture in [18]. In the surface case the construction is based on the classical Bogomolov inequality for Chern characters of semistable vector bundles. By analogy a generalized Bogomolov inequality for threefolds involving the third Chern character was conjectured in [18] that allows to construct Bridgeland stability. In [103] it was shown that this conjectural inequality needs to be modified, since it does not hold for the blow up of $\mathbb{P}^3$ in a point. There are though many cases in which the original inequality is true. The first case was $\mathbb{P}^3$ in [18, 82]. A similar argument was then successfully applied to the smooth quadric hypersurface in $\mathbb{P}^4$ in [101]. The case of abelian threefolds was independently proved in [79, 80] and [19]. Moreover, as pointed out in [19], this also implies the case of étale quotients of abelian threefolds and gives the existence of Bridgeland stability condition on orbifold quotients of abelian threefolds (this includes examples of Calabi-Yau threefolds which are simply-connected). The latest progress is the proof of the conjecture for all Fano threefolds of Picard rank one in [73] and a proof of a modified version for all Fano threefolds independently in [23] and [95].

Once stability conditions exist on threefolds, it is interesting to study moduli spaces therein and which geometric information one can get by varying stability. For projective space this approach has led to first results in [47, 102, 116].

Structure of the Notes In Sect. 2 we give a very light introduction to stability of vector bundles on curves. The chapter serves mainly as motivation and is logically independent of the remaining notes. Therefore, it can safely be skipped if the reader wishes to do so. In Sect. 3, we discuss classical generalizations of stability from curves to higher dimensional varieties. Moduli spaces appearing out of those are often times of classical interest and connecting these to Bridgeland stability is usually key. In Sects. 4 and 5 we give a full definition of what a Bridgeland stability condition is and prove or point out important basic properties. In Sect. 6, we demonstrate the construction of stability on smooth projective surfaces. Sections 7 and 8 are about concrete examples. We show how to compute the nef cone for Hilbert schemes of points in some cases, how one can use Bridgeland stability to prove Kodaira vanishing on surfaces and discuss some further questions on possible applications to projective normality for surfaces. Section 9 is about threefolds. We explain the construction of stability conditions on $\mathrm{D}^{\mathrm{b}}(\mathbb{P}^3)$. As an application we point out how Castelnuovo's classical genus bound for non degenerate curves turns out to be a simple consequence of the theory. In Appendix we give some background on derived categories.

These notes contain plenty of exercises and we encourage the reader to do as many of them as possible. A lot of aspects of the theory that seem obscure and abstract at first turn out to be fairly simple in practice.

What is Not in These Notes One of the main topics of current interest we do not cover in these notes is the "local case" (i.e., stability conditions on CY3 triangulated categories defined using quivers with potential; see, e.g., [29, 34, 68]) or more generally stability conditions on Fukaya categories (see, e.g., [40, 53, 65]). Another fundamental topic is the connection with counting invariants; for a survey we refer to [111–113]. Connections to string theory are described for example in [9, 28, 46]. Connections to Representation Theory are in [5].

There is also a recent survey [60] focusing more on both the classical theory of semistable sheaves and concrete examples still involving Bridgeland stability. The note [12] focuses instead on deep geometric applications of the theory (the classical Brill-Noether theorem for curves). The survey [62] focuses more on K3 surfaces and applications therein. Finally, Bridgeland's deformation theorem is the topic of the excellent survey [11], with a short proof recently appearing in [13].

Notation

G_k	The k-vector space $G \otimes k$ for a field k and abelian group G
X	A smooth projective variety over $\mathbb{C}$
$\mathcal{I}_Z$	The ideal sheaf of a closed subscheme $Z \subset X$
$\mathrm{D^b}(X)$	The bounded derived category of coherent sheaves on X
$\mathrm{ch}(E)$	The Chern character of an object $E \in D^b(X)$
$K_0(X)$	The Grothendieck group of X
$K_{\mathrm{num}}(X)$	The numerical Grothendieck group of X
$\mathrm{NS}(X)$	The Néron-Severi group of X
$N^1(X)$	$\mathrm{NS}(X)_{\mathbb{R}}$
$\mathrm{Amp}(X)$	The ample cone inside $N^1(X)$
$\mathrm{Pic}^d(C)$	The Picard variety of lines bundles of degree d on a smooth curve C

2 Stability on Curves

The theory of Bridgeland stability conditions builds on the usual notions of stability for sheaves. In this section we review the basic definitions and properties of stability for vector bundles on curves. Basic references for this section are [63, 72, 90, 104]. Throughout this section C will denote a smooth projective complex curve of genus $g \geq 0$.

2.1 The Projective Line

The starting point is the projective line $\mathbb{P}^1$. In this case, vector bundles can be fully classified, by using the following decomposition theorem, which is generally attributed to Grothendieck, who proved it in modern language. It was known way before in the work of Dedekind-Weber, Birkhoff, Hilbert, among others (see [63, Theorem 1.3.1]).

Theorem 2.1 *Let E be a vector bundle on $\mathbb{P}^1$. Then there exist unique integers $a_1, \ldots, a_n \in \mathbb{Z}$ satisfying $a_1 > \ldots > a_n$ and unique non-zero vector spaces $V_1, \ldots, V_n$ such that E is isomorphic to*

$$E \cong (\mathcal{O}_{\mathbb{P}^1}(a_1) \otimes_{\mathbb{C}} V_1) \oplus \ldots \oplus (\mathcal{O}_{\mathbb{P}^1}(a_n) \otimes_C V_n).$$

Proof For the existence of the decomposition we proceed by induction on the rank $r(E)$. If $r(E) = 1$ there is nothing to prove. Hence, we assume $r(E) > 1$.

By Serre duality and Serre vanishing, for all $a \gg 0$ we have

$$\mathrm{Hom}(\mathcal{O}_{\mathbb{P}^1}(a), E) = H^1(\mathbb{P}^1, E^\vee(a-2))^\vee = 0.$$

We pick the largest integer $a \in \mathbb{Z}$ such that $\mathrm{Hom}(\mathcal{O}_{\mathbb{P}^1}(a), E) \neq 0$ and a non-zero morphism $\phi \in \mathrm{Hom}(\mathcal{O}_{\mathbb{P}^1}(a), E)$. Since $\mathcal{O}_{\mathbb{P}^1}(a)$ is torsion-free of rank 1, ϕ is injective. Consider the exact sequence

$$0 \to \mathcal{O}_{\mathbb{P}^1}(a) \xrightarrow{\phi} E \to F = \mathrm{Cok}(\phi) \to 0.$$

We claim that F is a vector bundle of rank $r(E) - 1$. Indeed, by Exercise 2.2 below, if this is not the case, there is a subsheaf $T_F \hookrightarrow F$ supported in dimension 0. In particular, we have a morphism from the skyscraper sheaf $\mathbb{C}(x) \hookrightarrow F$. But this gives a non-zero map $\mathcal{O}_{\mathbb{P}^1}(a+1) \to E$, contradicting the maximality of a.

By induction F splits as a direct sum

$$F \cong \bigoplus_j \mathcal{O}_{\mathbb{P}^1}(b_j)^{\oplus r_j}.$$

The second claim is that $b_j \leq a$, for all j. If not, there is a non-zero morphism $\mathcal{O}_{\mathbb{P}^1}(a+1) \hookrightarrow F$. Since $\mathrm{Ext}^1(\mathcal{O}_{\mathbb{P}^1}(a+1), \mathcal{O}_{\mathbb{P}^1}(a)) = H^1(\mathbb{P}^1, \mathcal{O}_{\mathbb{P}^1}(-1)) = 0$, this morphism lifts to a non-zero map $\mathcal{O}_{\mathbb{P}^1}(a+1) \to E$, contradicting again the maximality of a.

Finally, we obtain a decomposition of E as direct sum of line bundles since

$$\mathrm{Ext}^1(\oplus_j \mathcal{O}_{\mathbb{P}^1}(b_j)^{\oplus r_j}, \mathcal{O}_{\mathbb{P}^1}(a)) = \oplus_j H^0(\mathbb{P}^1, \mathcal{O}_{\mathbb{P}^1}(b_j - a - 2))^{\oplus r_j} = 0.$$

The uniqueness now follows from

$$\mathrm{Hom}(\mathcal{O}_{\mathbb{P}^1}(a), \mathcal{O}_{\mathbb{P}^1}(b)) = 0$$

for all $a > b$. □

In higher genus, Theorem 2.1 fails, but some of its features still hold. The direct sum decomposition will be replaced by a filtration (the *Harder-Narasimhan filtration*) and the blocks $\mathcal{O}_{\mathbb{P}^1}(a) \otimes_{\mathbb{C}} V$ by *semistable* sheaves. The ordered sequence $a_1 > \ldots > a_n$ will be replaced by an analogously ordered sequence of *slopes*. Semistable sheaves (of fixed rank and degree) can then be classified as points of a projective variety, the *moduli space*. Finally, the uniqueness part of the proof will follow by an analogous vanishing for morphisms of semistable sheaves.

Exercise 2.2 Show that any coherent sheaf E on a smooth projective curve C fits into a unique canonical exact sequence

$$0 \to T_E \to E \to F_E \to 0,$$

where T_E is a torsion sheaf supported and F_E is a vector bundle. Show additionally that this sequence splits (non-canonically).

2.2 *Stability*

We start with the basic definition of slope stability.

Definition 2.3

(1) The *degree* $d(E)$ of a vector bundle E on C is defined to be the degree of the line bundle $\bigwedge^{r(E)} E$.
(2) The *degree* $d(T)$ for a torsion sheaf T on C is defined to be the length of its scheme theoretic support.
(3) The *rank* of an arbitrary coherent sheaf E on C is defined as $r(E) = r(F_E)$, while its *degree* is defined as $d(E) = d(T_E) + d(F_E)$.
(4) The *slope* of a coherent sheaf E on C is defined as

$$\mu(E) = \frac{d(E)}{r(E)},$$

where dividing by 0 is interpreted as $+\infty$.
(5) A coherent sheaf E on C is called *(semi)stable* if for any proper non-trivial subsheaf $F \subset E$ the inequality $\mu(F) < (\leq)\mu(E)$ holds.

The terms in the previous definition are often times only defined for vector bundles. The reason for this is the following exercise.

Exercise 2.4

(1) Show that the degree is additive in short exact sequences, i.e., for any short exact sequence

$$0 \to F \to E \to G \to 0$$

in $\mathrm{Coh}(X)$ the equality $d(E) = d(F) + d(G)$ holds.

(2) If $E \in \mathrm{Coh}(C)$ is semistable, then it is either a vector bundle or a torsion sheaf.
(3) A vector bundle E on C is (semi)stable if and only if for all non trivial subbundles $F \subset E$ with $r(F) < r(E)$ the inequality $\mu(F) < (\leq)\mu(E)$ holds.

Exercise 2.5 Let $g = 1$ and let $p \in C$ be a point. Then

$$\mathrm{Ext}^1(\mathcal{O}_C, \mathcal{O}_C) = \mathbb{C} \text{ and } \mathrm{Ext}^1(\mathcal{O}_C(p), \mathcal{O}_C) \cong \mathbb{C}.$$

Hence, we have two non-trivial extensions:

$$0 \to \mathcal{O}_C \to V_0 \to \mathcal{O}_C \to 0$$
$$0 \to \mathcal{O}_C \to V_1 \to \mathcal{O}_C(p) \to 0$$

Show that V_0 is semistable but not stable and that V_1 is stable.

A useful fact to notice is that whenever a non-zero sheaf $E \in \mathrm{Coh}(C)$ has rank 0, its degree is strictly positive. This is one of the key properties we want to generalize to higher dimensions. It also turns out to be useful for proving the following result, which generalizes Theorem 2.1 to any curve.

Theorem 2.6 ([55]) *Let E be a non-zero coherent sheaf on C. Then there is a unique filtration (called* Harder-Narasimhan filtration*)*

$$0 = E_0 \subset E_1 \subset \ldots \subset E_{n-1} \subset E_n = E$$

of coherent sheaves such that $A_i = E_i/E_{i-1}$ is semistable for all $i = 1, \ldots, n$ and $\mu(A_1) > \ldots > \mu(A_n)$.

We will give a proof of this statement in a more general setting in Proposition 4.10. In the case of $\mathbb{P}^1$ semistable vector bundles are simply direct sums of line bundles, and stable vector bundles are simply line bundles. Vector bundles on elliptic curves can also be classified (see [10] for the original source and [97] for a modern proof):

Example 2.7 Let $g = 1$, fix a pair $(r, d) \in \mathbb{Z}_{\geq 1} \times \mathbb{Z}$, let E be a vector bundle with $r(E) = r$ and $d(E) = d$. Then

- $\gcd(r, d) = 1$: E is semistable if and only if E is stable if and only if E is indecomposable (i.e., it cannot be decomposed as a non-trivial direct sum of vector bundles) if and only if E is simple (i.e., $\mathrm{End}(E) = \mathbb{C}$). Moreover, all such E are of the form $L \otimes E_0$, where $L \in \mathrm{Pic}^0(C)$ and E_0 can be constructed as an iterated extension similarly as in Exercise 2.5.
- $\gcd(r, d) \neq 1$: There exist no stable vector bundles. Semistable vector bundles can be classified via torsion sheaves. More precisely, there is a one-to-one correspondence between semistable vector bundles of rank r and degree d and torsion sheaves of length $\gcd(r, d)$. Under this correspondence, indecomposable vector bundles are mapped onto torsion sheaves topologically supported on a point.

Remark 2.8 Let E be a semistable coherent sheaf on C. Then there is a (non-unique!) filtration (called the *Jordan-Hölder filtration*)

$$0 = E_0 \subset E_1 \subset \ldots \subset E_{n-1} \subset E_n = E$$

of coherent sheaves such that $A_i = E_i/E_{i-1}$ is stable for all $i = 1, \ldots, n$ and their slopes are all equal. The factors A_i are unique up to reordering. We say that two semistable sheaves are *S-equivalent* if they have the same stable factors up to reordering. More abstractly, this follows from the fact that the category of semistable sheaves on C with fixed slope is an abelian category of finite length (we will return to this in Exercise 5.9). Simple objects in that category agree with stable sheaves.

2.3 Moduli Spaces

In this section, let $g \geq 2$ and denote by C a curve of genus g. We fix integers $r \in \mathbb{Z}_{\geq 1}$, $d \in \mathbb{Z}$, and a line bundle $L \in \mathrm{Pic}^d(C)$ of degree d.

Definition 2.9 We define two functors $\underline{\mathrm{Sch}}_{\mathbb{C}} \to \underline{\mathrm{Set}}$ as follows:

$$\underline{M}_C(r,d)(B) := \left\{ \mathcal{E} \text{ vector bundle on } C \times B \ : \ \forall b \in B, \begin{array}{r} \mathcal{E}|_{C\times\{b\}} \text{ is semistable} \\ r(\mathcal{E}|_{C\times\{b\}}) = r \\ d(\mathcal{E}|_{C\times\{b\}}) = d \end{array} \right\} / \cong,$$

$$\underline{M}_C(r,L)(B) := \left\{ \mathcal{E} \in \underline{M}_C(r,d)(B) \ : \ \det(\mathcal{E}|_{C\times\{b\}}) \cong L \right\},$$

where B is a scheme (locally) of finite-type over $\mathbb{C}$.

We will denote the (open) subfunctors consisting of stable vector bundles by $\underline{M}^s_C(r,d)$ and $\underline{M}^s_C(r,d)$. The following result summarizes the work of many people, including Drézet, Mumford, Narasimhan, Ramanan, Seshadri, among others (see [42, 63, 72, 90, 104]). It gives a further motivation to study stable vector bundles, besides classification reasons: the moduli spaces are interesting algebraic varieties.

Theorem 2.10

(i) There exists a coarse moduli space $M_C(r,d)$ for the functor $\underline{M}_C(r,d)$ parameterizing S-equivalence classes of semistable vector bundles on C. It has the following properties:

- *It is non-empty.*
- *It is an integral, normal, factorial, projective variety over $\mathbb{C}$ of dimension $r^2(g-1)+1$.*
- *It has a non-empty smooth open subset $M^s_C(r,d)$ parameterizing stable vector bundles.*

- *Except when $g = 2$, $r = 2$, and d is even,*[1] $M_C(r,d) \setminus M^s_C(r,d)$ *consists of all singular points of* $M_C(r,d)$*, and has codimension at least* 2.
- $\mathrm{Pic}(M_C(r,d)) \cong \mathrm{Pic}(\mathrm{Pic}^d(C)) \times \mathbb{Z}$.
- *If* $\gcd(r,d) = 1$*, then* $M_C(r,d) = M^s_C(r,d)$ *is a fine moduli space.*[2]

(ii) A similar statement holds for $\underline{M}_C(r,L)$*,*[3] *with the following additional properties:*

- *Its dimension is* $(r^2-1)(g-1)$.
- $\mathrm{Pic}(M_C(r,L) = \mathbb{Z}\cdot\theta$*, where* θ *is ample.*
- *If* $u := \gcd(r,d)$*, then the canonical line bundle is* $K_{M_C(r,L)} = -2u\theta$.

Ideas of the proof • Boundedness: Fix a very ample line bundle $\mathcal{O}_C(1)$ on C. Then, for any semistable vector bundle E on C, Serre duality implies

$$H^1(C, E\otimes\mathcal{O}_C(m-1)) \cong \mathrm{Hom}(E, \mathcal{O}_C(-m+1)\otimes K_C) = 0,$$

for all $m \geq m_0$, where m_0 only depends on $\mu(E)$. Hence, by replacing E with $E\otimes\mathcal{O}_C(m_0)$, we obtain a surjective map

$$\mathcal{O}_C^{\oplus\chi(C,E)} \twoheadrightarrow E.$$

- Quot scheme: Let $\gamma = \chi(C,E)$, where E is as in the previous step. We consider the subscheme R of the Quot scheme on C parameterizing quotients $\phi : \mathcal{O}_C^{\oplus\gamma} \twoheadrightarrow U$, where U is locally-free and ϕ induces an isomorphism $H^0(C,\mathcal{O}_C^{\oplus\gamma}) \cong H^0(C,U)$. Then it can be proved that R is smooth and integral. Moreover, the group $\mathrm{PGL}(\gamma)$ acts on R.
- Geometric Invariant Theory: We consider the set of semistable (resp., stable) points R^{ss} (resp., R^s) with respect to the group action given by $\mathrm{PGL}(\gamma)$ (and a certain multiple of the natural polarization on R). Then $M_C(r,d) = R^{ss}\,\mathrm{PGL}(\gamma)$ and $M^s_C(r,d) = R^s/\,\mathrm{PGL}(\gamma)$.

□

Remark 2.11 While the construction of the line bundle θ in Theorem 2.10 and its ampleness can be directly obtained from GIT, we can give a more explicit description by using purely homological algebra techniques; see Exercise 2.12 below.

In fact, the "positivity" of θ can be split into two statements: the first is the property of θ to be strictly nef [i.e., strictly positive when intersecting all curves in $M_C(r,L)$]; the second is semi-ampleness (or, in this case, to show that $\mathrm{Pic}(M_C(r,L) \cong \mathbb{Z})$. The second part is more subtle and it requires a deeper

[1]See Example 2.18.

[2]To be precise, we need to modify the equivalence relation for $\underline{M}^s_C(r,d)(B)$: $\mathcal{E}\sim\mathcal{E}'$ if and only if $\mathcal{E}\cong\mathcal{E}'\otimes p_B^*\mathcal{L}$, where $\mathcal{L}\in\mathrm{Pic}(B)$.

[3]By using the morphism det: $M_C(r,d)\to\mathrm{Pic}^d(C)$, and observing that $M_C(r,L) = \det^{-1}(L)$.

understanding of stable vector bundles or of the geometry of $M_C(r, L)$. But the first one turns out to be a fairly general concept, as we will see in Sect. 8 in the Positivity Lemma.

The following exercise foreshadows how we will construct nef divisors in the case of Bridgeland stability in Sect. 8.

Exercise 2.12 Let $Z : \mathbb{Z}^{\oplus 2} \to \mathbb{C}$ be the group homomorphism given by

$$Z(r, d) = -d + \sqrt{-1}r.$$

Let $r_0, d_0 \in \mathbb{Z}$ be such that $r_0 \geq 1$ and $\gcd(r_0, d_0) = 1$, and let $L_0 \in \mathrm{Pic}^{d_0}(C)$. We consider the moduli space $M := M_C(r_0, L_0)$ and we let $\mathcal{E}_0$ be a universal family on $C \times M$.

For an integral curve $\gamma \subset M$, we define

$$\ell_Z.\gamma := -\Im \frac{Z(r(\Phi_{\mathcal{E}_0}(\mathcal{O}_\gamma)), d(\Phi_{\mathcal{E}_0}(\mathcal{O}_\gamma)))}{Z(r_0, d_0)} \in \mathbb{R},$$

where $\mathcal{O}_\gamma \in \mathrm{Coh}(M)$ is the structure sheaf of γ in M, $\Phi_{\mathcal{E}_0} : \mathrm{D^b}(M) \to \mathrm{D^b}(C)$, $\Phi_{\mathcal{E}_0}(-) := (p_C)_*(\mathcal{E}_0 \otimes p_M^*(-))$ is the Fourier-Mukai transform with kernel $\mathcal{E}_0$, and $\Im$ denotes the imaginary part of a complex number.

(1) Show that by extending ℓ_Z by linearity we get the numerical class of a real divisor $\ell_Z \in N^1(M) = N_1(M)^\vee$.
(2) Show that

$$\Im \frac{Z(r(\Phi_{\mathcal{E}_0}(\mathcal{O}_\gamma)), d(\Phi_{\mathcal{E}_0}(\mathcal{O}_\gamma)))}{Z(r_0, d_0)} = \Im \frac{Z(r(\Phi_{\mathcal{E}_0}(\mathcal{O}_\gamma(m))), d(\Phi_{\mathcal{E}_0}(\mathcal{O}_\gamma(m))))}{Z(r_0, d_0)},$$

for any $m \in \mathbb{Z}$ and for any ample line bundle $\mathcal{O}_\gamma(1)$ on γ.
(3) Show that $\Phi_{\mathcal{E}_0}(\mathcal{O}_\gamma(m))$ is a sheaf, for $m \gg 0$.
(4) By using the existence of relative Harder-Narasimhan filtration (see, e.g., [63, Theorem 2.3.2]), show that, for $m \gg 0$,

$$\mu(\Phi_{\mathcal{E}_0}(\mathcal{O}_\gamma(m))) \leq \frac{d_0}{r_0}.$$

(5) Deduce that ℓ_Z is nef, namely $\ell_Z.\gamma \geq 0$ for all $\gamma \subset M$ integral curve.

2.4 Equivalent Definitions for Curves

In this section we briefly mention other equivalent notions of stability for curves.

Let C be a curve. We have four equivalent definition of stability for vector bundles:

(1) Slope stability (Definition 2.3); this is the easiest to handle.
(2) GIT stability (sketched in the proof of Theorem 2.10); this is useful in the construction of moduli spaces as algebraic varieties.
(3) Faltings-Seshadri definition (see below); this is useful to find sections of line bundles on moduli spaces, and more generally to construct moduli spaces without using GIT.
(4) Differential geometry definition (see below); this is the easiest for proving results and deep properties of stable objects.

Faltings' Approach The main result is the following.

Theorem 2.13 (Faltings-Seshadri) *Let E be a vector bundle on a curve C. Then E is semistable if and only if there exists a vector bundle F such that* $\mathrm{Hom}(F, E) = \mathrm{Ext}^1(F, E) = 0$.

Proof "$\Rightarrow$": This is very hard. We refer to [105].

"$\Leftarrow$": Suppose that E is not semistable. Then, by definition, there exists a subbundle $G \subset E$ such that $\mu(G) > \mu(E)$. By assumption, $\chi(F, E) = 0$. Due to the Riemann-Roch Theorem, we have

$$1 - g + \mu(E) - \mu(F) = 0.$$

Hence,

$$1 - g + \mu(G) - \mu(F) > 0.$$

By applying the Riemann-Roch Theorem again, we have $\chi(F, G) > 0$, and so $\mathrm{Hom}(F, G) \neq 0$. But this implies $\mathrm{Hom}(F, E) \neq 0$, which is a contradiction. □

An example for Theorem 2.13 is in Sect. 2.5.

Differential Geometry Approach To simplify, since this is not the main topic of this survey, we will treat only the degree 0 case. We refer to [89, Theorem 2] for the general result and for all details.

Theorem 2.14 (Narasimhan-Seshadri) *There is a one-to-one correspondence*

$$\{\text{Stable vector bundles on } C \text{ of degree } 0\} \xleftrightarrow{1:1} \{\text{Irreducible unitary representations of } \pi_1(C)\}$$

Idea of the proof We only give a very brief idea on how to associate a stable vector bundle to an irreducible unitary representation. Let C be a curve of genus $g \geq 2$. Consider its universal cover $p : \mathbb{H} \to C$. The group of deck transformation is $\pi_1(C)$. Consider the trivial bundle $\mathbb{H} \times \mathbb{C}^r$ on $\mathbb{H}$ and a representation $\rho \colon \pi_1(C) \to \mathrm{GL}(r, \mathbb{C})$. Then we get a $\pi_1(C)$-bundle via the action

$$\begin{aligned} a_\tau \colon \mathbb{H} \times \mathbb{C}^r &\to \mathbb{H} \times \mathbb{C}^r, \\ (y, v) &\mapsto (\tau \cdot y, \rho(\tau) \cdot v), \end{aligned}$$

for $\tau \in \pi_1(C)$. This induces a vector bundle E_ρ of rank r on C with degree 0.

If ρ is a unitary representation, then the semistability of E can be shown as follows. Assume there is a sub vector bundle $W \subset E$ with $d(W) > 0$. By taking $\bigwedge^{r(W)}$ we can reduce to the case $r(W) = 1$, namely W is a line bundle of positive degree.

By tensoring with a line bundle of degree 0, we can assume W has a section. But it can be shown (see [88, Proposition 4.1]) that a vector bundle corresponding to a non-trivial unitary irreducible representation does not have any section. By splitting ρ as direct sum of irreducible representation, this implies that W has to be a sub line bundle of the trivial bundle, which is impossible.

If the representation is irreducible, the bundle is stable. □

As mentioned before, an important remark about the interpretation of stability of vector bundles in terms of irreducible representations is that it is easier to prove results on stable sheaves in this language. The first example is the preservation of stability by the tensor product [63, Theorem 3.1.4]: the tensor product of two semistable vector bundles is semistable. Example 2.17 is an application of Theorem 2.14 and preservation of stability by taking the symmetric product.

Gieseker's Ample Bundle Characterization We also mention that for geometric applications (e.g., in the algebraic proof of preservation of stability by tensor product mentioned above), the following result by Gieseker [51] and [63, Theorem 3.2.7] is of fundamental importance.

Theorem 2.15 *Let E be a semistable vector bundle on C. Consider the projective bundle $\pi : \mathbb{P}_C(E) \to C$ and denote the tautological line bundle by $\mathcal{O}_\pi(1)$. Then:*

(1) *$d(E) \geq 0$ if and only if $\mathcal{O}_\pi(1)$ is nef.*
(2) *$d(E) > 0$ if and only if $\mathcal{O}_\pi(1)$ is ample.*

Idea of the proof We just sketch how to prove the non-trivial implication of (1). Assume that $d(E) \geq 0$. To prove that $\mathcal{O}_\pi(1)$ is nef, since it is relatively ample, we only need to prove that $\mathcal{O}_\pi(1)\cdot\gamma \geq 0$ for a curve $\gamma \subset \mathbb{P}_C(E)$ which maps surjectively onto C via π.

By taking the normalization, we can assume that γ is smooth. Denote by $f : \gamma \to C$ the induced finite morphism. Then we have a surjection $f^*E \to \mathcal{O}_\pi(1)|_\gamma$. But f^*E is semistable, by Exercise 2.16 below. Hence, $d(\mathcal{O}_\pi(1)|_\gamma) \geq 0$, which is what we wanted. □

Exercise 2.16 Let $f : C' \to C$ be a finite morphism between curves. Let E be a vector bundle on C. Show that E is semistable if and only if f^*E is semistable. *Hint: One implication ("$\Leftarrow$") is easy. For the other ("$\Rightarrow$"), by using the first implication, we can reduce to a Galois cover. Then use uniqueness of Harder-Narasimhan filtrations to show that if f^*E is not semistable. Then all its semistable factors must be invariant. Then use that all factors are vector bundles to show that they descend to achieve a contradiction.*

2.5 Applications and Examples

In this section we present a few examples and applications of stability of vector bundles on curves. We will mostly skip proofs, but give references to the literature.

Example 2.17 (Mumford's Example; [56, Theorem 10.5]) Let C be a curve of genus $g \geq 2$. Then there exists a stable rank 2 vector bundle U on C such that

$$H^0(C, \mathrm{Sym}^m U) = 0, \tag{1}$$

for all $m > 0$.[4]

Indeed, by using Theorem 2.14, since

$$\pi_1(C) = \langle a_1, b_1, \ldots, a_g, b_g | a_1 b_1 a_1^{-1} b_1^{-1} \cdot \ldots \cdot a_g b_g a_g^{-1} b_g^{-1} \rangle,$$

we only need to find two matrices $A, B \in \mathrm{U}(2)$ such that $\mathrm{Sym}^m A$ and $\mathrm{Sym}^m B$ have no common fixed subspace, for all $m > 0$. Indeed, by letting them be the matrices corresponding to a_1 and b_1, the matrices corresponding to the other a_i's and b_i's can be chosen to satisfy the relation.

We can choose

$$A = \begin{pmatrix} \lambda_1 & 0 \\ 0 & \lambda_2 \end{pmatrix}$$

where $|\lambda_i| = 1$ and λ_2/λ_1 is not a root of unity. Then

$$\mathrm{Sym}^m A = \begin{pmatrix} \lambda_1^m & 0 & \ldots & 0 \\ 0 & \lambda_1^{m-1}\lambda_2 & \ldots & 0 \\ \vdots & \vdots & \ddots & \vdots \\ 0 & 0 & \ldots & \lambda_2^m \end{pmatrix}$$

has all eigenvalues distinct, and so the only fixed subspaces are subspaces generated by the standard basis vectors.

If we let

$$B = \begin{pmatrix} \mu_{11} & \mu_{12} \\ \mu_{21} & \mu_{22} \end{pmatrix},$$

then all entries of $\mathrm{Sym}^m B$ are certain polynomials in the μ_{ij} not identically zero. Since $\mathrm{U}(2) \subset \mathrm{GL}(2, \mathbb{C})$ is not contained in any analytic hypersurface, we can find

---

[4] If we let $\pi\colon X := \mathbb{P}_C(U) \to C$ be the corresponding ruled surface, then (1) implies that the nef divisor $\mathcal{O}_\pi(1)$ is not ample, although it has the property that $\mathcal{O}_\pi(1) \cdot \gamma > 0$, for all curves $\gamma \subset X$.

μ_{ij} such that all entries of $\mathrm{Sym}^m B$ are non-zero, for all $m > 0$. This concludes the proof.

The two classical examples of moduli spaces are the following (see [39, 87]).

Example 2.18 Let C be a curve of genus $g = 2$. Then $M_C(2, \mathcal{O}_C) \cong \mathbb{P}^3$.

Very roughly, the basic idea behind this example is the following. We can identify $\mathbb{P}^3$ with $\mathbb{P}(H^0(\mathrm{Pic}^1(C), 2\Theta))$, where Θ denotes the theta-divisor on $\mathrm{Pic}^1(C)$. The isomorphism in the statement is then given by associating to any rank 2 vector bundle W the subset

$$\{\xi \in \mathrm{Pic}^1(C) : H^0(C, W \otimes \xi) \neq 0\}.$$

It is interesting to notice that the locus of strictly semistable vector bundles corresponds to the S-equivalence class of vector bundles of the form $M \oplus M^*$, where $M \in \mathrm{Pic}^0(C)$. Geometrically, this is the Kummer surface associated to $\mathrm{Pic}^0(C)$ embedded in $\mathbb{P}^3$.

Example 2.19 Let C be a curve of genus $g = 2$, and let L be a line bundle of degree 1. Then $M_C(2, L) \cong Q_1 \cap Q_2 \subset \mathbb{P}^5$, where Q_1 and Q_2 are two quadric hypersurfaces.

The space $M_2(2, L)$ is one of the first instances of derived category theory, semi-orthogonal decompositions, and their connection with birational geometry. Indeed, the result above can be enhanced by saying that there exists a *semi-orthogonal decomposition* of the derived category of $M_C(2, L)$ in which the non-trivial component is equivalent to the derived category of the curve C. We will not add more details here, since it is outside the scope of these lecture notes; we refer instead to [26, 69] and references therein. There is also an interpretation in terms of Bridgeland stability conditions (see [20]).

Example 2.20 Let C be a curve of genus 2. The moduli space $M_C(3, \mathcal{O}_C)$ also has a very nice geometric interpretation, and it is very closely related to the Coble cubic hypersurface in $\mathbb{P}^8$. We refer to [94] for the precise statement and for all the details.

Finally, we mention another classical application of vector bundles techniques to birational geometry. In principle, by using Bridgeland stability, this could be generalized to higher dimensions. We will talk briefly about this in Sect. 7.

Example 2.21 Let C be a curve of genus $g \geq 1$, and let L be an ample divisor on C of degree $d(L) \geq 2g + 1$. Then the vector bundle

$$M_L := \mathrm{Ker}\left(\mathrm{ev}\colon \mathcal{O}_C \otimes_{\mathbb{C}} H^0(C, L) \twoheadrightarrow L\right)$$

is semistable. As an application this implies that the embedding of C in $\mathbb{P}(H^0(C, L)^\vee)$ induced by L is projectively normal, a result by Castelnuovo, Mattuck, and Mumford. We refer to [71] for a proof of this result, and for more applications.

We will content ourselves to prove a somehow related statement for the canonical line bundle, by following [48, Theorem 1.6], to illustrate Faltings' definition of stability: the vector bundle M_{K_C} is semistable.

To prove this last statement we let L be a general line bundle of degree $g+1$ on C. Then L is base-point free and $H^1(C,L)=0$.

Start with the exact sequence

$$0 \to M_{K_C} \to \mathcal{O}_C \otimes_{\mathbb{C}} H^0(C,K_C) \to K_C \to 0,$$

tensor by L and take cohomology. We have a long exact sequence

$$\begin{aligned} 0 \to H^0(C, M_{K_C} \otimes L) \to H^0(C,K_C) \otimes H^0(C,L) \xrightarrow{\alpha} H^0(C, K_C \otimes L) \\ \to H^1(C, M_{K_C} \otimes L) \to 0. \end{aligned} \tag{2}$$

By the Riemann-Roch Theorem, $h^0(C,L)=2$. Hence, we get an exact sequence

$$0 \to L^\vee \to \mathcal{O}_C \otimes H^0(C,L) \to L \to 0.$$

By tensoring by K_C and taking cohomology, we get

$$0 \to H^0(C, K_C \otimes L^\vee) \to H^0(C,K_C) \otimes H^0(C,L) \xrightarrow{\alpha} H^0(C, K_C \otimes L),$$

and so $\mathrm{Ker}(\alpha) \cong H^0(C, K_C \otimes L^\vee) = H^1(C,L)^\vee = 0$.

By using (2), this gives $H^0(C, M_{K_C} \otimes L) = H^1(C, M_{K_C} \otimes L) = 0$, and so M_{K_C} is semistable, by Theorem 2.13.

3 Generalizations to Higher Dimensions

The definitions of stability from Sect. 2.4 generalize in different ways from curves to higher dimensional varieties. In this section we give a quick overview on this. We will not talk about how to generalize the Narasimhan-Seshadri approach. There is a beautiful theory (for which we refer for example to [67]), but it is outside the scope of this survey.

Let X be a smooth projective variety over $\mathbb{C}$ of dimension $n \geq 2$. We fix an ample divisor class $\omega \in \mathrm{N}^1(X)$ and another divisor class $B \in N^1(X)$.

Definition 3.1 We define the twisted Chern character as

$$\mathrm{ch}^B := \mathrm{ch} \cdot e^{-B}.$$

If the reader is unfamiliar with Chern characters we recommend either a quick look at Appendix A of [57] or a long look at [45]. By expanding Definition 3.1, we have for example

$$\begin{aligned}\mathrm{ch}_0^B &= \mathrm{ch}_0 = \mathrm{rk},\\ \mathrm{ch}_1^B &= \mathrm{ch}_1 - B\cdot \mathrm{ch}_0,\\ \mathrm{ch}_2^B &= \mathrm{ch}_2 - B\cdot \mathrm{ch}_1 + \frac{B^2}{2}\cdot \mathrm{ch}_0 .\end{aligned}$$

Gieseker-Maruyama-Simpson Stability GIT stability does generalize to higher dimensions. The corresponding "numerical notion" is the one of (twisted) Gieseker-Maruyama-Simpson stability (see [50, 83–85, 107]).

Definition 3.2 Let $E \in \mathrm{Coh}(X)$ be a pure sheaf of dimension d.

(1) The B-twisted Hilbert polynomial is

$$P(E,B,t) := \int_X \mathrm{ch}^B(E)\cdot e^{t\omega}\cdot \mathrm{td}_X = a_d(E,B)t^d + a_{d-1}(E,B)t^{d-1} + \ldots + a_0(E,B)$$

(2) We say that E is B-twisted Gieseker (semi)stable if, for any proper non-trivial subsheaf $F \subset E$, the inequality

$$\frac{P(F,B,t)}{a_d(F,B)} < (\leq) \frac{P(E,B,t)}{a_d(E,B)}$$

holds, for $t \gg 0$.

If E is torsion-free, then $d = n$ and $a_d(E,B) = \mathrm{ch}_0(E)$. Theorem 2.6 and the existence part in Theorem 2.10 generalize for (twisted) Gieseker stability (the second when ω and B are rational classes), but singularities of moduli spaces become very hard to understand.

Slope Stability We can also define a twisted version of the standard slope stability function for $E \in \mathrm{Coh}(X)$ by

$$\mu_{\omega,B}(E) := \frac{\omega^{n-1}\cdot \mathrm{ch}_1^B(E)}{\omega^n \cdot \mathrm{ch}_0^B(E)} = \frac{\omega^{n-1}\cdot \mathrm{ch}_1(E)}{\omega^n\cdot \mathrm{ch}_0(E)} - \frac{\omega^{n-1}\cdot B}{\omega^n},$$

where dividing by 0 is interpreted as $+\infty$.

Definition 3.3 A sheaf $E \in \mathrm{Coh}(X)$ is called slope (semi)stable if for all subsheaves $F \subset E$ the inequality $\mu_{\omega,B}(F) < (\leq) \mu_{\omega,B}(E/F)$ holds.

Notice that this definition is independent of B and classically one sets $B = 0$. However, this more general notation will turn out useful in later parts of these notes. Also, torsion sheaves are all slope semistable of slope $+\infty$, and if a sheaf with non-zero rank is semistable then it is torsion-free.

This stability behaves well with respect to certain operations, like restriction to (general and high degree) hypersurfaces, pull-backs by finite morphisms, tensor products, and we will see that Theorem 2.6 holds for slope stability as well. The main problem is that moduli spaces (satisfying a universal property) are harder to construct, even in the case of surfaces.

Bridgeland Stability We finally come to the main topic of this survey, Bridgeland stability. This is a direct generalization of slope stability. The main difference is that we will need to change the category we are working with. Coherent sheaves will never work in dimension ≥ 2. Instead, we will look for other abelian categories inside the bounded derived category of coherent sheaves on X.

To this end, we first have to treat the general notion of slope stability for abelian categories. Then we will introduce the notion of a bounded t-structure and the key operation of tilting. Finally, we will be able to define Bridgeland stability on surfaces and sketch how to conjecturally construct Bridgeland stability conditions in higher dimensions.

The key advantage of Bridgeland stability is the very nice behavior with respect to change of stability. We will see that moduli spaces of Gieseker semistable and slope semistable sheaves are both particular cases of Bridgeland semistable objects, and one can pass from one to the other by varying stability. This will give us a technique to study those moduli spaces as well. Finally, we remark that even Faltings' approach to stability on curves, which despite many efforts still lacks a generalization to higher dimensions as strong as for the case of curves, seems to require as well to look at stability for complexes of sheaves (see, e.g., [4, 58, 59]).

As drawbacks, moduli spaces in Bridgeland stability are not associated naturally to a GIT problem and therefore harder to construct. Also, the categories to work with are not "local", in the sense that many geometric constructions for coherent sheaves will not work for these more general categories, since we do not have a good notion of "restricting a morphism to an open subset".

4 Stability in Abelian Categories

The goal of the theory of stability conditions is to generalize the situation from curves to higher dimensional varieties with similar nice properties. In order to do so it turns out that the category of coherent sheaves is not good enough. In this chapter, we will lay out some general foundation for stability in a more general abelian category. Throughout this section we let $\mathcal{A}$ be an abelian category with Grothendieck group $K_0(\mathcal{A})$.

Definition 4.1 Let $Z : K_0(\mathcal{A}) \to \mathbb{C}$ be an additive homomorphism. We say that Z is a *stability function* if, for all non-zero $E \in \mathcal{A}$, we have

$$\Im Z(E) \geq 0, \text{ and } \Im Z(E) = 0 \Rightarrow \Re Z(E) < 0.$$

Given a stability function, we will often denote $R := \Im Z$, the *generalized rank*, $D := -\Re Z$, the *generalized degree*, and $M := \frac{D}{R}$, the *generalized slope*, where as before dividing by zero is interpreted as $+\infty$.

Definition 4.2 Let $Z : K_0(\mathcal{A}) \to \mathbb{C}$ be a stability function. A non-zero object $E \in \mathcal{A}$ is called *(semi)stable* if, for all proper non trivial subobjects $F \subset E$, the inequality $M(F) < (\leq) M(E)$ holds.

When we want to specify the function Z, we will say that objects are Z-semistable, and add the suffix Z to all the notation.

Example 4.3

(1) Let C be a curve, and let $\mathcal{A} := \mathrm{Coh}(C)$. Then the group homomorphism

$$Z : K_0(C) \to \mathbb{C},\ Z(E) = -d(E) + \sqrt{-1}\, r(E)$$

is a stability function on $\mathcal{A}$. Semistable objects coincide with semistable sheaves in the sense of Definition 2.3.

(2) Let X be a smooth projective variety of dimension $n \geq 2$, $\omega \in N^1(X)$ be an ample divisor class and $B \in N^1(X)$. Let $\mathcal{A} := \mathrm{Coh}(X)$. Then the group homomorphism

$$\overline{Z}_{\omega,B} : K_0(X) \to \mathbb{C},\ \overline{Z}_{\omega,B}(E) = -\omega^{n-1} \cdot \mathrm{ch}_1^B(E) + \sqrt{-1}\, \omega^n \cdot \mathrm{ch}_0^B(E)$$

is **not** a stability function on $\mathcal{A}$. Indeed, for a torsion sheaf T supported in codimension ≥ 2, we have $\overline{Z}_{\omega,B}(T) = 0$. More generally, by Toda [110, Lemma 2.7] there is no stability function whose central charge factors through the Chern character for $\mathcal{A} = \mathrm{Coh}(X)$.

(3) Let X, ω, and B be as before. Consider the quotient category

$$\mathcal{A} := \mathrm{Coh}_{n,n-1}(X) := \mathrm{Coh}(X)/\,\mathrm{Coh}(X)_{\leq n-2}$$

of coherent sheaves modulo the (Serre) subcategory $\mathrm{Coh}(X)_{\leq n-2}$ of coherent sheaves supported in dimension $\leq n-2$ (see [63, Definition 1.6.2]). Then $\overline{Z}_{\omega,B}$ is a stability function on $\mathcal{A}$. Semistable objects coincide with slope semistable sheaves in the sense of Definition 3.3.

Exercise 4.4 Show that an object $E \in \mathcal{A}$ is (semi)stable if and only for all quotient $E \twoheadrightarrow G$ the inequality $M(E) < (\leq) M(G)$ holds.

Stability for abelian categories still has Schur's property:

Lemma 4.5 *Let $Z : K_0(\mathcal{A}) \to \mathbb{C}$ be a stability function. Let $A, B \in \mathcal{A}$ be non-zero objects which are semistable with $M(A) > M(B)$. Then* $\mathrm{Hom}(A, B) = 0$.

Proof Let $f : A \to B$ be a morphism, and let $Q := \mathrm{im}(f) \subset B$. If f is non-zero, then $Q \neq 0$.

Since B is semistable, we have $M(Q) \leq M(B)$. Since A is semistable, by Exercise 4.4, we have $M(A) \leq M(Q)$. This is a contradiction to $M(A) > M(B)$. □

Definition 4.6 Let $Z : K_0(\mathcal{A}) \to \mathbb{C}$ be an additive homomorphism. We call the pair $(\mathcal{A}, Z)$ a *stability condition* if

- Z is a stability function, and
- Any non-zero $E \in \mathcal{A}$ has a filtration, called the *Harder-Narasimhan filtration*,

$$0 = E_0 \subset E_1 \subset \ldots \subset E_{n-1} \subset E_n = E$$

of objects $E_i \in \mathcal{A}$ such that $A_i = E_i/E_{i-1}$ is semistable for all $i = 1, \ldots, n$ and $M(F_1) > \ldots > M(F_n)$.

Exercise 4.7 Show that for any stability condition $(\mathcal{A}, Z)$ the Harder-Narasimhan filtration for any object $E \in \mathcal{A}$ is unique up to isomorphism.

In concrete examples we will need a criterion for the existence of Harder-Narasimhan filtrations. This is the content of Proposition 4.10. It requires the following notion.

Definition 4.8 An abelian category $\mathcal{A}$ is called *noetherian*, if the *descending chain condition* holds, i.e. for any descending chain of objects in $\mathcal{A}$

$$A_0 \supset A_1 \supset \ldots \supset A_i \supset \ldots$$

we have $A_i = A_j$ for $i, j \gg 0$.

Standard examples of noetherian abelian categories are the category of modules over a noetherian ring or the category of coherent sheaves on a variety. Before dealing with Harder-Narasimhan filtrations, we start with a preliminary result.

Lemma 4.9 *Let $Z : K_0(\mathcal{A}) \to \mathbb{C}$ be a stability function. Assume that*

- *$\mathcal{A}$ is noetherian, and*
- *the image of R is discrete in $\mathbb{R}$.*

Then, for any object $E \in \mathcal{A}$, there is a number $D_E \in \mathbb{R}$ such that for any $F \subset E$ the inequality $D(F) \leq D_E$ holds.

Proof Since the image of R is discrete in $\mathbb{R}$, we can do induction on $R(E)$. If $R(E) = 0$ holds, then $R(F) = 0$. In particular, $0 < D(F) \leq D(E)$.

Let $R(E) > 0$. Assume there is a sequence $F_n \subset E$ for $n \in \mathbb{N}$ such that

$$\lim_{n \to \infty} D(F_n) = \infty.$$

If the equality $R(F_n) = R(E)$ holds for any $n \in \mathbb{N}$, then the quotient satisfies $R(E/F_n) = 0$. In particular, $D(E/F_n) \geq 0$ implies $D(F_n) \leq D(E_n)$. Therefore, we can assume $R(F_n) < R(E)$ for all n.

We will construct an increasing sequence of positive integers n_k such that $D(F_{n_1} + \ldots + F_{n_k}) > k$ and $R(F_{n_1} + \ldots + F_{n_k}) < R(E)$. We are done after that, because this contradicts $\mathcal{A}$ being noetherian. By assumption, we can choose n_1 such that $D(F_{n_1}) \geq 1$. Assume we have constructed $n_1, \ldots, n_{k-1}$. There is an exact sequence

$$0 \to (F_{n_1}+\ldots+F_{n_{k-1}})\cap F_n \to (F_{n_1}+\ldots+F_{n_{k-1}})\oplus F_n \to F_{n_1}+\ldots+F_{n_{k-1}}+F_n \to 0$$

for any $n > n_{k-1}$, where intersection and sum are taken inside of E. Therefore,

$$D(F_{n_1}+\ldots+F_{n_{k-1}}+F_n) = D(F_{n_1}+\ldots+F_{n_{k-1}})+D(F_n)-D((F_{n_1}+\ldots+F_{n_{k-1}})\cap F_n).$$

By induction $D((F_{n_1} + \ldots + F_{n_{k-1}}) \cap F_n)$ is bounded from above and we obtain

$$\lim_{n\to\infty} D(F_{n_1} + \ldots + F_{n_{k-1}} + F_n) = \infty.$$

As before this is only possible if $R(F_{n_1} + \ldots + F_{n_{k-1}} + F_n) < R(E)$ for $n \gg 0$. Therefore, we can choose $n_k > n_{k-1}$ as claimed. □

We will follow a proof from [52] (the main idea goes back at least to [106]) for the existence of Harder-Narasimhan filtrations, as reviewed in [11].

Proposition 4.10 *Let $Z : K_0(\mathcal{A}) \to \mathbb{C}$ be a stability function. Assume that*

- *$\mathcal{A}$ is noetherian, and*
- *the image of R is discrete in $\mathbb{R}$.*

Then Harder-Narasimhan filtrations exist in $\mathcal{A}$ with respect to Z.

Proof We will give a proof when the image of D is discrete in $\mathbb{R}$ and leave the full proof as an exercise.

Let $E \in \mathcal{A}$. The goal is to construct a Harder-Narasimhan filtration for E. By $\mathcal{H}(E)$ we denote the convex hull of the set of all $Z(F)$ for $F \subset E$. By Lemma 4.9 this set is bounded from the left. By $\mathcal{H}_l$ we denote the half plane to the left of the line between $Z(E)$ and 0. If E is semistable, we are done. Otherwise the set $\mathcal{P}(E) = \mathcal{H}(E) \cap \mathcal{H}_L$ is a convex polygon (see Fig. 1).

Let $v_0 = 0, v_1, \ldots, v_n = Z(E)$ be the extremal vertices of $\mathcal{P}(E)$ in order of ascending imaginary part. Choose $F_i \subset E$ with $Z(F_i) = v_i$ for $i = 1, \ldots, n-1$. We will now prove the following three claims to conclude the proof.

(1) The inclusion $F_{i-1} \subset F_i$ holds for all $i = 1, \ldots, n$.
(2) The object $G_i = F_i/F_{i-1}$ is semistable for all $i = 1, \ldots, n$.
(3) The inequalities $M(G_1) > M(G_2) > \ldots > M(G_n)$ hold.

By definition of $\mathcal{H}(E)$ both $Z(F_{i-1} \cap F_i)$ and $Z(F_{i-1} + F_i)$ have to lie in $\mathcal{H}(E)$. Moreover, we have $R(F_{i-1} \cap F_i) \le R(F_{i-1}) < R(F_i) \le R(F_{i-1} + F_i))$. To see part (1) observe that

$$Z(F_{i-1} + F_i) - Z(F_{i-1} \cap F_i) = (v_{i-1} - Z(F_{i-1} \cap F_i)) + (v_i - Z(F_{i-1} \cap F_i)).$$

This is only possible if $Z(F_{i-1} \cap F_i) = Z(F_{i-1})$ and $Z(F_{i-1} + F_i)) = Z(F_i)$ (see Fig. 2).

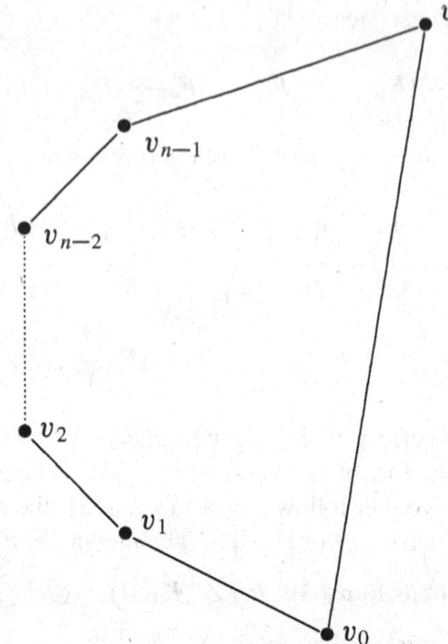

Fig. 1 The polygon $\mathcal{P}(E)$

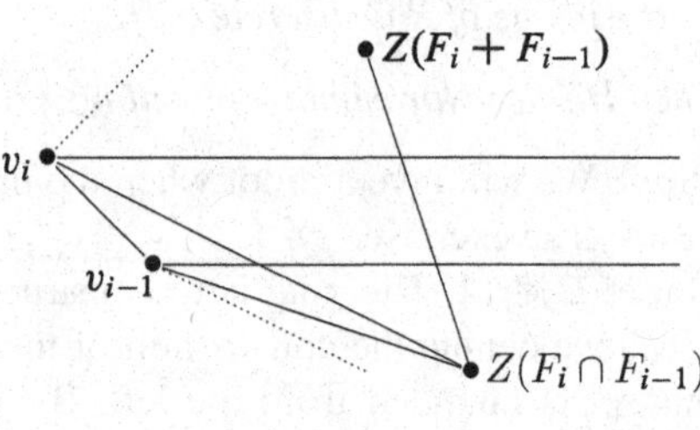

Fig. 2 Vectors adding up within a convex set

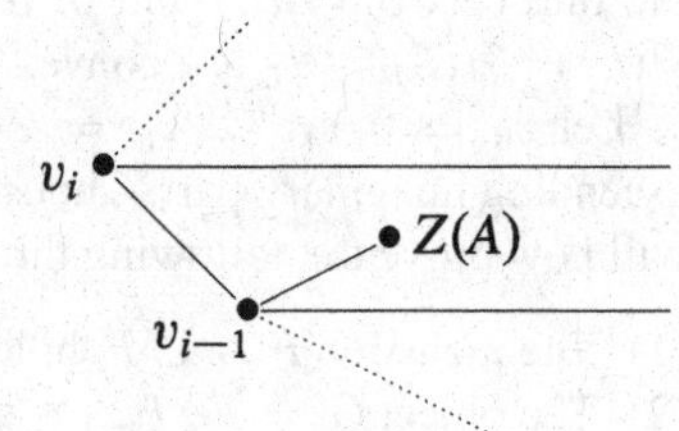

Fig. 3 Subobjects of G_i have smaller or equal slope

This implies $Z(F_{i-1}/(F_{i-1} \cap F_i)) = 0$ and the fact that Z is a stability function shows $F_{i-1} = F_{i-1} \cap F_i$. This directly leads to (1).

Let $\overline{A} \subset G_i$ be a non zero subobject with preimage $A \subset F_i$. Then A has to be in $\mathcal{H}(E)$ and $R(F_{i-1}) \leq R(A) \leq R(F_i)$. That implies $Z(A) - Z(F_{i-1})$ has smaller or equal slope than $Z(F_i) - Z(F_{i-1})$. But this is the same as $M(\overline{A}) \leq M(G_i)$, proving (2) (Fig. 3).

The slope $M(G_i)$ is the slope of the line through v_i and v_{i-1}. Convexity of $\mathcal{P}(E)$ implies (3). □

Exercise 4.11 The goal of this exercise is to remove the condition that the image of D is discrete in the proof of Proposition 4.10. Show that there are objects $F_i \subset E$ with $Z(F_i) = v_i$ for $i = 1, \ldots, n-1$. *Hint: By definition of $\mathcal{P}(E)$ there is a sequence of objects $F_{j,i} \subset E$ such that*

$$\lim_{j \to \infty} Z(F_{j,i}) = Z(F_i).$$

Since R has discrete image, we can assume $R(F_{j,i}) = \Im v_i$ for all j. Replace the $F_{j,i}$ by an ascending chain of objects similarly to the proof of Lemma 4.9.

Example 4.12 For the stability functions in Example 4.3 (1) and (3) Harder-Narasimhan filtrations exist.

We conclude the section with an example coming from representation theory.

Exercise 4.13 Let A be a finite-dimensional algebra over $\mathbb{C}$ and $\mathcal{A} = \text{mod-}A$ be the category of finitely dimensional (right) A-modules. We denote the simple modules by $S_1, \ldots, S_m$. Pick $Z_i \in \mathbb{C}$ for $i = 1, \ldots, m$, where $\Im(Z_i) \geq 0$ holds and if $\Im Z_i = 0$ holds, then we have $\Re Z_i < 0$.

(1) Show that there is a unique homomorphism $Z : K_0(\mathcal{A}) \to \mathbb{C}$ with $Z(S_i) = Z_i$.
(2) Show that $(\mathcal{A}, Z)$ is a stability condition.

The corresponding projective moduli spaces of semistable A-modules were constructed in [66].

Remark 4.14 There is also a weaker notion of stability on abelian categories, with similar properties. We say that a homomorphism Z is a *weak stability function* if, for all non-zero $E \in \mathcal{A}$, we have

$$\Im Z(E) \geq 0, \text{ and } \Im Z(E) = 0 \Rightarrow \Re Z(E) \leq 0.$$

Semistable objects can be defined similarly, and Proposition 4.10 still holds. The proof can be extended to this more general case by demanding that the F_i are maximal among those objects satisfying $Z(F_i) = v_i$. A *weak stability condition* is then defined accordingly.

For example, slope stability of sheaves is coming from a weak stability condition on $\text{Coh}(X)$. This is going to be useful for Bridgeland stability on higher dimensional varieties (see Sect. 9, or [18, 19, 96]).

5 Bridgeland Stability

In this section we review the basic definitions and results in the general theory of Bridgeland stability conditions. The key notions are the heart of a bounded t-structure and slicings recalled in Sect. 5.1. We then introduce the two equivalent definitions in Sect. 5.2. Bridgeland's Deformation Theorem, together with a sketch of its proof, is in Sect. 5.3. Finally, we discuss moduli spaces in Sect. 5.4 and the fundamental wall and chamber structure of the space of stability conditions in Sect. 5.5. Some of the proofs we sketch here will be fully proved in the case of surfaces in the next section.

Throughout this section, we let X be a smooth projective variety over $\mathbb{C}$. We also denote the bounded derived category of coherent sheaves on X by $\mathrm{D}^{\mathrm{b}}(X) := \mathrm{D}^{\mathrm{b}}(\mathrm{Coh}(X))$. A quick introduction to bounded derived categories can be found in Appendix. The results in this section are still true in the more general setting of triangulated categories.

5.1 Heart of a Bounded t-Structure and Slicing

For the general theory of t-structures we refer to [22]. In this section we content ourselves to speak about the heart of a bounded t-structure, which in this case is equivalent.

Definition 5.1 The *heart of a bounded t-structure* on $\mathrm{D}^{\mathrm{b}}(X)$ is a full additive subcategory $\mathcal{A} \subset \mathrm{D}^{\mathrm{b}}(X)$ such that

- for integers $i > j$ and $A, B \in \mathcal{A}$, the vanishing $\mathrm{Hom}(A[i], B[j]) = \mathrm{Hom}(A, B[j-i]) = 0$ holds, and
- for all $E \in \mathrm{D}^{\mathrm{b}}(X)$ there are integers $k_1 > \ldots > k_m$, objects $E_i \in \mathrm{D}^{\mathrm{b}}(X)$, $A_i \in \mathcal{A}$ for $i = 1, \ldots, m$ and a collection of triangles

$$\begin{array}{ccccccccc} 0 = E_0 & \longrightarrow & E_1 & \longrightarrow & E_2 & \longrightarrow \ldots \longrightarrow & E_{m-1} & \longrightarrow & E_m = E \\ & \nwarrow & \downarrow & \nwarrow & \downarrow & \nwarrow & \downarrow & \nwarrow & \downarrow \\ & & A_1[k_1] & & A_2[k_2] & & A_{m-1}[k_{m-1}] & & A_m[k_m]. \end{array}$$

Lemma 5.2 *The heart $\mathcal{A}$ of a bounded t-structure in $\mathrm{D}^{\mathrm{b}}(X)$ is abelian.*

Sketch of the proof Exact sequences in $\mathcal{A}$ are nothing but exact triangles in $\mathrm{D}^{\mathrm{b}}(X)$ all of whose objects are in $\mathcal{A}$. To define kernel and cokernel of a morphism $f : A \to B$, for $A, B \in \mathcal{A}$ we can proceed as follows. Complete the morphism f to an exact triangle

$$A \xrightarrow{f} B \to C \to A[1].$$

By the definition of a heart, we have a triangle

$$C_{>0} \to C \to C_{\leq 0} \to C_{>0}[1],$$

where $C_{>0}$ belongs to the category generated by extensions of $\mathcal{A}[i]$, with $i > 0$, and $C_{\leq 0}$ to the category generated by extensions of $\mathcal{A}[j]$, with $j \leq 0$.

Consider the composite map $B \to C \to C_{\leq 0}$. Then an easy diagram chase shows that $C_{\leq 0} \in \mathcal{A}$. Similarly, the composite map $C_{>0} \to C \to A[1]$ gives $C_{>0} \in \mathcal{A}[1]$. Then $\mathrm{Ker}(f) = C_{>0}[-1]$ and $\mathrm{Cok}(f) = C_{\leq 0}$. □

The standard example of a heart of a bounded t-structure on $\mathrm{D^b}(X)$ is given by $\mathrm{Coh}(X)$. If $\mathrm{D^b}(\mathcal{A}) \cong \mathrm{D^b}(X)$, then $\mathcal{A}$ is the heart of a bounded t-structure. The converse does not hold (see Exercise 5.3 below), but this is still one of the most important examples for having an intuition about this notion. In particular, given $E \in \mathrm{D^b}(X)$, the objects $A_1, \ldots, A_n \in \mathcal{A}$ in the definition of a heart are its *cohomology objects* with respect to $\mathcal{A}$. They are denoted $A_1 =: \mathcal{H}_{\mathcal{A}}^{-k_1}(E), \ldots, A_n =: \mathcal{H}_{\mathcal{A}}^{-k_n}(E)$. In the case $\mathcal{A} = \mathrm{Coh}(X)$, these are nothing but the cohomology sheaves of a complex. These cohomology objects have long exact sequences, just as cohomology sheaves of a complexes.

Exercise 5.3 Let $X = \mathbb{P}^1$. Let $\mathcal{A} \subset \mathrm{D^b}(\mathbb{P}^1)$ be the additive subcategory generated by extensions by $\mathcal{O}_{\mathbb{P}^1}[2]$ and $\mathcal{O}_{\mathbb{P}^1}(1)$.

(1) Let $A \in \mathcal{A}$. Show that there exists $a_0, a_1 \geq 0$ such that

$$A \cong \mathcal{O}_{\mathbb{P}^1}^{\oplus a_0}[2] \oplus \mathcal{O}_{\mathbb{P}^1}(1)^{\oplus a_1}.$$

(2) Show that $\mathcal{A}$ is the heart of a bounded t-structure on $\mathrm{D^b}(\mathbb{P}^1)$. *Hint: Use Theorem 2.1.*
(3) Show that, as an abelian category, $\mathcal{A}$ is equivalent to the direct sum of two copies of the category of finite-dimensional vector spaces.
(4) Show that $\mathrm{D^b}(\mathcal{A})$ is not equivalent to $\mathrm{D^b}(\mathbb{P}^1)$.

Exercise 5.4 Let $\mathcal{A} \subset \mathrm{D^b}(X)$ be the heart of a bounded t-structure. Show that there is a natural identification between Grothendieck groups

$$K_0(\mathcal{A}) = K_0(X) = K_0(\mathrm{D^b}(X)).$$

A slicing further refines the heart of a bounded t-structure.

Definition 5.5 ([30]) A *slicing* $\mathcal{P}$ of $\mathrm{D^b}(X)$ is a collection of subcategories $\mathcal{P}(\phi) \subset \mathrm{D^b}(X)$ for all $\phi \in \mathbb{R}$ such that

- $\mathcal{P}(\phi)[1] = \mathcal{P}(\phi + 1)$,
- if $\phi_1 > \phi_2$ and $A \in \mathcal{P}(\phi_1)$, $B \in \mathcal{P}(\phi_2)$ then $\mathrm{Hom}(A, B) = 0$,
- for all $E \in \mathrm{D^b}(X)$ there are real numbers $\phi_1 > \ldots > \phi_m$, objects $E_i \in \mathrm{D^b}(X)$, $A_i \in \mathcal{P}(\phi_i)$ for $i = 1, \ldots, m$ and a collection of triangles

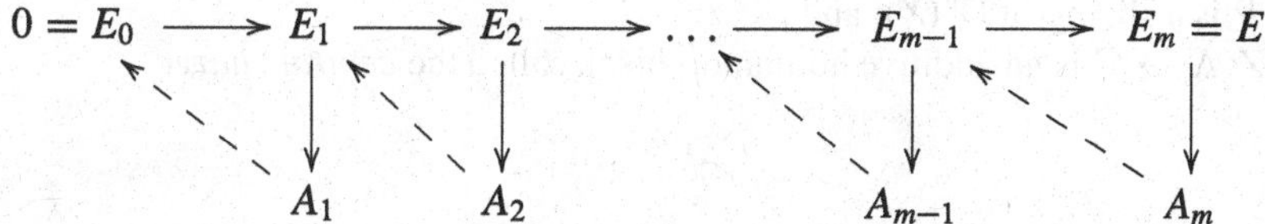

where $A_i \in P(\phi_i)$.

For this filtration of an element $E \in \mathrm{D^b}(X)$ we write $\phi^-(E) := \phi_m$ and $\phi^+(E) := \phi_1$. Moreover, for $E \in P(\phi)$ we call $\phi(E) := \phi$ the *phase* of E.

The last property is called the *Harder-Narasimhan filtration*. By setting $\mathcal{A} := \mathcal{P}((0,1])$ to be the extension closure of the subcategories $\{\mathcal{P}(\phi) : \phi \in (0,1]\}$ one gets the heart of a bounded t-structure from a slicing.[5] In both cases of a slicing and the heart of a bounded t-structure the Harder-Narasimhan filtration is unique similarly to Exercise 4.7.

Exercise 5.6 Let $\mathcal{A}, \mathcal{B} \subset \mathrm{D^b}(X)$ be hearts of bounded t-structures. Show that if $\mathcal{A} \subset \mathcal{B}$, then $\mathcal{A} = \mathcal{B}$. Similarly, if $\mathcal{P}$ and $\mathcal{P}'$ are two slicings and $\mathcal{P}'(\phi) \subset \mathcal{P}(\phi)$ for all $\phi \in \mathbb{R}$, then $\mathcal{P}$ and $\mathcal{P}'$ are identical.

5.2 Bridgeland Stability Conditions: Definition

The definition of a Bridgeland stability condition will depend on some additional data. More precisely, we fix a finite rank lattice Λ and a surjective group homomorphism

$$v\colon K_0(X) \twoheadrightarrow \Lambda.$$

We also fix a norm $\|\cdot\|$ on $\Lambda_{\mathbb{R}}$. Recall that all choices of norms are equivalent here and subsequent definitions will not depend on it.

Example 5.7 Let T be the set of $v \in K_0(X)$ such that $\chi(v, w) = 0$ for all $w \in K_0(X)$. Then the *numerical Grothendieck group* $K_{\mathrm{num}}(X)$ is defined as $K_0(X)/T$. We have that $K_{\mathrm{num}}(X)$ is a finitely generated $\mathbb{Z}$-lattice. The choice of $\Lambda = K_{\mathrm{num}}(X)$ together with the natural projection is an example of a pair (Λ, v) as before.

If X is a surface, then $K_{\mathrm{num}}(X)$ is nothing but the image of the Chern character map

$$\mathrm{ch}\colon K_0(X) \to H^*(X, \mathbb{Q}),$$

and the map v is the Chern character. For K3 or abelian surfaces, $K_{\mathrm{num}}(X) = H^*_{\mathrm{alg}}(X, \mathbb{Z}) = H^0(X, \mathbb{Z}) \oplus \mathrm{NS}(X) \oplus H^4(X, \mathbb{Z})$.

Definition 5.8 ([30]) A *Bridgeland stability condition* on $\mathrm{D^b}(X)$ is a pair $\sigma = (\mathcal{P}, Z)$ where:

- $\mathcal{P}$ is a slicing of $\mathrm{D^b}(X)$, and
- $Z\colon \Lambda \to \mathbb{C}$ is an additive homomorphism, called the *central charge*,

[5]More generally, by fixing $\phi_0 \in \mathbb{R}$, the category $\mathcal{P}((\phi_0, \phi_0 + 1])$ is also a heart of a bounded t-structure. A slicing is a family of hearts, parameterized by $\mathbb{R}$.

satisfying the following properties:

(1) For any non-zero $E \in \mathcal{P}(\phi)$, we have

$$Z(v(E)) \in \mathbb{R}_{>0} \cdot e^{\sqrt{-1}\pi\phi}.$$

(2) (*support property*)

$$C_\sigma := \inf\left\{ \frac{|Z(v(E))|}{\|v(E)\|} : 0 \neq E \in \mathcal{P}(\phi), \phi \in \mathbb{R} \right\} > 0.$$

As before, the heart of a bounded t-structure can be defined by $\mathcal{A} := P((0, 1])$. Objects in $\mathcal{P}(\phi)$ are called σ-semistable of phase ϕ. The *mass* of an object $E \in \mathrm{D}^{\mathrm{b}}(X)$ is defined as $m_\sigma(E) = \sum |Z(A_j)|$, where $A_1, \ldots, A_m$ are the Harder-Narasimhan factors of E.

Exercise 5.9 Let $\sigma = (\mathcal{P}, Z)$ be a stability condition. Show that the category $\mathcal{P}(\phi)$ is abelian and of finite length (i.e., it is noetherian and artinian). *Hint: Use the support property to show there are only finitely many classes of subobjects in $\mathcal{P}(\phi)$ of a given object in $\mathcal{P}(\phi)$.*

The simple object in $\mathcal{P}(\phi)$ are called *σ-stable*. As in the case of stability of sheaves on curves (see Remark 2.8), since the category $\mathcal{P}(\phi)$ is of finite length, σ-semistable objects admit finite *Jordan-Hölder filtrations* in σ-stable ones of the same phase. The notion of *S-equivalent* σ-semistable objects is defined analogously as well.

The support property was introduced in [68]. It is equivalent to Bridgeland's notion of a full locally-finite stability condition in [31, Definition 4.2] (see [15, Proposition B.4]). There is an equivalent formulation: There is a symmetric bilinear form Q on $\Lambda_{\mathbb{R}}$ such that

(1) all semistable objects $E \in \mathcal{P}$ satisfy the inequality $Q(v(E), v(E)) \geq 0$ and
(2) all non zero vectors $v \in \Lambda_{\mathbb{R}}$ with $Z(v) = 0$ satisfy $Q(v, v) < 0$.

The inequality $Q(v(E), v(E)) \geq 0$ can be viewed as some generalization of the classical Bogomolov inequality for vector bundles; we will see the precise relation in Sect. 6. By abuse of notation we will often forget v from the notation. For example, we will write $Q(E, F)$ instead of $Q(v(E), v(F))$. We will also use the notation $Q(E) = Q(E, E)$.

Exercise 5.10 Show that the previous two definitions of the support property are equivalent. *Hint: Use* $Q(w) = C^2|Z(w)|^2 - ||w||^2$.

Definition 5.8 is short and good for abstract argumentation, but it is not very practical for finding concrete examples. The following lemma shows that this definition of a stability condition on $\mathrm{D}^{\mathrm{b}}(X)$ and the one given in the previous section for an arbitrary abelian category $\mathcal{A}$ are closely related.

Lemma 5.11 ([30, Proposition 5.3]) *Giving a stability condition $(\mathcal{P}, Z)$ on $\mathrm{D}^{\mathrm{b}}(X)$ is equivalent to giving a stability condition $(\mathcal{A}, Z)$ in the sense of Definition 4.6,* *where $\mathcal{A}$ is the heart of a bounded t-structure on $\mathrm{D}^b(X)$ together with the support property*

$$\inf\left\{\frac{|Z(v(E))|}{\|v(E)\|} : 0 \neq E \in \mathcal{A} \text{ semistable}\right\} > 0.$$

Proof Assume we have a stability conditions $(\mathcal{P}, Z)$ on $\mathrm{D}^b(X)$. Then we can define a heart $\mathcal{A} = \mathcal{P}((0,1])$. Then $(\mathcal{A}, Z)$ is a stability conditions in the sense of Definition 4.6 satisfying the support property. The other way around, we can define $\mathcal{P}(\phi)$ to be the category of semistable objects of phase ϕ in $\mathcal{A}$ whenever $\phi \in (0,1]$. This definition can be extended to any $\phi \in \mathbb{R}$ via the property $\mathcal{P}(\phi)[1] = \mathcal{P}(\phi+1)$. □

From here on we will interchangeably use $(\mathcal{P}, Z)$ and $(\mathcal{A}, Z)$ to denote a stability condition.

Example 5.12 Let C be a curve. Then the stability condition in Example 4.3 (1) gives a Bridgeland stability condition on $\mathrm{D}^b(C)$. The lattice Λ is nothing but $H^0(C,\mathbb{Z}) \oplus H^2(C,\mathbb{Z})$, the map v is nothing but (r,d), and we can choose $Q = 0$.

Exercise 5.13 Show that the stability condition in Exercise 4.13 gives a Bridgeland stability condition on $\mathrm{D}^b(A) = \mathrm{D}^b(\text{mod-}A)$. What is Q?

5.3 Bridgeland's Deformation Theorem

The main theorem in [30] is the fact that the set of stability conditions can be given the structure of a complex manifold.

Let $\mathrm{Stab}(X)$ be the set of stability conditions on $\mathrm{D}^b(X)$ (with respect to Λ and v). This set can be given a topology as the coarsest topology such that for any $E \in \mathrm{D}^b(X)$ the maps $(\mathcal{A}, Z) \mapsto Z$, $(\mathcal{A}, Z) \mapsto \phi^+(E)$ and $(\mathcal{A}, Z) \mapsto \phi^-(E)$ are continuous. Equivalently, the topology is induced by the generalized (i.e., with values in $[0, +\infty]$) metric

$$d(\sigma_1, \sigma_2) = \sup_{0 \neq E \in \mathrm{D}^b(X)} \left\{ |\phi^+_{\sigma_1}(E) - \phi^+_{\sigma_2}(E)|, |\phi^-_{\sigma_1}(E) - \phi^-_{\sigma_2}(E)|, \|Z_1 - Z_2\| \right\},$$

for $\sigma_i = (\mathcal{P}_i, Z_i) \in \mathrm{Stab}(X)$, $i = 1, 2$; here, by abuse of notation, $\| \cdot \|$ also denotes the induced operator norm on $\mathrm{Hom}(\Lambda, \mathbb{C})$.

Remark 5.14 There are two group actions on the space of stability conditions.

(1) The universal cover $\widetilde{\mathrm{GL}}^+(2,\mathbb{R})$ of $\mathrm{GL}^+(2,\mathbb{R})$, the 2×2 matrices with real entries and positive determinant, acts on the right of $\mathrm{Stab}(X)$ as follows. We

first recall the presentation

$$\widetilde{\mathrm{GL}}^+(2,\mathbb{R}) = \left\{ (T,f) : \begin{array}{l} f\colon \mathbb{R} \to \mathbb{R} \text{ increasing, } f(\phi+1) = f(\phi)+1 \\ T \in \mathrm{GL}^+(2,\mathbb{R}) \\ f|_{\mathbb{R}/2\mathbb{Z}} = T|_{(\mathbb{R}^2\setminus\{0\})/\mathbb{R}_{>0}} \end{array} \right\}$$

Then (T,f) acts on $(\mathcal{P},Z)$ by $(T,f)\cdot(\mathcal{P},Z) = (\mathcal{P}',Z')$, where $Z' = T^{-1}\circ Z$ and $\mathcal{P}'(\phi) = \mathcal{P}(f(\phi))$.

(2) The group of exact autoequivalences $\mathrm{Aut}_\Lambda(\mathrm{D^b}(X))$ of $\mathrm{D^b}(X)$, whose action Φ_* on $K_0(X)$ is compatible with the map $v\colon K_0(X)\to\Lambda$, acts on the left of $\mathrm{Stab}(X)$ by $\Phi\cdot(\mathcal{P},Z) = (\Phi(\mathcal{P}), Z\circ\Phi_*)$. In [14], Bayer and Bridgeland use this action and a description of the geometry of the space of stability conditions on K3 surfaces of Picard rank 1 to describe the full group of derived autoequivalences. This idea should work for all K3 surfaces, as envisioned by Bridgeland in [31, Conjecture 1.2].

Theorem 5.15 ([30]) *The map* $\mathcal{Z}\colon \mathrm{Stab}(X) \to \mathrm{Hom}(\Lambda,\mathbb{C})$ *given by* $(\mathcal{A},Z)\mapsto Z$ *is a local homeomorphism. In particular,* $\mathrm{Stab}(X)$ *is a complex manifold of dimension* $\mathrm{rk}(\Lambda)$.

Ideas from the proof We follow the presentation in [11, Sect. 5.5] and we refer to it for the complete argument. Let $\sigma = (\mathcal{A},Z)\in\mathrm{Stab}(X)$. We need to prove that any group homomorphism W near Z extends to a stability condition near σ.

The first key property is the following [11, Lemma 5.5.4]: The function $C:\mathrm{Stab}(X)\to\mathbb{R}_{>0}$, $\sigma\mapsto C_\sigma$ is continuous, where C_σ is the constant appearing in the support property in Definition 5.8.

By using this, and the $\widetilde{\mathrm{GL}}^+(2,\mathbb{R})$-action described in Remark 5.14 (by just rotating of $\pi/2$), it is possible to reduce to the following case: We let $W\colon\Lambda\to\mathbb{Z}$ satisfy the assumptions

- $\Im W = \Im Z$;
- $\|W - Z\| < \epsilon C_\sigma$, with $\epsilon < \frac{1}{8}$.

The claim is that $(\mathcal{A},W)\in\mathrm{Stab}(X)$. To prove this, under our assumptions, the only thing needed is to show that HN filtrations exist with respect to W. We proceed as in the proof of Proposition 4.10. We need an analogue of Lemma 4.9 and to show that there are only finitely many vertices in the Harder-Narasimhan polygon. The idea for these is the following. Let $E\in\mathcal{A}$.

- The existence of HN filtrations for Z implies there exists a constant Γ_E such that

$$\Re Z(F) \geq \Gamma_E + m_\sigma(F),$$

for all $F\subset E$.

- By taking the HN filtration of F with respect to Z, and by using the support property for σ, we get

$$|\Re W(F_i) - \Re Z(F_i)| \leq \epsilon |Z(F_i)|,$$

where the F_i's are the HN factors of F with respect to Z.
- By summing up, we get

$$\Re W(F) \geq \Re Z(F) - \epsilon m_\sigma(F) \geq \Gamma_E + \underbrace{(1-\epsilon)}_{>0} m_\sigma(F) > \Gamma_E,$$

namely Lemma 4.9 holds.
- If $F \subset E$ is an extremal point of the polygon for W, then

$$\max\{0, \Re W(E)\} > \Re W(F),$$

and so

$$m_\sigma(F) \leq \frac{\max\{0, \Re W(E)\} - \Gamma_E}{1-\epsilon} =: \Gamma'_E.$$

- Again, by taking the HN factors of F with respect to Z, we get $|Z(F_i)| < \Gamma'_E$, and so by using the support property,

$$\|F_i\| < \frac{\Gamma'_E}{C_\sigma}.$$

- From this we deduce there are only finitely many classes, and so finitely many vertices in the polygon. The proof of Proposition 4.10 applies now and this gives the existence of Harder-Narasimhan filtrations with respect to W.
- The support property follows now easily.

□

We can think about Bridgeland's main theorem more explicitly in terms of the quadratic form Q appearing in the support property. This approach actually gives an "ϵ-free proof" of Theorem 2.15 and appears in [13]. The main idea is contained in the following proposition, which is [19, Proposition A.5]. We will not prove this, but we will see it explicitly in the case of surfaces.

Proposition 5.16 *Assume that* $\sigma = (\mathcal{P}, Z) \in \mathrm{Stab}(X)$ *satisfies the support property with respect to a quadratic form* Q *on* $\Lambda_{\mathbb{R}}$. *Consider the open subset of* $\mathrm{Hom}(\Lambda, \mathbb{C})$ *consisting of central charges on whose kernel* Q *is negative definite, and let* U *be the connected component containing* Z. *Let* $\mathcal{U} \subset \mathrm{Stab}(X)$ *be the connected component of the preimage* $\mathcal{Z}^{-1}(U)$ *containing* σ. *Then:*

(1) *The restriction* $\mathcal{Z}|_{\mathcal{U}}: \mathcal{U} \to U$ *is a covering map.*

(2) *Any stability condition $\sigma' \in \mathcal{U}$ satisfies the support property with respect to the same quadratic form Q.*

Example 5.17 Spaces of stability conditions are harder to study in general. For curves everything is known. As before, we keep Λ and v as in Example 5.12.

(1) $\mathrm{Stab}(\mathbb{P}^1) \cong \mathbb{C}^2$ (see [92]).
(2) Let C be a curve of genus $g \geq 1$. Then $\mathrm{Stab}(C) \cong \mathbb{H} \times \mathbb{C}\,(= \sigma_0 \cdot \widetilde{\mathrm{GL}}^+(2, \mathbb{R}))$ (see [30, 81]), where σ_0 is the stability $(\mathrm{Coh}(C), -d + \sqrt{-1}\,r)$ in Example 5.12.

Let $\sigma = (\mathcal{P}, Z) \in \mathrm{Stab}(C)$. We will examine only the case $g \geq 1$. The key point of the proof is to show that the skyscraper sheaves $\mathbb{C}(x)$, $x \in C$, and all line bundles L are all in the category $\mathcal{P}((\phi_0, \phi_0 + 1])$, for some $\phi_0 \in \mathbb{R}$, and they are all σ-stable.

Assume first that $\mathbb{C}(x)$ is not σ-semistable, for some $x \in C$. By taking its HN filtration, we obtain an exact triangle

$$A \to \mathbb{C}(x) \to B \to A[1]$$

with $\mathrm{Hom}^{\leq 0}(A, B) = 0$.[6] By taking cohomology sheaves, we obtain a long exact sequence of coherent sheaves on C

$$0 \to \mathcal{H}^{-1}(B) \to \mathcal{H}^0(A) \to \mathbb{C}(x) \xrightarrow{f} \mathcal{H}^0(B) \to \mathcal{H}^1(A) \to 0,$$

and $\mathcal{H}^{i-1}(B) \cong \mathcal{H}^i(A)$, for all $i \neq 0, 1$.

But, since C is smooth of dimension 1, an object F in $\mathrm{D}^{\mathrm{b}}(C)$ is isomorphic to the direct sum of its cohomology sheaves: $F \cong \oplus_i \mathcal{H}^i(F)[-i]$. Since $\mathrm{Hom}^{\leq 0}(A, B) = 0$, this gives $\mathcal{H}^{i-1}(B) \cong \mathcal{H}^i(A) = 0$, for all $i \neq 0, 1$.

We look at the case $f = 0$. The case in which $f \neq 0$ (and so, f is injective) can be dealt with similarly. In this case, $\mathcal{H}^0(B) \cong \mathcal{H}^1(A)$ and so they are both trivial as well, since otherwise $\mathrm{Hom}^{-1}(A, B) \neq 0$. Therefore, $A \cong \mathcal{H}^0(A)$ and $B \cong \mathcal{H}^{-1}(B)[1]$. But, by using Serre duality, we have

$$\begin{aligned} 0 \neq \mathrm{Hom}(\mathcal{H}^{-1}(B), \mathcal{H}^0(A)) \hookrightarrow \mathrm{Hom}(\mathcal{H}^{-1}(B), \mathcal{H}^0(A) \otimes K_C) &\cong \mathrm{Hom}(\mathcal{H}^0(A), \mathcal{H}^{-1}(B)[1]) \\ &\cong \mathrm{Hom}(A, B), \end{aligned}$$

a contradiction.

We now claim that $\mathbb{C}(x)$ is actually stable. Indeed, a similar argument as before shows that if $\mathbb{C}(x)$ is not σ-stable, then the only possibility is that all its stable factors are isomorphic to a single object K. But then, by looking at its cohomology sheaves, K must be a sheaf as well, which is impossible. Set ϕ_x as the phase of $\mathbb{C}(x)$ in σ.

In the same way, all line bundles are σ-stable as well. For a line bundle A on C, we set ϕ_A as the phase of A in σ.

[6] More precisely, $A \in \mathcal{P}(< \phi_0)$ and $B \in \mathcal{P}(\geq \phi_0)$, for some $\phi_0 \in \mathbb{R}$.

The existence of maps $A \to \mathbb{C}(x)$ and $\mathbb{C}(x) \to A[1]$ gives the inequalities

$$\phi_x - 1 \leq \phi_A \leq \phi_x.$$

Since A and $\mathbb{C}(x)$ are both σ-stable, the strict inequalities hold. Hence, Z is an isomorphism as a map $\mathbb{R}^2 \to \mathbb{R}^2$ and orientation-preserving. By acting with an element of $\widetilde{\mathrm{GL}}^+(2,\mathbb{R})$, we can therefore assume that $Z = Z_0 = -d + \sqrt{-1}\, r$ and that, for some $x \in C$, the skyscraper sheaf $\mathbb{C}(x)$ has phase 1. But then all line bundles on C have phases in the interval $(0, 1)$, and so all skyscraper sheaves have phase 1 as well. But this implies that $\mathcal{P}((0, 1]) = \mathrm{Coh}(C)$, and so $\sigma = \sigma_0$.

Example 5.18 One of the motivations for the introduction of Bridgeland stability conditions is coming from mirror symmetry (see [41]). In case of elliptic curves, the mirror symmetry picture is particularly easy to explain (see [30, Sect. 9] and [32]). Indeed, let C be an elliptic curve. We can look at the action of the subgroup $\mathbb{C} \subset \widetilde{\mathrm{GL}}^+(2,\mathbb{R})$ and of $\mathrm{Aut}(\mathrm{D}^{\mathrm{b}}(C))$ on $\mathrm{Stab}(C)$, and one deduces

$$\mathrm{Aut}(\mathrm{D}^{\mathrm{b}}(C))\backslash \mathrm{Stab}(C)/\mathbb{C} \cong \mathbb{H}/\,\mathrm{PSL}(2,\mathbb{Z}),$$

coherently with the idea that spaces of stability conditions should be related to variations of complex structures on the mirror variety (in this case, a torus itself).

5.4 Moduli Spaces

The main motivation for stability conditions is the study of moduli spaces of semistable objects. In this section we recall the general theory of moduli spaces of complexes, and then define in general moduli spaces of Bridgeland semistable objects.

To define the notion of a *family of complexes*, we recall first the notion of a *perfect complex*. We follow [64, 76]. Let B be a scheme locally of finite type over $\mathbb{C}$. We denote the unbounded derived category of quasi-coherent sheaves by $\mathrm{D}(\mathrm{Qcoh}(X \times B))$. A complex $E \in \mathrm{D}(\mathrm{Qcoh}(X \times B))$ is called *B-perfect* if it is, locally over B, isomorphic to a bounded complex of flat sheaves over B of finite presentation. The triangulated subcategory of B-perfect complexes is denoted by $\mathrm{D}_{B\text{-perf}}(X \times B)$.

Definition 5.19 We let $\mathfrak{M}\colon \underline{\mathrm{Sch}} \to \underline{\mathrm{Grp}}$ be the 2-functor which maps a scheme B locally of finite type over $\mathbb{C}$ to the groupoid $\mathfrak{M}(B)$ given by

$$\left\{\mathcal{E} \in \mathrm{D}_{B\text{-perf}}(X \times B) \,:\, \mathrm{Ext}^i(\mathcal{E}_b, \mathcal{E}_b) = 0, \text{ for all } i < 0 \text{ and all geometric pts } b \in B\right\},$$

where $\mathcal{E}_b$ is the restriction $\mathcal{E}|_{X\times\{b\}}$

Theorem 5.20 ([76, Theorem 4.2.1]) *$\mathfrak{M}$ is an Artin stack, locally of finite type, locally quasi separated, and with separated diagonal.*

We will not discuss Artin stacks in detail any further in this survey; we refer to the introductory book [93] for the basic terminology (or [108]). We will mostly use the following special case. Let $\mathfrak{M}_{\mathrm{Spl}}$ be the open substack of $\mathfrak{M}$ parameterizing simple objects (as in the sheaf case, these are complexes with only scalar endomorphisms). This is also an Artin stack with the same properties as $\mathfrak{M}$. Define also a functor $\underline{M}_{\mathrm{Spl}}\colon \underline{\mathrm{Sch}} \to \underline{\mathrm{Set}}$ by forgetting the groupoid structure on $\mathfrak{M}_{\mathrm{Spl}}$ and by sheafifying it (namely, as in the stable sheaves case, through quotienting by the equivalence relation given by tensoring with line bundles from the base B).

Theorem 5.21 ([64, Theorem 0.2]) *The functor $\underline{M}_{\mathrm{Spl}}$ is representable by an algebraic space locally of finite type over $\mathbb{C}$. Moreover, the natural morphism $\mathfrak{M}_{\mathrm{Spl}} \to \underline{M}_{\mathrm{Spl}}$ is a $\mathbb{G}_m$-gerbe.*

We can now define the moduli functor for Bridgeland semistable objects.

Definition 5.22 Let $\sigma = (\mathcal{P}, Z) \in \mathrm{Stab}(X)$. Fix a class $v_0 \in \Lambda$ and a phase $\phi \in \mathbb{R}$ such that $Z(v_0) \in \mathbb{R}_{>0} \cdot e^{\sqrt{-1}\pi\phi}$. We define $\widehat{M}_\sigma(v, \phi)$ to be the set of σ-semistable objects in $\mathcal{P}(\phi)$ of numerical class v_0 and $\mathfrak{M}_\sigma(v, \phi) \subset \mathfrak{M}$ the substack of objects in $M_\sigma(v, \phi)$. Similarly, we define $\mathfrak{M}^s_\sigma(v, \phi)$ as the substack parameterizing stable objects.

Question 5.23 Are the inclusions $\mathfrak{M}^s_\sigma(v, \phi) \subset \mathfrak{M}_\sigma(v, \phi) \subset \mathfrak{M}$ open? Is the set $\widehat{M}_\sigma(v, \phi)$ bounded?

If the answers to Question 5.23 are both affirmative, then by Theorem 5.20 and [109, Lemma 3.4], $\mathfrak{M}_\sigma(v, \phi)$ and $\mathfrak{M}^s_\sigma(v, \phi)$ are both Artin stack of finite type over $\mathbb{C}$. Moreover, by Theorem 5.21, the associated functor $\underline{M}^s_\sigma(v, \phi)$ would be represented by an algebraic space of finite type over $\mathbb{C}$.

We will see in Sect. 6.6 that Question 5.23 has indeed an affirmative answer in the case of certain stability conditions on surfaces [109] (or more generally, when Bridgeland stability conditions are constructed via an iterated tilting procedure [96]).

A second fundamental question is about the existence of a moduli space:

Question 5.24 Is there a coarse moduli space $M_\sigma(v, \phi)$ parameterizing S-equivalence classes of σ-semistable objects? Is $M_\sigma(v, \phi)$ a projective scheme?

Question 5.24 has an affirmative complete answer only for the projective plane $\mathbb{P}^2$ (see [8]), $\mathbb{P}^1 \times \mathbb{P}^1$ and the blow up of a point in $\mathbb{P}^2$ (see [7]), and partial answers for other surfaces, including abelian surfaces (see [86, 118]), K3 surfaces (see [17]), and Enriques surfaces (see [91, 117]). In Sect. 6.6 we show how the projectivity is shown in case of $\mathbb{P}^2$ via quiver representations. The other two del Pezzo cases were proved with the same method, but turn out more technically involved. As remarked in Sect. 3, since Bridgeland stability is not a priori associated to a GIT problem, it is harder to answer Question 5.24 in general. The recent works [2, 3] on good quotients for Artin stacks may lead to a general answer to this question, though. Once a coarse moduli space exists, then separatedness and properness of the moduli space is a general result by Abramovich and Polishchuk [1], which we will review in Sect. 8.

It is interesting to note that the technique in [1] is also very useful in the study of the geometry of the moduli space itself. This will also be reviewed in Sect. 8.

Exercise 5.25 Show that Questions 5.23 and 5.24 have both affirmative answers for curves (you can assume, for simplicity, that the genus is ≥ 1). Show also that moduli spaces in this case are exactly moduli spaces of semistable sheaves as reviewed in Sect. 2.

5.5 *Wall and Chamber Structure*

We conclude the section by explaining how stable object change when the stability condition is varied within $\mathrm{Stab}(X)$ itself. It turns out that the topology on $\mathrm{Stab}(X)$ is defined in such as way that this happens in a controlled way. This will be one of the key properties in studying the geometry of moduli spaces of semistable objects.

Definition 5.26 Let $v_0, w \in \Lambda \setminus \{0\}$ be two non-parallel vectors. A *numerical wall* $W_w(v_0)$ for v_0 with respect to w is a non empty subset of $\mathrm{Stab}(X)$ given by

$$W_w(v_0) := \{\sigma = (\mathcal{P}, Z) \in \mathrm{Stab}(X) \,:\, \Re Z(v_0) \cdot \Im Z(w) = \Re Z(w) \cdot \Im Z(v_0)\}.$$

We denote the set of numerical walls for v_0 by $\mathcal{W}(v_0)$. By Theorem 5.15, a numerical wall is a real submanifold of $\mathrm{Stab}(X)$ of codimension 1.

We will use walls only for special subsets of the space of stability conditions (the (α, β)-plane in Sect. 6.4). We will give a complete proof in this case in Sect. 6. The general behavior is explained in [31, Sect. 9]. We direct the reader to [15, Proposition 3.3] for the complete proof of the following result.

Proposition 5.27 *Let $v_0 \in \Lambda$ be a primitive class, and let $S \subset \mathrm{D}^{\mathrm{b}}(X)$ be an arbitrary set of objects of class v_0. Then there exists a collection of walls $W_w^S(v_0)$, $w \in \Lambda$, with the following properties:*

(1) *Every wall $W_w^S(v_0)$ is a closed submanifold with boundary of real codimension one.*
(2) *The collection $W_w^S(v_0)$ is locally finite (i.e., every compact subset $K \subset \mathrm{Stab}(X)$ intersects only a finite number of walls).*
(3) *For every stability condition $(\mathcal{P}, Z)$ on a wall in $W_w^S(v_0)$, there exists a phase $\phi \in \mathbb{R}$ and an inclusion $F_w \hookrightarrow E_{v_0}$ in $\mathcal{P}(\phi)$ with $v(F_w) = w$ and $E_{v_0} \in S$.*
(4) *If $C \subset \mathrm{Stab}(X)$ is a connected component of the complement of $\bigcup_{w \in \Lambda} W_w^S(v_0)$ and $\sigma_1, \sigma_2 \in C$, then an object $E_{v_0} \in S$ is σ_1-stable if and only if it is σ_2-stable.*

In particular, the property for an object to be stable is open in $\mathrm{Stab}(X)$.

The walls in Proposition 5.27 will be called *actual walls*. A *chamber* is defined to be a connected component of the complement of the set of actual walls, as in (4) above.

Idea of the proof For a class $w \in \Lambda$, we let V_w^S be the set of stability conditions for which there exists an inclusion as in (3). Clearly V_w^S is contained in the numerical wall $W_w(v_0)$.

The first step is to show that there only finitely many w for which V_w^S intersects a small neighborhood around a stability condition. This is easy to see by using the support property: see also Lemma 5.29 below.

We then let $W_w^S(v_0)$ be the codimension one component of V_w^S. It remains to show (4). The idea is the following: higher codimension components of V_w^S always come from objects E_{v_0} that are semistable on this component and unstable at any nearby point. □

Remark 5.28 We notice that if v_0 is not primitive and we ask only for semistability on chambers, then part (3) cannot be true: namely actual walls may be of higher codimension. On the other hand, the following holds (see [31, Sect. 9]): Let $v_0 \in \Lambda$. Then there exists a locally finite collection of numerical walls $W_w(v_0)$ such that on each chamber the set of semistable objects $\widehat{M}_\sigma(v_0, \phi)$ is constant.

The following lemma clarifies how walls behave with respect to the quadratic form appearing in the support property. It was first observed in [19, Appendix A].

Lemma 5.29 *Let Q be a quadratic form on a real vector space V and $Z : V \to \mathbb{C}$ a linear map such that the kernel of Z is negative semi-definite with respect to Q. If ρ is a ray in $\mathbb{C}$ starting at the origin, we define*

$$C_\rho^+ = Z^{-1}(\rho) \cap \{Q \geq 0\}.$$

(1) *If $w_1, w_2 \in C_\rho^+$, then $Q(w_1, w_2) \geq 0$.*
(2) *The set C_ρ^+ is a convex cone.*
(3) *Let $w, w_1, w_2 \in C_\rho^+$ with $w = w_1 + w_2$. Then $0 \leq Q(w_1) + Q(w_2) \leq Q(w)$. Moreover, $Q(w_1) = Q(w)$ implies $Q(w) = Q(w_1) = Q(w_2) = Q(w_1, w_2) = 0$.*
(4) *If the kernel of Z is negative definite with respect to Q, then any vector $w \in C_\rho^+$ with $Q(w) = 0$ generates an extremal ray of C^+.*

Proof *For any non trivial $w_1, w_2 \in C_\rho^+$ there is $\lambda > 0$ such that $Z(w_1 - \lambda w_2) = 0$. Therefore, we get*

$$0 \geq Q(w_1 - \lambda w_2) = Q(w_1) + \lambda^2 Q(w_2) - 2\lambda Q(w_1, w_2).$$

The inequalities $Q(w_1) \geq 0$ and $Q(w_2) \geq 0$ lead to $Q(w_1, w_2) \geq 0$. Part (2) follows directly from (1). The first claim in (3) also follows immediately from (1) by using

$$Q(w) = Q(w_1) + Q(w_2) + 2Q(w_1, w_2) \geq 0.$$

Observe that $Q(w_1) = Q(w)$ implies $Q(w_2) = Q(w_1, w_2) = 0$ and therefore,

$$0 = 2\lambda Q(w_1, w_2) \geq Q(w_1) + \lambda^2 Q(w_2) = Q(w_1) \geq 0.$$

Let $w \in \mathcal{C}_\rho^+$ with $Q(w) = 0$. Assume that w is not extremal, i.e., there are linearly independent $w_1, w_2 \in \mathcal{C}_\rho^+$ such that $w = w_1 + w_2$. By (3) we get $Q(w_1) = Q(w_2) = Q(w_1, w_2) = 0$. As before $Z(w_1 - \lambda w_2) = 0$, but this time $w_1 - \lambda w_2 \neq 0$. That means

$$0 > Q(w_1 - \lambda w_2) = Q(w_1) + \lambda^2 Q(w_2) - 2\lambda Q(w_1, w_2) = 0.$$

□

Corollary 5.30 *[19, Proposition A.8] Assume that U is a path-connected open subset of* Stab(X) *such that all $\sigma \in U$ satisfy the support property with respect to the quadratic form Q. If $E \in \mathrm{D}^{\mathrm{b}}(X)$ with $Q(E) = 0$ is σ-stable for some $\sigma \in U$ then it is σ'-stable for all $\sigma' \in U$.*

Proof *If there is a point in U at which E is unstable, then there is a point $\sigma = (\mathcal{A}, Z)$ at which E is strictly semistable. If ρ is the ray that $Z(E)$ lies on, then the previous lemma implies that $v(E)$ is extremal in $\mathcal{C}_\rho^+$. That is contradiction to E being strictly semistable.* □

6 Stability Conditions on Surfaces

This is one of the key sections of these notes. We introduce the fundamental operation of tilting for hearts of bounded t-structures and use it to give examples of Bridgeland stability conditions on surfaces. Another key ingredient is the Bogomolov inequality for slope semistable sheaves, which we will recall as well, and show its close relation with the support property for Bridgeland stability conditions introduced in the previous section. We then conclude with a few examples of stable objects, explicit description of walls, and the behavior at the large volume limit point.

6.1 Tilting of t-Structures

Given the heart of a bounded t-structure, the process of tilting is used to obtain a new heart. For a detailed account of the general theory of tilting we refer to [54] and [27, Sect. 5]. The idea of using this operation to construct Bridgeland stability conditions is due to Bridgeland in [31], in the case of K3 surfaces. The extension to general smooth projective surfaces is in [6].

Definition 6.1 Let $\mathcal{A}$ be an abelian category. A *torsion pair* on $\mathcal{A}$ consists of a pair of full additive subcategories $(\mathcal{F}, \mathcal{T})$ of $\mathcal{A}$ such that the following two properties hold:

- For any $F \in \mathcal{F}$ and $T \in \mathcal{T}$, $\mathrm{Hom}(T, F) = 0$.
- For any $E \in \mathcal{A}$ there are $F \in \mathcal{F}$, $T \in \mathcal{T}$ together with an exact sequence

$$0 \to T \to E \to F \to 0.$$

By the vanishing property, the exact sequence in the definition of a torsion pair is unique.

Exercise 6.2 Let X be a smooth projective variety. Show that the pair of subcategories

$$\mathcal{T} := \{\text{Torsion sheaves on } X\}$$
$$\mathcal{F} := \{\text{Torsion-free sheaves on } X\}$$

gives a torsion pair in $\mathrm{Coh}(X)$.

Lemma 6.3 (Happel-Reiten-Smalø) *Let X be a smooth projective variety. Let $\mathcal{A} \subset \mathrm{D}^{\mathrm{b}}(X)$ be the heart of a bounded t-structure, and let $(\mathcal{F}, \mathcal{T})$ be a torsion pair in $\mathcal{A}$. Then the category*

$$\mathcal{A}^\sharp := \left\{ E \in \mathrm{D}^{\mathrm{b}}(X) : \begin{array}{l} \mathcal{H}^i_{\mathcal{A}}(E) = 0, \text{ for all } i \neq 0, -1 \\ \mathcal{H}^0_{\mathcal{A}}(E) \in \mathcal{T} \\ \mathcal{H}^{-1}_{\mathcal{A}}(E) \in \mathcal{F} \end{array} \right\}$$

is the heart of a bounded t-structure on $\mathrm{D}^{\mathrm{b}}(X)$, where $\mathcal{H}^\bullet_{\mathcal{A}}$ denotes the cohomology object with respect to the t-structure $\mathcal{A}$.

Proof We only need to check the conditions in Definition 5.1. We check the first one and leave the second as an exercise.

Given $E, E' \in \mathcal{A}^\sharp$, we need to show that $\mathrm{Hom}^{<0}(E, F) = 0$. By definition of $\mathcal{A}^\sharp$, we have exact triangles

$$F_E[1] \to E \to T_E \quad \text{and} \quad F_{E'}[1] \to E' \to T_{E'},$$

where $F_E, F_{E'} \in \mathcal{F}$ and $T_E, T_{E'} \in \mathcal{T}$. By taking the functor Hom, looking at the induced long exact sequences, and using the fact that negative Hom's vanish for objects in $\mathcal{A}$, the required vanishing amounts to showing that $\mathrm{Hom}(T_E, F_{E'}) = 0$. This is exactly the first condition in the definition of a torsion pair. □

Exercise 6.4 Show that $\mathcal{A}^\sharp$ can also be defined as the smallest extension closed full subcategory of $\mathrm{D}^{\mathrm{b}}(X)$ containing both $\mathcal{T}$ and $\mathcal{F}[1]$. We will use the notation $\mathcal{A}^\sharp = \langle \mathcal{F}[1], \mathcal{T} \rangle$.

Exercise 6.5 Let $\mathcal{A}$ and $\mathcal{B}$ be two hearts of bounded t-structures on $\mathrm{D}^{\mathrm{b}}(X)$ such that $\mathcal{A} \subset \langle \mathcal{B}, \mathcal{B}[1] \rangle$. Show that $\mathcal{A}$ is a tilt of $\mathcal{B}$. *Hint: Set $\mathcal{T} = \mathcal{B} \cap \mathcal{A}$ and $\mathcal{F} = \mathcal{B} \cap \mathcal{A}[-1]$ and show that $(\mathcal{F}, \mathcal{T})$ is a torsion pair on $\mathcal{B}$.*

6.2 Construction of Bridgeland Stability Conditions on Surfaces

Let X be a smooth projective surface over $\mathbb{C}$. As in Sect. 3, we fix an ample divisor class $\omega \in \mathrm{N}^1(X)$ and a divisor class $B \in N^1(X)$. We will construct a family of Bridgeland stability conditions that depends on these two parameters.

As remarked in Example 4.3 (2), $\mathrm{Coh}(X)$ will never be the heart of a Bridgeland stability condition. The idea is to use tilting, by starting with coherent sheaves, and use slope stability to define a torsion pair.

Let $\mathcal{A} = \mathrm{Coh}(X)$. We define a pair of subcategories

$$\mathcal{T}_{\omega,B} = \{E \in \mathrm{Coh}(X) : \text{any semistable factor } F \text{ of } E \text{ satisfies } \mu_{\omega,B}(F) > 0\},$$
$$\mathcal{F}_{\omega,B} = \{E \in \mathrm{Coh}(X) : \text{any semistable factor } F \text{ of } E \text{ satisfies } \mu_{\omega,B}(F) \leq 0\}.$$

The existence of Harder-Narasimhan filtrations for slope stability (see Example 4.12) and Lemma 4.5 show that this is indeed a torsion pair on $\mathrm{Coh}(X)$.

Definition 6.6 We define the tilted heart

$$\mathrm{Coh}^{\omega,B}(X) := \mathcal{A}^\sharp = \langle \mathcal{F}_{\omega,B}[1], \mathcal{T}_{\omega,B} \rangle.$$

Exercise 6.7 Show that the categories $\mathcal{T}_{\omega,B}$ and $\mathcal{F}_{\omega,B}$ can also be defined as follows:

$$\mathcal{T}_{\omega,B} = \{E \in \mathrm{Coh}(X) : \forall\, E \twoheadrightarrow Q \neq 0,\ \mu_{\omega,B}(Q) \leq 0\},$$
$$\mathcal{F}_{\omega,B} = \{E \in \mathrm{Coh}(X) : \forall\, 0 \neq F \subset E,\ \mu_{\omega,B}(F) > 0\}.$$

Exercise 6.8 Show that the category $\mathrm{Coh}^{\omega,B}(X)$ depends only on $\frac{\omega}{\omega^2}$ and $\frac{\omega \cdot B}{\omega^2}$.

Exercise 6.9 We say that an object S in an abelian category is *minimal*[7] if S does not have any non-trivial subobjects (or quotients). Show that skyscraper sheaves are minimal objects in the category $\mathrm{Coh}^{\omega,B}(X)$. Let E be a $\mu_{\omega,B}$-stable vector bundle on X with $\mu_{\omega,B}(E) = 0$. Show that $E[1]$ is a minimal object in $\mathrm{Coh}^{\omega,B}(X)$.

We now need to define a stability function on the tilted heart. We set

$$Z_{\omega,B} := -\int_X e^{\sqrt{-1}\,\omega} \cdot \mathrm{ch}^B.$$

[7] This is commonly called a *simple* object. Unfortunately, in the theory of semistable sheaves, the word "simple" is used to indicate $\mathrm{Hom}(S, S) = \mathbb{C}$; this is why we use this slightly non-standard notation.

Explicitly, for $E \in \mathrm{D}^{\mathrm{b}}(X)$,

$$Z_{\omega,B}(E) = \left(-\mathrm{ch}_2^B(E) + \frac{\omega^2}{2} \cdot \mathrm{ch}_0^B(E)\right) + \sqrt{-1}\,\omega \cdot \mathrm{ch}_1^B(E).$$

The corresponding slope function is

$$\nu_{\omega,B}(E) = \frac{\mathrm{ch}_2^B(E) - \frac{\omega^2}{2} \cdot \mathrm{ch}_0^B(E)}{\omega \cdot \mathrm{ch}_1^B(E)}.$$

The main result is the following (see [6, 15, 31]). We will choose $\Lambda = K_{\mathrm{num}}(X)$ and v as the Chern character map as in Example 5.7.

Theorem 6.10 *Let X be a smooth projective surface. The pair $\sigma_{\omega,B} = (\mathrm{Coh}^{\omega,B}(X), Z_{\omega,B})$ gives a Bridgeland stability condition on X. Moreover, the map*

$$\mathrm{Amp}(X) \times N^1(X) \to \mathrm{Stab}(X), \;\; (\omega, B) \mapsto \sigma_{\omega,B}$$

is a continuous embedding.

Unfortunately the proof of this result is not so direct, even in the case of K3 surfaces. We will give a sketch in Sect. 6.3 below. The idea is to first prove the case in which ω and B are rational classes. The non trivial ingredient in this part of the proof is the classical Bogomolov inequality for slope semistable torsion-free sheaves. Then we show we can deform by keeping B fixed and letting ω vary. This, together with the behavior for ω "large", will give a Bogomolov inequality for Bridgeland stable objects. This will allow to deform B as well and to show a general Bogomolov inequality for Bridgeland semistable objects, which will finally imply the support property and conclude the proof of the theorem.

The key result in the proof of Theorem 6.10 can be summarized as follows. It is one of the main theorems from [18] (see also [19, Theorem 3.5]). We first need to introduce three notions of discriminant for an object in $\mathrm{D}^{\mathrm{b}}(X)$.

Exercise 6.11 Let $\omega \in N^1(X)$ be an ample real divisor class. Then there exists a constant $C_\omega \geq 0$ such that, for every effective divisor $D \subset X$, we have

$$C_\omega(\omega \cdot D)^2 + D^2 \geq 0.$$

Definition 6.12 Let $\omega, B \in N^1(X)$ with ω ample. We define the *discriminant* function as

$$\Delta := \left(\mathrm{ch}_1^B\right)^2 - 2\,\mathrm{ch}_0^B \cdot \mathrm{ch}_2^B = (\mathrm{ch}_1)^2 - 2\,\mathrm{ch}_0 \cdot \mathrm{ch}_2\,.$$

We define the *ω-discriminant* as

$$\overline{\Delta}_\omega^B := \left(\omega \cdot \mathrm{ch}_1^B\right)^2 - 2\omega^2 \cdot \mathrm{ch}_0^B \cdot \mathrm{ch}_2^B\,.$$

Choose a rational non-negative constant C_ω as in Exercise 6.11 above. Then we define the *(ω, B, C_ω)-discriminant* as

$$\Delta^C_{\omega,B} := \Delta + C_\omega(\omega \cdot \mathrm{ch}_1^B)^2.$$

Theorem 6.13 *Let X be a smooth projective surface over $\mathbb{C}$. Let $\omega, B \in N^1(X)$ with ω ample. Assume that E is $\sigma_{\omega,B}$-semistable. Then*

$$\Delta^C_{\omega,B}(E) \geq 0 \ \ and \ \ \overline{\Delta}^B_\omega(E) \geq 0.$$

The quadratic form $\Delta^C_{\omega,B}$ will give the support property for $\sigma_{\omega,B}$. The quadratic form $\overline{\Delta}^B_\omega$ will give the support property on the (α, β)-plane (see Sect. 6.4).

6.3 Sketch of the Proof of Theorem 6.10 and Theorem 6.13

We keep the notation as in the previous section.

Bogomolov Inequality Our first goal is to show that $Z_{\omega,B}$ is a stability function on $\mathrm{Coh}^{\omega,B}(X)$. The key result we need is the following [24, 51, 99], [72, Theorem 12.1.1], [63, Theorem 3.4.1].

Theorem 6.14 (Bogomolov Inequality) *Let X be a smooth projective surface. Let $\omega, B \in N^1(X)$ with ω ample, and let E be a $\mu_{\omega,B}$-semistable torsion-free sheaf. Then*

$$\Delta(E) = \mathrm{ch}_1^B(E)^2 - 2\,\mathrm{ch}_0^B(E) \cdot \mathrm{ch}_2^B(E) \geq 0.$$

Sketch of the proof Since neither slope stability nor Δ depend on B, we can assume $B = 0$. Also, by Huybrechts and Lehn [63, Lemma 4.C.5], slope stability with respect to an ample divisor changes only at integral classes. Hence, we can assume $\omega = H \in \mathrm{NS}(X)$ is the divisor class of a very ample integral divisor (by abuse of notation, we will still denote it by H).

Since Δ is invariant both by tensor products of line bundles and by pull-backs via finite surjective morphisms, and since these two operations preserve slope stability of torsion-free sheaves (see e.g., Exercise 2.16 for the case of curves, and [63, Lemma 3.2.2] in general), by taking a $\mathrm{rk}(E)$-cyclic cover (see, e.g., [63, Theorem 3.2.9]) and by tensoring by a $\mathrm{rk}(E)$-root of $\mathrm{ch}_1(E)$, we can assume $\mathrm{ch}_1(E) = 0$. Finally, by taking the double-dual, Δ can only become worse. Hence, we can assume E is actually a vector bundle on X.

Hence, we are reduced to show $\mathrm{ch}_2(E) \leq 0$. We use the restriction theorem for slope stability (see [63, Sect. 7.1 or 7.2] or [70]): up to replacing H with a multiple uH, for $u \gg 0$ fixed, there exists a smooth projective curve $C \in |H|$ such that $E|_C$ is again slope semistable. Also, as remarked in Sect. 2.4, the tensor product $E|_C \otimes \ldots \otimes E|_C$ is still semistable on C.

Now the theorem follows by estimating Euler characteristics. Indeed, on the one hand, by the Riemann-Roch Theorem we have

$$\chi(X, \underbrace{E \otimes \ldots \otimes E}_{m\text{-times}}) = \mathrm{rk}(E)^m \chi(X, \mathcal{O}_X) + m\,\mathrm{rk}(E)^{m-1}\,\mathrm{ch}_2(E),$$

for all $m > 0$.

On the other hand, we can use the exact sequence

$$0 \to (E \otimes \ldots \otimes E) \otimes \mathcal{O}_X(-H) \to E \otimes \ldots \otimes E \to E|_C \otimes \ldots \otimes E|_C \to 0,$$

and, by using the remark above, deduce that

$$h^0(X, E \otimes \ldots \otimes E) \leq h^0(C, E|_C \otimes \ldots \otimes E|_C) \leq \gamma_E\,\mathrm{rk}(E)^m,$$

for a constant $\gamma_E > 0$ which is independent on m. Similarly, by Serre duality, and by using an analogous argument, we have

$$h^2(X, E \otimes \ldots \otimes E) \leq \delta_E\,\mathrm{rk}(E)^m,$$

for another constant δ_E which depends only on E.

By putting all together, since

$$\chi(X, E \otimes \ldots \otimes E) \leq h^0(X, E \otimes \ldots \otimes E) + h^2(X, E \otimes \ldots \otimes E) \leq (\gamma_E + \delta_E)\,\mathrm{rk}(E)^m,$$

we deduce

$$\chi(X, \mathcal{O}_X) + m\frac{\mathrm{ch}_2(E)}{\mathrm{rk}(E)} \leq \gamma_E + \delta_E,$$

for any $m > 0$, namely $\mathrm{ch}_2(E) \leq 0$, as we wanted. □

Remark 6.15 Let E be a torsion sheaf. Then

$$\Delta^C_{\omega,B}(E) \geq 0.$$

Indeed, $\mathrm{ch}_1^B(E) = \mathrm{ch}^1(E)$ is an effective divisor. Then the inequality follows directly by the definition of C_ω in Exercise 6.11.

We can now prove our result:

Proposition 6.16 *The group homomorphism $Z_{\omega,B}$ is a stability function on $\mathrm{Coh}^{\omega,B}(X)$.*

Proof By definition, it is immediate to see that $\Im Z_{\omega,B} \geq 0$ on $\mathrm{Coh}^{\omega,B}(X)$. Moreover, if $E \in \mathrm{Coh}^{\omega,B}(X)$ is such that $\Im Z_{\omega,B}(E) = 0$, then E fits in an exact triangle

$$\mathcal{H}^{-1}(E)[1] \to E \to \mathcal{H}^0(E),$$

where $\mathcal{H}^{-1}(E)$ is a $\mu_{\omega,B}$-semistable torsion-free sheaf with $\mu_{\omega,B}(\mathcal{H}^{-1}(E)) = 0$ and $\mathcal{H}^0(E)$ is a torsion sheaf with zero-dimensional support. Here, as usual, $\mathcal{H}^\bullet$ denotes the cohomology sheaves of a complex.

We need to show $\Re Z_{\omega,B}(E) < 0$. Since $\Re Z_{\omega,B}$ is additive, we only need to show this for its cohomology sheaves. But, on the one hand, since $\mathcal{H}^0(E)$ is a torsion sheaf with support a zero-dimensional subscheme, we have

$$\Re Z_{\omega,B}(\mathcal{H}^0(E)) = -\operatorname{ch}_2(\mathcal{H}^0(E)) < 0.$$

On the other hand, by the Hodge Index Theorem, since $\omega \cdot \operatorname{ch}_1^B(\mathcal{H}^{-1}(E)) = 0$, we have $\operatorname{ch}_1^B(\mathcal{H}^{-1}(E))^2 \leq 0$. Therefore, by the Bogomolov inequality, we have $\operatorname{ch}_2^B(\mathcal{H}^{-1}(E)) \leq 0$. Hence,

$$\Re Z_{\omega,B}(\mathcal{H}^{-1}(E)[1]) = -\Re Z_{\omega,B}(\mathcal{H}^{-1}(E)) = \underbrace{\operatorname{ch}_2^B(\mathcal{H}^{-1}(E))}_{\leq 0} - \underbrace{\frac{\omega^2}{2}\operatorname{ch}_0^B(\mathcal{H}^{-1}(E))}_{>0} < 0,$$

thus concluding the proof. □

The Rational Case We now let $B \in \mathrm{NS}_{\mathbb{Q}}$ be a rational divisor class, and we let $\omega = \alpha H$, where $\alpha \in \mathbb{R}$ and $H \in \mathrm{NS}(X)$ is an integral ample divisor class.

We want to show that the group homomorphism $Z_{\omega,B}$ gives a stability condition in the abelian category $\mathrm{Coh}^{\omega,B}(X)$, in the sense of Definition 4.6. By using Proposition 6.16 above, we only need to show that HN filtrations exist. To this end, we use Proposition 4.10, and since $\Im Z_{\omega,B}$ is discrete under our assumptions, we only need to show the following.

Lemma 6.17 *Under the previous assumption, the tilted category* $\mathrm{Coh}^{\omega,B}(X)$ *is noetherian.*

Proof This is a general fact about tilted categories. It was first observed in the case of K3 surfaces in [31, Proposition 7.1]. We refer to [96, Lemma 2.17] for all details and just give a sketch of the proof. As observed above, under our assumptions, $\Im Z_{\omega,B}$ is discrete and non-negative on $\mathrm{Coh}^{\omega,B}(X)$. This implies that it is enough to show that, for any $M \in \mathrm{Coh}^{\omega,B}(X)$, there exists no infinite filtration

$$0 = A_0 \subsetneq A_1 \subsetneq \ldots \subsetneq A_l \subsetneq \ldots \subset M$$

with $\Im Z_{\omega,B}(A_l) = 0$.

Write $Q_l := M/A_l$. By definition, the exact sequence

$$0 \to A_l \to M \to Q_l \to 0$$

in $\mathrm{Coh}^{\omega,B}(X)$ induces a long exact sequence of sheaves

$$0 \to \mathcal{H}^{-1}(A_l) \to \mathcal{H}^{-1}(M) \to \mathcal{H}^{-1}(Q_l) \to \mathcal{H}^0(A_l) \to \mathcal{H}^0(M) \to \mathcal{H}^0(Q_l) \to 0.$$

Since $\mathrm{Coh}(X)$ is noetherian, we have that both sequences

$$\begin{aligned}\mathcal{H}^0(M) = \mathcal{H}^0(Q_0) \twoheadrightarrow \mathcal{H}^0(Q_1) \twoheadrightarrow \mathcal{H}^0(Q_2) \twoheadrightarrow \dots \\ 0 = \mathcal{H}^{-1}(A_0) \hookrightarrow \mathcal{H}^{-1}(A_1) \hookrightarrow \mathcal{H}^{-1}(A_2) \hookrightarrow \dots \hookrightarrow \mathcal{H}^{-1}(M)\end{aligned}$$

stabilize. Hence, from now on we will assume $\mathcal{H}^0(Q_l) = \mathcal{H}^0(Q_{l+1})$ and $\mathcal{H}^{-1}(A_l) = \mathcal{H}^{-1}(A_{l+1})$ for all l. Both $U := \mathcal{H}^{-1}(M)/\mathcal{H}^{-1}(A_l)$ and $V := \mathrm{Ker}(\mathcal{H}^0(M) \twoheadrightarrow \mathcal{H}^0(Q_l))$ are constant and we have an exact sequence

$$0 \to U \to \mathcal{H}^{-1}(Q_l) \to \mathcal{H}^0(A_l) \to V \to 0, \tag{3}$$

again by definition of $\mathrm{Coh}^{\omega,B}(X)$, since $\Im Z_{\omega,B}(A_l) = 0$, $\mathcal{H}^0(A_l)$ is a torsion sheaf supported in dimension 0. Write $B_l := A_l/A_{l-1}$. Again, by looking at the induced long exact sequence of sheaves and the previous observation, we deduce that

$$0 \to \mathcal{H}^{-1}(A_{l-1}) \xrightarrow{\cong} \mathcal{H}^{-1}(A_l) \xrightarrow{0} \underbrace{\mathcal{H}^{-1}(B_l)}_{\text{torsion-free}\Rightarrow=0} \to \underbrace{\mathcal{H}^0(A_{l-1})}_{\text{torsion}} \to \mathcal{H}^0(A_l) \to \mathcal{H}^0(B_l) \to 0.$$

Hence, the only thing we need is to show that $\mathcal{H}^0(B_l) = 0$, for all $l \gg 0$, or equivalently to bound the length of $\mathcal{H}^0(A_l)$. But, by letting $K_l := \mathrm{Ker}(\mathcal{H}^0(A_l) \twoheadrightarrow V)$, (3) gives an exact sequence of sheaves

$$0 \to U \to \mathcal{H}^{-1}(Q_l) \to K_l \to 0.$$

where K_l is a torsion sheaf supported in dimension 0 as well and $\mathcal{H}^{-1}(Q_l)$ is torsion-free. This gives a bound on the length of K_l and therefore a bound on the length of $\mathcal{H}^0(A_l)$, as we wanted. □

To finish the proof that, under our rationality assumptions, $\sigma_{\omega,B}$ is a Bridgeland stability condition, we still need to prove the support property, namely Theorem 6.13 in our case. The idea is to let $\alpha \to \infty$ and use the Bogomolov theorem for sheaves (together with Remark 6.15). First we use the following lemma (we will give a more precise statement in Sect. 6.4). It first appeared in the K3 surface case as a particular case of [31, Proposition 14.2].

Lemma 6.18 *Let $\omega, B \in N^1(X)$ with ω ample. If $E \in \mathrm{Coh}^{\omega,B}(X)$ is $\sigma_{\alpha\cdot\omega,B}$-semistable for all $\alpha \gg 0$, then it satisfies one of the following conditions:*

(1) *$\mathcal{H}^{-1}(E) = 0$ and $\mathcal{H}^0(E)$ is a $\mu_{\omega,B}$-semistable torsion-free sheaf.*
(2) *$\mathcal{H}^{-1}(E) = 0$ and $\mathcal{H}^0(E)$ is a torsion sheaf.*
(3) *$\mathcal{H}^{-1}(E)$ is a $\mu_{\omega,B}$-semistable torsion-free sheaf and $\mathcal{H}^0(E)$ is either 0 or a torsion sheaf supported in dimension zero.*

Proof One can compute $\sigma_{\alpha\omega,B}$-stability with slope $\frac{2\nu_{\alpha\omega,B}}{\alpha}$ instead of $\nu_{\alpha\omega,B}$. This is convenient in the present argument because

$$\lim_{\alpha\to\infty} \frac{2\nu_{\alpha\omega,B}}{\alpha}(E) = -\mu^{-1}_{\omega,B}(E).$$

By definition of $\mathrm{Coh}^{\omega,B}(X)$ the object E is an extension $0 \to F[1] \to E \to T \to 0$, where $F \in \mathcal{F}_{\omega,B}$ and $T \in \mathcal{T}_{\omega,B}$.

Assume that $\omega \cdot \mathrm{ch}_1^B(E) = 0$. Then both $\omega \cdot \mathrm{ch}_1^B(F) = 0$ and $\omega \cdot \mathrm{ch}_1^B(T) = 0$. By definition of $\mathcal{F}_{\omega,B}$ and $\mathcal{T}_{\omega,B}$ this means T is 0 or has be supported in dimension 0 and F is 0 or a $\mu_{\omega,B}$-semistable torsion free sheaf. Therefore, for the rest of the proof we can assume $\omega \cdot \mathrm{ch}_1^B(E) > 0$.

Assume that $\mathrm{ch}_0^B(E) \geq 0$. By definition $-\mu_{\omega,B}^{-1}(F[1]) \geq 0$. The inequality $\omega \cdot \mathrm{ch}_1^B(E) > 0$ implies $-\mu_{\omega,B}^{-1}(E) < 0$. Since E is $\sigma_{\alpha\omega,B}$-semistable for $\alpha \gg 0$, we get $F = 0$ and $E \in \mathcal{T}_{\omega,B}$ is a sheaf. If E is torsion, we are in case (2). Assume E is neither torsion nor slope semistable. Then by definition of $\mathcal{T}_{\omega,B}$ there is an exact sequence

$$0 \to A \to E \to B \to 0$$

in $\mathcal{T}_{\omega,B} = \mathrm{Coh}^{\omega,B}(X) \cap \mathrm{Coh}(X)$ such that $\mu_{\omega,B}(A) > \mu_{\omega,B}(E) > 0$. But then $-\mu_{\omega,B}^{-1}(A) > -\mu_{\omega,B}^{-1}(E)$ contradicts the fact that E is $\sigma_{\alpha\omega,B}$-semistable for $\alpha \gg 0$.

Assume that $\mathrm{ch}_0^B(E) < 0$. If $\omega \cdot \mathrm{ch}_1^B(T) = 0$, then $T \in \mathcal{T}_{\omega,B}$ implies $\mathrm{ch}_0^B(T) = 0$ and up to semistability of F we are in case (3). Let $\omega \cdot \mathrm{ch}_1^B(T) > 0$. By definition $-\mu_{\omega,B}^{-1}(T) < 0$. As before we have $-\mu_{\omega,B}^{-1}(T) > 0$ and the fact that E is $\sigma_{\alpha\omega,B}$-semistable for $\alpha \gg 0$ implies $T = 0$. In either case we are left to show that F is $\mu_{\omega,B}$-semistable. If not, there is an exact sequence

$$0 \to A \to F \to B \to 0$$

in $\mathcal{F}_{\omega,B} = \mathrm{Coh}^{\omega,B}(X)[-1] \cap \mathrm{Coh}(X)$ such that $\mu_{\omega,B}(A) > \mu_{\omega,B}(F)$. Therefore, there is an injective map $A[1] \hookrightarrow E$ in $\mathrm{Coh}^{\omega,B}(X)$ such that $-\mu_{\omega,B}^{-1}(A[1]) > -\mu_{\omega,B}^{-1}(E)$ in contradiction to the fact that E is $\sigma_{\alpha\omega,B}$-semistable for $\alpha \gg 0$. □

Exercise 6.19 Let $B \in \mathrm{NS}(X)_{\mathbb{Q}}$ be a rational divisor class and let $\omega = \alpha H$, where $\alpha \in \mathbb{R}_{>0}$ and $H \in \mathrm{NS}(X)_{\mathbb{Q}}$ is an integral ample divisor. Show that

$$c := \min \left\{ H \cdot \mathrm{ch}_1^B(F) : \begin{array}{c} F \in \mathrm{Coh}^{\omega,B}(X) \\ H \cdot \mathrm{ch}_1^B(F) > 0 \end{array} \right\} > 0.$$

exists. Let $E \in \mathrm{Coh}^{\omega,B}(X)$ satisfy $H \cdot \mathrm{ch}_1^B(E) = c$ and $\mathrm{Hom}(A, E) = 0$, for all $A \in \mathrm{Coh}^{\omega,B}(X)$ with $H \cdot \mathrm{ch}_1^B(A) = 0$. Show that E is $\sigma_{\omega,B}$-stable.

We can now prove Theorem 6.13 (and so Theorem 6.10) in the rational case.

Proof of Theorem 6.13, rational case Since B is rational and $\omega = \alpha H$, with H being an integral divisor, the imaginary part is a discrete function. We proceed by induction on $H \cdot \mathrm{ch}_1^B$.

Let $E \in \mathrm{Coh}^{H,B}(X)$ be $\sigma_{\alpha_0 H,B}$-semistable for some $\alpha_0 > 0$ such that $H \cdot \mathrm{ch}_1^B > 0$ is minimal. Then by Exercise 6.19, E is stable for all $\alpha \gg 0$. Therefore, by Lemma 6.18, both $\Delta_{\omega,B}^C(E) \geq 0$ and $\overline{\Delta}_{\omega}^B(E) \geq 0$ hold, since they are true for semistable sheaves.

The induction step is subtle: we would like to use the wall and chamber structure. If an object E is stable for all $\alpha \gg 0$, then again by Lemma 6.18 we are done. If not, it is destabilized at a certain point. By Lemma 5.29, we can then proceed by induction on $H \cdot \mathrm{ch}_1^B$. The issue is that we do not know the support property yet and therefore, that walls are indeed well behaved. Fortunately, as α increases, all possible destabilizing subobjects and quotients have strictly smaller $H \cdot \mathrm{ch}_1^B$, which satisfy the desired inequality by our induction assumption. This is enough to ensure that E satisfies well-behaved wall-crossing. By following the same argument as in Proposition 5.27 (and Remark 5.28), it is enough to know a support property type statement for all potentially destabilizing classes. □

The General Case We only sketch the argument for the general case. This is explained in detail in the case of K3 surfaces in [31], and can be directly deduced from [30, Sects. 6 and 7].

The idea is to use the support property we just proved in the rational case to deform B and ω in all possible directions, by using Bridgeland's Deformation Theorem 5.15 (and Proposition 5.16). The only thing we have to make sure is that once we have the correct central charge, the category is indeed $\mathrm{Coh}^{\omega,B}(X)$. The intuition behind this is in the following result.

Lemma 6.20 *Let $\sigma = (\mathcal{A}, Z_{\omega,B})$ be a stability condition satisfying the support property such that all skyscraper sheaves are stable of phase one. Then $\mathcal{A} = \mathrm{Coh}^{\omega,B}(X)$ holds.*

Proof We start by showing that all objects $E \in \mathcal{A}$ have $H^i(E) = 0$ whenever $i \neq 0, -1$ and that $H^{-1}(E)$ is torsion free. Notice that E is an iterated extension of stable objects, so we can assume that E is stable. This is clearly true for skyscraper sheaves, so we can assume that E is not a skyscraper sheaf. Then Serre duality implies that for $i \neq 0, 1$ and any $x \in X$ we have

$$\mathrm{Ext}^i(E, \mathbb{C}(x)) = \mathrm{Ext}^{2-i}(\mathbb{C}(x), E) = 0.$$

By Proposition 10 we get that E is isomorphic to a two term complex of locally free sheaves and the statement follows. This also implies that $H^{-1}(E)$ is torsion free.

Therefore, the inclusion $\mathcal{A} \subset \langle \mathrm{Coh}(X), \mathrm{Coh}(X)[1] \rangle$ holds. Set $\mathcal{T} = \mathrm{Coh}(X) \cap \mathcal{A}$ and $\mathcal{F} = \mathrm{Coh}(X) \cap \mathcal{A}[-1]$. By Exercise 6.5 we get $\mathcal{A} = \langle \mathcal{T}, \mathcal{F}[1] \rangle$ and $\mathrm{Coh}(X) = \langle \mathcal{T}, \mathcal{F} \rangle$. We need to show $\mathcal{T} = \mathcal{T}_{\omega,B}$ and $\mathcal{F} = \mathcal{F}_{\omega,B}$. In fact, it will be enough to show $\mathcal{T}_{\omega,B} \subset \mathcal{T}$ and $\mathcal{F}_{\omega,B} \subset \mathcal{F}$.

Let $E \in \mathrm{Coh}(X)$ be slope semistable. There is an exact sequence $0 \to T \to E \to F \to 0$ with $T \in \mathcal{T}$ and $F \in \mathcal{F}$. We already showed that $F = H^{-1}(F)$ is torsion free. If E is torsion this implies $F = 0$ and $E = T \in \mathcal{T}$ as claimed.

Assume that E is torsion free. Since $F[1] \in \mathcal{A}$ and $T \in \mathcal{A}$, we get $\mu_{\omega,B}(F) \leq 0$ and $\mu_{\omega,B}(T) \geq 0$. This is a contradiction to E being stable unless either $F = 0$ or $T = 0$. Therefore, we showed $E \in \mathcal{F}$ or $E \in \mathcal{T}$.

If $\omega \cdot \mathrm{ch}_1^B(E) > 0$, then $E \in \mathcal{T}$ and if $\omega \cdot \mathrm{ch}_1^B(E) < 0$, then $E \in \mathcal{F}$. Assume that $\omega \cdot \mathrm{ch}_1^B(E) = 0$, but $E \in \mathcal{T}$. Then by definition of a stability condition we have

$Z_{\omega,B}(E) \in \mathbb{R}_{<0}$ and E is σ-semistable. Since E is a sheaf, there is a skyscraper sheaf $\mathbb{C}(x)$ together with a surjective morphism of coherent sheaves $E \twoheadrightarrow \mathbb{C}(x)$. Since $\mathbb{C}(x)$ is stable of slope ∞ this morphism is also a surjection in $\mathcal{A}$. Let $F \in \mathcal{A} \cap \mathrm{Coh}(X) = \mathcal{T}$ be the kernel of this map. Then $Z(F) = Z(E) + 1$. Iterating this procedure will lead to an object F with $Z(F) \in \mathbb{R}_{\geq 0}$, a contradiction. □

6.4 The Wall and Chamber Structure in the (α, β)-Plane

If we consider a certain slice of the space of Bridgeland stability conditions, the structure of the walls turns out rather simple.

Definition 6.21 Let $H \in \mathrm{NS}(X)$ be an ample integral divisor class and let $B_0 \in \mathrm{NS}_{\mathbb{Q}}$. We define the (α, β)-plane as the set of stability conditions of the form $\sigma_{\alpha H, B_0 + \beta H}$, for $\alpha, \beta \in \mathbb{R}$, $\alpha > 0$.

When it is clear from context which (α, β) plane we choose (for example, if the Picard number is one), we will abuse notation and drop H and B_0 from the notation; for example, we denote stability conditions by $\sigma_{\alpha,\beta}$, the twisted Chern character by ch^{β}, etc.

The following proposition describes all walls in the (α, β)-plane. It was originally observed by Bertram and completely proved in [77].

Proposition 6.22 (Structure Theorem for Walls on Surfaces) *Fix a class* $v \in K_{\mathrm{num}}(X)$.

(1) *All numerical walls are either semicircles with center on the* β*-axis or vertical rays.*
(2) *Two different numerical walls for* v *cannot intersect.*
(3) *For a given class* $v \in K_{\mathrm{num}}(X)$ *the hyperbola* $\Re Z_{\alpha,\beta}(v) = 0$ *intersects all numerical semicircular walls at their top points.*
(4) *If* $\mathrm{ch}_0^{B_0}(v) \neq 0$, *then there is a unique numerical vertical wall defined by the equation*

$$\beta = \frac{H \, \mathrm{ch}_1^{B_0}(v)}{H^2 \, \mathrm{ch}_0^{B_0}(v)}.$$

(5) *If* $\mathrm{ch}_0^{B_0}(v) \neq 0$, *then all semicircular walls to either side of the unique numerical vertical wall are strictly nested semicircles.*
(6) *If* $\mathrm{ch}_0^{B_0}(v) = 0$, *then there are only semicircular walls that are strictly nested.*
(7) *If a wall is an actual wall at a single point, it is an actual wall everywhere along the numerical wall.*

Exercise 6.23 Prove Proposition 6.22. *(Hint: For (2) ignore the slope and rephrase everything with just* Z *using linear algebra. For (7) show that a destabilizing subobject or quotient would have to destabilize itself at some point of the wall.)*

As application, it is easy to show that a "largest wall" exists and we will also be able to prove that walls are locally finite in the surface case without using Proposition 5.27. Both will directly follow from the following statement that is taken from [102, Lemma 5.6].

Lemma 6.24 *Let $v \in K_{\mathrm{num}}(X)$ be a non-zero class such that $\overline{\Delta}_H^{B_0}(v) \geq 0$. For fixed $\beta_0 \in \mathbb{Q}$ there are only finitely many walls intersecting the vertical line $\beta = \beta_0$.*

Proof Any wall has to come from an exact sequence $0 \to F \to E \to G \to 0$ in $\mathrm{Coh}^\beta(X)$. Let $(H^2 \cdot \mathrm{ch}_0^\beta(E), H \cdot \mathrm{ch}_1^\beta(E), \mathrm{ch}_2^\beta(E)) = (R, C, D)$ and $(H^2 \cdot \mathrm{ch}_0^\beta(F), H \cdot \mathrm{ch}_1^\beta(F), \mathrm{ch}_2^\beta(F)) = (r, c, d)$. Notice that due to the fact that $\beta \in \mathbb{Q}$ the possible values of r, c and d are discrete in $\mathbb{R}$. Therefore, it will be enough to bound those values to obtain finiteness.

By definition of $\mathrm{Coh}^\beta(X)$ one has $0 \leq c \leq C$. If $C = 0$, then $c = 0$ and we are dealing with the unique vertical wall. Therefore, we may assume $C \neq 0$. Let $\Delta := C^2 - 2RD$. The Bogomolov inequality together with Lemma 5.29 implies $0 \leq c^2 - 2rd \leq \Delta$. Therefore, we get

$$\frac{c^2}{2} \geq rd \geq \frac{c^2 - \Delta}{2}.$$

Since the possible values of r and d are discrete in $\mathbb{R}$, there are finitely many possible values unless $r = 0$ or $d = 0$.

Assume $R = r = 0$. Then the equality $\nu_{\alpha,\beta}(F) = \nu_{\alpha,\beta}(E)$ holds if and only if $Cd - Dc = 0$. In particular, it is independent of (α, β). Therefore, the sequence does not define a wall.

If $r = 0$, $R \neq 0$, and $D - d \neq 0$, then using the same type of inequality for G instead of E will finish the proof. If $r = 0$ and $D - d = 0$, then $d = D$ and there are only finitely many walls like this, because we already bounded c.

Assume $D = d = 0$. Then the equality $\nu_{\alpha,\beta}(F) = \nu_{\alpha,\beta}(E)$ holds if and only if $Rc - Cr = 0$. Again this cannot define a wall.

If $d = 0$, $D \neq 0$, and $R - r \neq 0$, then using the same type of inequality for G instead of E will finish the proof. If $d = 0$ and $R - r = 0$, then $r = R$ and there are only finitely many walls like this, because we already bounded c. □

Corollary 6.25 *Let $v \in K_{\mathrm{num}}(X)$ be a non-zero class such that $\overline{\Delta}_H^{B_0}(v) \geq 0$. Then semicircular walls in the (α, β)-plane with respect to v are bounded from above.*

Corollary 6.26 *Let $v \in K_{\mathrm{num}}(X)$ be a non-zero class such that $\overline{\Delta}_H^{B_0}(v) \geq 0$. Walls in the (α, β)-plane with respect to v are locally finite, i.e., any compact subset of the upper half plane intersects only finitely many walls.*

We will compute the largest wall in examples: see Sect. 7.1. An immediate consequence of Proposition 6.25 is the following precise version of Lemma 6.18. This was proved in the case of K3 surfaces in [31, Proposition 14.2]. The general proof is essentially the same if the statement is correctly adjusted.

Exercise 6.27 Let $v \in K_{\mathrm{num}}(X)$ be a non-zero class with positive rank and let $\beta \in \mathbb{R}$ such that $H \cdot \mathrm{ch}_1^{\beta}(v) > 0$. Then there exists $\alpha_0 > 0$ such that for any $\alpha > \alpha_0$ the set of $\sigma_{\alpha,\beta}$-semistable objects with class v is the same as the set of twisted $(\omega, B_0 - \frac{1}{2}K_X)$-Gieseker semistable sheaves with class v. Moreover, $\sigma_{\alpha,\beta}$-stable objects of class v are the same as twisted $(\omega, B_0 - \frac{1}{2}K_X)$-Gieseker stable sheaves with class v. *Hint: Follow the proof of Lemma 6.18 and compare lower terms.*

6.5 Examples of Semistable Objects

We already saw a few easy examples of stable objects with respect to $\sigma_{\omega,B}$: skyscraper sheaves and objects with minimal $H \cdot \mathrm{ch}_1^B$ (or with $H \cdot \mathrm{ch}_1^B = 0$). See Exercises 6.9 and 6.19.

The key example of Bridgeland semistable objects are those with trivial discriminant.

Lemma 6.28 *Let E be a $\mu_{\omega,B}$-stable vector bundle. Assume that either $\Delta^C_{\omega,B}(E) = 0$, or $\overline{\Delta}^B_H(E) = 0$. Then E is $\sigma_{\omega,B}$-stable.*

Proof We can assume ω and B to be rational. Consider the (α, β)-plane. By Exercise 6.27, E (or $E[1]$) is stable for $\alpha \gg 0$. The statement now follows directly from Corollary 5.30. □

In particular, all line bundles are stable everywhere only if the Néron-Severi group is of rank one, or if the constant C_ω of Exercise 6.11 is zero (e.g., for abelian surfaces).

Exercise 6.29 Let $\sigma: X \to \mathbb{P}^2$ be the blow up of $\mathbb{P}^2$ at one point. Let h be the pull-back $\sigma^*\mathcal{O}_{\mathbb{P}^2}(1)$ and let f denote the fiber of the $\mathbb{P}^1$-bundle $X \to \mathbb{P}^1$. Consider the anti-canonical line bundle $H := -K_X = 2h + f$, and consider the (α, β)-plane with respect to $\omega = \alpha H$, $B = \beta H$. Show that $\mathcal{O}_X(h)$ is $\sigma_{\alpha,\beta}$-stable for all α, β, while there exists (α_0, β_0) for which $\mathcal{O}(2h)$ is not $\sigma_{\alpha_0,\beta_0}$-semistable.

6.6 Moduli Spaces

We keep the notation as in the beginning of this section, namely $\omega, B \in N^1(X)$ with ω ample. We fix $v_0 \in K_{\mathrm{num}}(X)$ and $\phi \in \mathbb{R}$. Consider the stack $\mathfrak{M}_{\omega,B}(v_0, \phi) := \mathfrak{M}_{\sigma_{\omega,B}}(v_0, \phi)$ (and $\mathfrak{M}^s_{\omega,B}(v_0, \phi)$) as in Sect. 5.4.

Theorem 6.30 (Toda) *$\mathfrak{M}_{\omega,B}(v_0, \phi)$ is a universally closed Artin stack of finite type over $\mathbb{C}$. Moreover, $\mathfrak{M}^s_{\omega,B}(v_0, \phi)$ is a $\mathbb{G}_m$-gerbe over an algebraic space $M_{\omega,B}(v_0, \phi)$. Finally, if $\mathfrak{M}_{\omega,B}(v_0, \phi) = \mathfrak{M}^s_{\omega,B}(v_0, \phi)$, then $M_{\omega,B}(v_0, \phi)$ is a proper algebraic space over $\mathbb{C}$.*

Ideas from the proof As we observed after Question 5.23, we only need to show openness and boundedness. The idea to show boundedness is to reduce to semistable sheaves and use boundedness from them. For openness, the key technical result is a construction by Abramovich-Polishchuk [1], which we will recall in Theorem 8.2. Then openness for Bridgeland semistable objects follows from the existence of relative Harder-Narasimhan filtrations for sheaves. We refer to [109] for all the details. □

There are only few examples where we can be more precise on the existence of moduli spaces. We will present a few of them below. To simplify notation, we will drop from now on the phase ϕ from the notation for a moduli space.

The Projective Plane Let $X = \mathbb{P}^2$. We identify the lattice $K_0(\mathbb{P}^2) = K_{\mathrm{num}}(\mathbb{P}^2)$ with $\mathbb{Z}^{\oplus 2} \oplus \frac{1}{2}\mathbb{Z}$, and the Chern character with a triple $\mathrm{ch} = (r, c, s)$, where r is the rank, c is the first Chern character, and s is the second Chern character. We fix $v_0 \in K_{\mathrm{num}}(\mathbb{P}^2)$.

Theorem 6.31 *For all $\alpha, \beta \in \mathbb{R}$, $\alpha > 0$, there exists a coarse moduli space $M_{\alpha,\beta}(v_0)$ parameterizing S-equivalence classes of $\sigma_{\alpha,\beta}$-semistable objects. It is a projective variety. Moreover, if v_0 is primitive and $\sigma_{\alpha,\beta}$ is outside a wall for v_0, then $M^s_{\alpha,\beta}(v_0) = M_{\alpha,\beta}(v_0)$ is a smooth irreducible projective variety.*

The projectivity was first observed in [8], while generic smoothness is proved in this generality in [75]. For the proof a GIT construction is used, but with a slightly different GIT problem. First of all, an immediate consequence of Proposition 6.22 is the following (we leave the details to the reader).

Exercise 6.32 Given $\alpha, \beta \in \mathbb{R}, \alpha > 0$, there exist $\alpha_0, \beta_0 \in \mathbb{R}$, such that $0 < \alpha_0 < \frac{1}{2}$ and $\widehat{M}_{\alpha,\beta}(v_0) = \widehat{M}_{\alpha_0,\beta_0}(v_0)$.

For $0 < \alpha < \frac{1}{2}$, Bridgeland stability on the projective plane is related to finite dimensional algebras, where moduli spaces are easy to construct via GIT (see Exercise 4.13). For $k \in \mathbb{Z}$, we consider the vector bundle $\mathcal{E} := \mathcal{O}_{\mathbb{P}^2}(k-1) \oplus \Omega_{\mathbb{P}^2}(k+1) \oplus \mathcal{O}_{\mathbb{P}^2}(k)$. Let A be the finite-dimensional associative algebra $\mathrm{End}(E)$. Then, by the Beilinson Theorem [21], the functor

$$\Phi_{\mathcal{E}} : \mathrm{D^b}(\mathbb{P}^2) \to \mathrm{D^b}(\text{mod-}A), \ \Phi_{\mathcal{E}}(F) := \mathbf{R}\mathrm{Hom}(\mathcal{E}, F)$$

is an equivalence of derived categories. The simple objects in the category $\Phi_{\mathcal{E}}^{-1}(\text{mod-}A)$ are $\mathcal{O}_{\mathcal{P}^2}(k-3)[2], \mathcal{O}_{\mathcal{P}^2}(k-2)[1], \mathcal{O}_{\mathcal{P}^2}(k-1)$.

Exercise 6.33 For $\alpha, \beta \in \mathbb{R}$, $0 < \alpha < \frac{1}{2}$, there exist $k \in \mathbb{Z}$ and an element $A \in \widetilde{\mathrm{GL}}^+(2, \mathbb{R})$ such that $A \cdot \sigma_{\alpha,\beta}$ is a stability condition as in Exercise 4.13.

It is not hard to prove that families of Bridgeland semistable objects coincide with families of modules over an algebra, as defined in [66]. Therefore, this proves the first part of Theorem 6.31. The proof of smoothness in the generic primitive case is more subtle. The complete argument is given in [75].

K3 Surfaces Let X be a K3 surface. We consider the algebraic Mukai lattice

$$H^*_{\mathrm{alg}}(X) := H^0(X,\mathbb{Z}) \oplus \mathrm{NS}(X) \oplus H^4(X,\mathbb{Z})$$

together with the Mukai pairing

$$\left((r,c,s),(r',c',s')\right) = c.c' - rs' - sr'.$$

We also consider the *Mukai vector*, a modified version of the Chern character by the square root of the Todd class:

$$v := \mathrm{ch}\cdot\sqrt{\mathrm{td}_X} = (\mathrm{ch}_0, \mathrm{ch}_1, \mathrm{ch}_2 + \mathrm{ch}_0)\,.$$

Our lattice Λ is $H^*_{\mathrm{alg}}(X)$ and the map v is nothing but the Mukai vector.

Bridgeland stability conditions for K3 surfaces can be described beyond the ones we just defined using ω, B. More precisely, the main result in [31] is that there exists a connected component $\mathrm{Stab}^\dagger(X)$, containing all stability conditions $\sigma_{\omega,B}$ such that the map

$$\eta\colon \mathrm{Stab}^\dagger(X) \xrightarrow{Z} \mathrm{Hom}(H^*_{\mathrm{alg}}(X),\mathbb{C}) \xrightarrow{(-,-)} H^*_{\mathrm{alg}}(X)_{\mathbb{C}}$$

is a covering onto its image, which can be described as a certain period domain.

Theorem 6.34 *Let $v_0 \in H^*_{\mathrm{alg}}(X)$. Then for all generic stability conditions $\sigma \in$ $\mathrm{Stab}^\dagger(X)$, there exists a coarse moduli space $M_\sigma(v_0)$ parameterizing S-equivalence classes of σ-semistable objects. It is a projective variety. Moreover, if v_0 is primitive, then $M^s_\sigma(v_0) = M_\sigma(v_0)$ is a smooth integral projective variety.*

We will not prove Theorem 6.34; we refer to [17]. The idea of the proof, based on [86], is to reduce to the case of semistable sheaves by using a Fourier-Mukai transform. The corresponding statement for non-generic stability conditions in $\mathrm{Stab}^\dagger(X)$ is still unknown.

When the vector is primitive, varieties appearing as moduli spaces of Bridgeland stable objects are so called *Irreducible Holomorphic Symplectic*. They are all deformation equivalent to Hilbert schemes of points on a K3 surface.

7 Applications and Examples

In this section we will give examples and applications for studying stability on surfaces.

7.1 The Largest Wall for Hilbert Schemes

Let X be a smooth complex projective surface of Picard rank one. We will deal with computing the largest wall for ideal sheaves of zero dimensional schemes $Z \subset X$. The moduli space of these ideal sheaves turns out to be the Hilbert scheme of points on X. The motivation for this problem lies in understanding its nef cone. We will explain in the next section how, given a stability condition σ, one constructs nef divisors on moduli spaces of σ-semistable objects. It turns out that stability conditions on walls will often times induce boundary divisors of the nef cone. If the number of points is large, this was done in [25].

Proposition 7.1 *Let X be a smooth complex projective surface with* $\mathrm{NS}(X) = \mathbb{Z}\cdot H$, *where H is ample. Moreover, let $a > 0$ be the smallest integer such that aH is effective and $n > a^2H^2$. Then the biggest wall for the Chern character $(1, 0, -n)$ to the left of the unique vertical wall is given by the equation $\nu_{\alpha,\beta}(\mathcal{O}(-aH)) = \nu_{\alpha,\beta}(1, 0, -n)$. Moreover, the ideal sheaves $\mathcal{I}_Z$ that destabilize at this wall are exactly those for which $Z \subset C$ for a curve $C \in |aH|$.*

We need the following version of a result from [38].

Lemma 7.2 *Let $0 \to F \to E \to G \to 0$ be an exact sequence in* $\mathrm{Coh}^{\beta}(X)$ *defining a non empty semicircular wall W. Assume further that* $\mathrm{ch}_0(F) > \mathrm{ch}_0(E) \geq 0$. *Then the radius ρ_W satisfies the inequality*

$$\rho_W^2 \leq \frac{\overline{\Delta}_H(E)}{4H^2 \cdot \mathrm{ch}_0(F)(H^2 \cdot \mathrm{ch}_0(F) - H^2 \cdot \mathrm{ch}_0(E))}.$$

Proof *Let $v, w \in K_0(X)$ be two classes such that the wall W given by $\nu_{\alpha,\beta}(v) = \nu_{\alpha,\beta}(w)$ is a non empty semicircle. Then a straightforward computation shows that the radius ρ_W and center s_W satisfy the equation*

$$(H^2 \cdot \mathrm{ch}_0(v))^2 \rho_W^2 + \overline{\Delta}_H(v) = (H^2 \cdot \mathrm{ch}_0(v) s_W - H \cdot \mathrm{ch}_1(v))^2. \tag{4}$$

For all $(\alpha, \beta) \in W$ we have the inequalities $H \cdot \mathrm{ch}_1^{\beta}(E) \geq H \cdot \mathrm{ch}_1^{\beta}(F) \geq 0$. This can be rewritten as

$$H \cdot \mathrm{ch}_1(E) + \beta(H^2 \cdot \mathrm{ch}_0(F) - H^2 \cdot \mathrm{ch}_0(E)) \geq H \cdot \mathrm{ch}_1(F) \geq \beta H^2 \cdot \mathrm{ch}_0(F).$$

Since $H \cdot \mathrm{ch}_1(F)$ is independent of β we can maximize the right hand side and minimize the left hand side individually in the full range of β between $s_W - \rho_W$ and $s_W + \rho_W$. By our assumptions this leads to

$$H \cdot \mathrm{ch}_1(E) + (s_W - \rho_W)(H^2 \cdot \mathrm{ch}_0(F) - H^2 \cdot \mathrm{ch}_0(E)) \geq (s_W + \rho_W) H^2 \cdot \mathrm{ch}_0(F).$$

By rearranging the terms and squaring we get

$$(2H^2\cdot \mathrm{ch}_0(F)-H^2\cdot \mathrm{ch}_0(E))^2\rho_W^2 \leq (H\cdot \mathrm{ch}_1(E)-H^2\cdot \mathrm{ch}_0(E)s_W)^2 = (H^2\cdot \mathrm{ch}_0(E))^2\rho_W^2+\overline{\Delta}_H(E).$$

The claim follows by simply solving for ρ_W^2. □

Proof of Proposition 7.1 We give the proof in the case where H is effective, i.e., $a = 1$. The general case is longer, but not substantially harder. The full argument can be found in [25].

The equation $\nu_{\alpha,\beta}(1, 0, -n) = \nu_{\alpha,\beta}(\mathcal{O}(-H))$ is equivalent to

$$\alpha^2 + \left(\beta + \frac{1}{2} + \frac{n}{H^2}\right)^2 = \left(\frac{n}{H^2} - \frac{1}{2}\right)^2. \tag{5}$$

This shows that every larger semicircle intersects the line $\beta = -1$. Moreover, the object $\mathcal{O}(-H)$ is in the category along the wall if and only if $n > \frac{H^2}{2}$.

We will first show that there is no bigger semicircular wall. Assume we have an exact sequence

$$0 \to F \to \mathcal{I}_Z \to G \to 0$$

where $Z \subset X$ has dimension 0 and length n. Moreover, assume the equation $\nu_{\alpha,-1}(F) = \nu_{\alpha,-1}(G)$ has a solution $\alpha > 0$. We have $\mathrm{ch}^{-1}(E) = (1, H, \frac{H^2}{2} - n)$. By definition of $\mathrm{Coh}^{-1}(X)$ we have $H \cdot \mathrm{ch}_1^{-1}(F), H \cdot \mathrm{ch}_1^{-1}(G) \geq 0$ and both those numbers add up to $H \cdot \mathrm{ch}_1^{-1}(E) = H^2$. Since H is the generator of the Picard group, this implies either $H \cdot \mathrm{ch}_1^{-1}(F) = 0$ or $H \cdot \mathrm{ch}_1^{-1}(G) = 0$. In particular, either F or G have slope infinity and it is impossible for F, E and G to have the same slope for $\beta = -1$ and $\alpha > 0$.

Next, assume that $0 \to F \to \mathcal{I}_Z \to G \to 0$ induces the wall W. By the long exact sequence in cohomology F is a torsion free sheaf. By Lemma 7.2 the inequality $\mathrm{ch}_0(F) \geq 2$ leads to

$$\rho^2 \leq \frac{2H^2 n}{8(H^2)^2} = \frac{n}{4H^2} < \left(\frac{n}{H^2} - \frac{1}{2}\right)^2.$$

Therefore, any such sequence giving the wall $\nu_{\alpha,\beta}(1, 0, -n) = \nu_{\alpha,\beta}(\mathcal{O}(-H))$ must satisfy $\mathrm{ch}_0(F) = 1$. Moreover, we must also have $H\cdot \mathrm{ch}_1^{-1}(F) = H\cdot \mathrm{ch}_1(F)+H^2 \geq 0$ and $H \cdot \mathrm{ch}_1^{-1}(G) = -H \cdot \mathrm{ch}_1(F) \geq 0$. A simple calculation shows that $\mathrm{ch}_1(F) = 0$ does not give the correct wall and therefore, $\mathrm{ch}_1(F) = -H$. Another straightforward computation implies that only $\mathrm{ch}_2(F) = \frac{H^2}{2}$ defines the right wall numerically. Since F is a torsion free sheaf, this means $F = \mathcal{O}(-H)$ implying the claim. □

Exercise 7.3 Any subscheme $Z \subset \mathbb{P}^n$ of dimension 0 and length 4 is contained in a quadric. Said differently there is a morphism $\mathcal{O}(-2) \hookrightarrow \mathcal{I}_Z$, i.e., no ideal sheaf is stable below the wall $\nu_{\alpha,\beta}(\mathcal{O}(-2)) = \nu_{\alpha,\beta}(\mathcal{I}_Z)$. Note that $\mathrm{ch}(\mathcal{I}_Z) = (1, 0, -4)$. The goal of this exercise, is to compute all bigger walls for this Chern character.

(1) Compute the equation of the wall $\nu_{\alpha,\beta}(\mathcal{O}(-2)) = \nu_{\alpha,\beta}(\mathcal{I}_Z)$. Why do all bigger walls intersect the line $\beta = -2$?
(2) Show that there are two walls bigger than $\nu_{\alpha,\beta}(\mathcal{O}(-2)) = \nu_{\alpha,\beta}(\mathcal{I}_Z)$. *Hint: Let* $0 \to F \to \mathcal{I}_Z \to G \to 0$ *define a wall for some* $\alpha > 0$ *and* $\beta = -2$. *Then* F *and* G *are semistable, i.e. they satisfy the Bogomolov inequality. Additionally show that* $0 < \mathrm{ch}_1^{\beta}(F) < \mathrm{ch}_1^{\beta}(E)$.
(3) Determine the Jordan-Hölder filtration of any ideal sheaf $\mathcal{I}_Z$ that destabilizes at any of these three walls. What do these filtrations imply about the geometry of Z? *Hint: Use the fact that a semistable sheaf on* $\mathbb{P}^2$ *with Chern character* $n \cdot \mathrm{ch}(\mathcal{O}(-m))$ *has to be* $\mathcal{O}(-m)^{\oplus n}$.

7.2 *Kodaira Vanishing*

As another application, we will give a proof of Kodaira vanishing for surfaces using tilt stability. The argument was first pointed out in [6]. While it is a well-known argument by Mumford that Kodaira vanishing in the surface case is a consequence of Bogomolov's inequality, this proof follows a slightly different approach.

Theorem 7.4 (Kodaira Vanishing for Surfaces) *Let* X *be a smooth projective complex surface and* H *an ample divisor on* X. *If* K_X *is the canonical divisor class of* X, *then the vanishing*

$$H^i(\mathcal{O}(H + K_X)) = 0$$

holds for all $i > 0$.

Proof By Serre duality $H^2(\mathcal{O}(H + K_X)) = H^0(\mathcal{O}(-H))$. Since anti-ample divisors are never effective we get $H^0(\mathcal{O}(-H)) = 0$.

The same way Serre duality implies $H^1(\mathcal{O}(H + K_X)) = H^1(\mathcal{O}(-H)) = \mathrm{Hom}(\mathcal{O}, \mathcal{O}(-H)[1])$. By Lemma 6.28 both $\mathcal{O}$ and $\mathcal{O}(-H)[1]$ are tilt semistable for all $\omega = \alpha H$, $B = \beta H$, where $\alpha > 0$ and $\beta \in (-1, 0)$. A straightforward computation shows

$$\nu_{\omega,B}(\mathcal{O}) > \nu_{\omega,B}(\mathcal{O}(-H)[1]) \Leftrightarrow \alpha^2 + \left(\beta - \frac{1}{2}\right)^2 > \frac{1}{4}.$$

Therefore, there is a region in which $\nu_{\omega,B}(\mathcal{O}) > \nu_{\omega,B}(\mathcal{O}(-H)[1])$ and both these objects are tilt stable, i.e., $\mathrm{Hom}(\mathcal{O}, \mathcal{O}(-H)[1]) = 0$. □

7.3 *Stability of Special Objects*

As in Sect. 2.5, we can look at Bridgeland stability for Lazarsfeld-Mukai bundles on a surface. The setting is the following. Let X be a smooth projective surface. Let L be an ample integral divisor on X. We assume the following: $L^2 \geq 8$ and $L.C \geq 2$, for all integral curves $C \subset X$. By Reider's Theorem [100], the divisor $L + K_X$ is globally generated. We define the *Lazarsfeld-Mukai* vector bundle as the kernel of the evaluation map:

$$M_{L+K_X} := \mathrm{Ker}\left(\mathcal{O}_X \otimes H^0(X, \mathcal{O}_X(L+K_X)) \twoheadrightarrow \mathcal{O}_X(L+K_X)\right).$$

We consider the ample divisor $H := 2L + K_X$ and $B := \frac{K_X}{2}$, and consider the (α, β)-plane with respect to H and B.

The following question would have interesting geometric applications:

Question 7.5 For which (α, β) is M_{L+K_X} tilt stable?

Exercise 7.6 Assume that M_{L+K_X} is $\nu_{\alpha,\beta}$-semistable for (α, β) inside the wall defined by $\nu_{\alpha,\beta}(M_{L+K_X}) = \nu_{\alpha,\beta}(\mathcal{O}_X(-L)[1])$. Show that $H^1(X, M_{L+K_X} \otimes \mathcal{O}_X(K_X + L)) = 0$.

If the assumption in Exercise 7.6 is satisfied, then we would prove that the multiplication map $H^0(X, \mathcal{O}_X(L+K_X)) \otimes H^0(X, \mathcal{O}_X(L+K_X)) \to H^0(X, \mathcal{O}_X(2L + 2K_X))$ is surjective, the first step toward projective normality for the embedding of X given by $L + K_X$.

Example 7.7 The assumption in Exercise 7.6 is satisfied when X is a K3 surface. Indeed, this is an explicit computation by using the fact that in that case, the vector bundle M_L is a *spherical* object (namely, $\mathrm{End}(M_L) = \mathbb{C}$ and $\mathrm{Ext}^1(M_L, M_L) = 0$).

8 Nef Divisors on Moduli Spaces of Bridgeland Stable Objects

So far we neglected another important aspect of moduli spaces of Bridgeland stable objects, namely the construction of divisors on them. Let X be an arbitrary smooth projective variety. Given a stability condition σ and a fixed set of invariants $v \in \Lambda$ we will demonstrate how to construct a nef divisor on moduli spaces of σ-semistable objects with class v. This was originally described in [17].

Assume that $\sigma = (Z, \mathcal{A})$ is a stability condition on $\mathrm{D^b}(X)$, S is a proper algebraic space of finite type over $\mathbb{C}$ and $v \in \Lambda$ a fixed set of invariants. Moreover, let $\mathcal{E} \in \mathrm{D^b}(X \times S)$ be a flat family of σ-semistable objects of class v, i.e., $\mathcal{E}$ is S-perfect (see Sect. 5.4 for a definition) and for every $\mathbb{C}$-point $P \in S$, the derived restriction $\mathcal{E}_{|X \times \{P\}}$ is σ-semistable of class v. The purpose of making the complex S-perfect is to make the derived restriction well defined. A divisor class $D_{\sigma,\mathcal{E}}$ can defined on S

by its intersection number with any projective integral curve $C \subset S$:

$$D_{\sigma,\mathcal{E}} \cdot C = \Im\left(-\frac{Z((p_X)_*\mathcal{E}_{|X\times C})}{Z(v)}\right).$$

We skipped over a detail in the definition that is handled in Sect. 4 of [17]. It is necessary to show that the number only depends on the numerical class of the curve C.

The motivation for this definition is the following Theorem due to [17].

Theorem 8.1 (Positivity Lemma)

(1) *The divisor $D_{\sigma,\mathcal{E}}$ is nef on S.*
(2) *A projective integral curve $C \subset X$ satisfies $D_{\sigma,\mathcal{E}} \cdot C = 0$ if and only if for two general elements $c, c' \in C$ the restrictions $\mathcal{E}_{|X\times\{c\}}$ and $\mathcal{E}_{|X\times\{c'\}}$ are S-equivalent, i.e., their Jordan-Hölder filtrations have the same stable factors up to order.*

We will give a proof that the divisor is nef and refer to [17] for the whole statement. Further background material is necessary. We will mostly follow the proof in [17]. Without loss of generality we can assume that $Z(v) = -1$ by scaling and rotating the stability condition. Indeed, assume that $Z(v) = r_0 e^{\sqrt{-1}\phi_0\pi}$ and let $\mathcal{P}$ be the slicing of our stability condition. We can then define a new stability condition by setting $\mathcal{P}'(\phi) := \mathcal{P}(\phi+1-\phi_0)$ and $Z' = Z \cdot \frac{1}{r_0} e^{\sqrt{-1}(1-\phi_0)\pi}$. Clearly, $Z'(v) = -1$. The definition of $D_{\sigma,\mathcal{E}}$ simplifies to

$$D_{\sigma,\mathcal{E}} \cdot C = \Im\left(Z((p_X)_*\mathcal{E}_{|X\times C})\right)$$

for any projective integral curve $C \subset S$. From this formula the motivation for the definition becomes more clear. If $(p_X)_*\mathcal{E}_{|X\times C} \in \mathcal{A}$ holds, the fact that the divisor is nef follows directly from the definition of a stability function. As always, things are more complicated. We can use Bridgeland's Deformation Theorem and assume further, without changing semistable objects (and the previous assumption), that the heart $\mathcal{A}$ is noetherian.

One of the key properties in the proof is an extension of a result by Abramovich and Polishchuk from [1] to the following statement by Polishchuk. We will not give a proof.

Theorem 8.2 *[98, Theorem 3.3.6] Let $\mathcal{A}$ be the heart of a noetherian bounded t-structure on $\mathrm{D}^{\mathrm{b}}(X)$. The category $\mathcal{A}^{qc} \subset D_{qc}(X)$ is the closure of $\mathcal{A}$ under infinite coproducts in the (unbounded) derived category of quasi-coherent sheaves on X. There is a noetherian heart $\mathcal{A}_S$ on $\mathrm{D}^{\mathrm{b}}(X \times S)$ for any finite type scheme S over the complex numbers satisfying the following three properties.*

(1) *The heart $\mathcal{A}_S$ is characterized by the property*

$$\mathcal{E} \in \mathcal{A}_S \Leftrightarrow (p_X)_*\mathcal{E}_{|X\times U} \in \mathcal{A}^{qc} \text{ for every open affine } U \subset S.$$

(2) *If* $S = \bigcup_i U_i$ *is an open covering of* S, *then*

$$\mathcal{E} \in \mathcal{A}_S \Leftrightarrow \mathcal{E}_{|X \times U_i} \in \mathcal{A}_{U_i} \text{ for all } i.$$

(3) *If* S *is projective and* $\mathcal{O}_S(1)$ *is ample, then*

$$\mathcal{E} \in \mathcal{A}_S \Leftrightarrow (p_X)_*(\mathcal{E} \otimes p_S^*\mathcal{O}_S(n)) \in \mathcal{A} \text{ for all } n \gg 0.$$

In order to apply this theorem to our problem we need the family $\mathcal{E}$ to be in $\mathcal{A}_S$. The proof of the following statement is somewhat technical and we refer to [17, Lemma 3.5].

Lemma 8.3 *Let* $\mathcal{E} \in \mathrm{D^b}(X \times S)$ *be a flat family of* σ*-semistable objects of class* v. *Then* $\mathcal{E} \in \mathcal{A}_S$.

Proof of Theorem 8.1 (1) Let $\mathcal{O}_C(1)$ be an ample line bundle on C. If we can show

$$D_{\sigma,\mathcal{E}} \cdot C = \Im\left(Z((p_X)_*(\mathcal{E} \otimes p_S^*\mathcal{O}_C(n)))\right)$$

for $n \gg 0$, then we are done. Indeed, Theorem 8.2 part (3) implies $(p_X)_*(\mathcal{E} \otimes p_S^*\mathcal{O}_C(n)) \in \mathcal{A}$ for $n \gg 0$ and the proof is concluded by the positivity properties of a stability function.

Choose $n \gg 0$ large enough such that $H^0(\mathcal{O}_C(n)) \neq 0$. Then there is a torsion sheaf T together with a short exact sequence

$$0 \to \mathcal{O}_C \to \mathcal{O}_C(n) \to T \to 0.$$

Since T has zero dimensional support and $\mathcal{E}$ is a family of objects with class v, we can show $Z((p_X)_*(\mathcal{E} \otimes p_S^*T)) \in \mathbb{R}$ by induction on the length of the support of T. But that shows

$$\Im\left(Z((p_X)_*(\mathcal{E} \otimes p_S^*\mathcal{O}_C(n)))\right) = \Im\left(Z((p_X)_*(\mathcal{E} \otimes p_S^*\mathcal{O}_C))\right).$$

□

8.1 The Donaldson Morphism

The definition of the divisor is made such that the proof of the Positivity Lemma is as clear as possible. However, it is hard to explicitly compute it in examples directly via the definition. Computation is often times done via the Donaldson morphism. This is originally explained in Sect. 4 of [17]. Recall that, for a proper scheme S, the Euler characteristic gives a well-defined pairing

$$\chi \colon K_0(\mathrm{D^b}(S)) \times K_0(\mathrm{D_{perf}}(S)) \to \mathbb{Z}$$

between the Grothendieck groups of the bounded derived categories of coherent sheaves $\mathrm{D}^{\mathrm{b}}(S)$ and of perfect complexes $\mathrm{D}_{\mathrm{perf}}(S)$. Taking the quotient with respect to the kernel of χ on each side we obtain numerical Grothendieck groups $K_{\mathrm{num}}(S)$ and $K_{\mathrm{num}}^{\mathrm{perf}}(S)$, respectively, with an induced perfect pairing

$$\chi : K_{\mathrm{num}}(S) \otimes K_{\mathrm{num}}^{\mathrm{perf}}(S) \to \mathbb{Z}.$$

Definition 8.4 We define the additive *Donaldson morphism* $\lambda_{\mathcal{E}} : v^{\#} \to N^1(S)$ by

$$w \mapsto \det((p_S)_*(p_X^* w \cdot [\mathcal{E}])).$$

Here,

$$v^{\#} = \{w \in K_{\mathrm{num}}(S)_{\mathbb{R}} : \chi(v \cdot w) = 0\}.$$

Let $w_\sigma \in v^{\#}$ be the unique vector such that

$$\chi(w_\sigma \cdot w') = \Im\left(-\frac{Z(w')}{Z(v)}\right)$$

for all $w' \in K_{\mathrm{num}}(X)_{\mathbb{R}}$.

Proposition 8.5 ([17, Theorem 4.4]) *We have* $\lambda_{\mathcal{E}}(w_\sigma) = D_{\sigma,\mathcal{E}}$.

Proof *Let* $\mathcal{L}_\sigma = (p_S)_*(p_X^* w_\sigma \cdot [\mathcal{E}])$. *For any* $s \in S$ *we can compute the rank*

$$r(\mathcal{L}_\sigma) = \chi(S, [\mathcal{O}_s] \cdot \mathcal{L}_\sigma) = \chi(S \times X, [\mathcal{O}_{\{s\}\times X}] \cdot [\mathcal{E}] \cdot p_X^* w_\sigma) = \chi(w_\sigma \cdot v) = 0.$$

Therefore, $\mathcal{L}_\sigma$ *has rank zero. This implies*

$$\lambda_{\mathcal{E}}(w_\sigma) \cdot C = \chi(\mathcal{L}_{\sigma|C})$$

for any projective integral curve $C \subset S$. *Let* $i_C : C \hookrightarrow S$ *be the embedding of* C *into* S. *Cohomology and base change implies*

$$\begin{aligned}
\mathcal{L}_{\sigma|C} &= i_C^*(p_S)_*(p_X^* w_\sigma \cdot [\mathcal{E}]) \\
&= (p_C)_*(i_C \times \mathrm{id}_X)^*(p_X^* w_\sigma \cdot [\mathcal{E}]) \\
&= (p_C)_*(p_X^* w_\sigma \cdot [\mathcal{E}_{|C\times X}]).
\end{aligned}$$

The proof can be finished by using the projection formula as follows

$$\begin{aligned}\chi(C, \mathcal{L}_{\sigma|C}) &= \chi(C, (p_C)_*(p_X^* w_\sigma \cdot [\mathcal{E}_{|C\times X}]))\\ &= \chi(X, w_\sigma \cdot (p_X)_*[\mathcal{E}_{|C\times X}])\\ &= \Im\left(-\frac{Z((p_X)_*\mathcal{E}_{|X\times C})}{Z(v)}\right).\end{aligned}$$

□

8.2 Applications to Hilbert Schemes of Points

Recall that for any positive integer $n \in \mathbb{N}$ the Hilbert scheme of n-points $X^{[n]}$ parameterizes subschemes $Z \subset X$ of dimension zero and length n. This scheme is closely connected to the symmetric product $X^{(n)}$ defined as the quotient X^n/S_n where S_n acts on X^n via permutation of the factors. By work of Fogarty in [43] the natural map $X^{[n]} \to X^{(n)}$ is a birational morphism that resolves the singularities of $X^{(n)}$.

We will recall the description of $\mathrm{Pic}(X^{[n]})$ in case the surface X has irregularity zero, i.e., $H^1(\mathcal{O}_X) = 0$. It is a further result in [44]. If D is any divisor on X, then there is an S_n invariant divisor $D^{\boxtimes n}$ on X^n. This induces a divisor $D^{(n)}$ on $X^{(n)}$, which we pull back to a divisor $D^{[n]}$ on $X^{[n]}$. If D is a prime divisor, then $D^{[n]}$ parameterizes those $Z \subset X$ that intersect D non trivially. Then

$$\mathrm{Pic}(X^{[n]}) \cong \mathrm{Pic}(X) \oplus \mathbb{Z} \cdot \frac{E}{2},$$

where E parameterizes the locus of non reduces subschemes Z. Moreover, the restriction of this isomorphism to $\mathrm{Pic}(X)$ is the embedding given by $D \mapsto D^{[n]}$. A direct consequence of this result is the description of the Néron-Severi group as

$$\mathrm{NS}(X^{[n]}) \cong \mathrm{NS}(X) \oplus \mathbb{Z} \cdot \frac{E}{2}.$$

The divisor $\frac{E}{2}$ is integral because it is given by $\det((p_{X^{[n]}})_*\mathcal{U}_n)$, where $\mathcal{U}_n \in \mathrm{Coh}(X \times X^{[n]})$ is the *universal ideal sheaf* of $X^{[n]}$.

Theorem 8.6 *Let X be a smooth complex projective surface with* $\mathrm{Pic}(X) = \mathbb{Z} \cdot H$, *where H is ample. Moreover, let $a > 0$ be the smallest integer such that aH is effective. If $n \geq a^2H^2$, then the divisor*

$$D = \frac{1}{2}K_X^{[n]} + \left(\frac{a}{2} + \frac{n}{aH^2}\right) H^{[n]} - \frac{1}{2}E$$

is nef. If g is the arithmetic genus of a curve $C \in |aH|$ and $n \geq g + 1$, then D is extremal. In particular, the nef cone of $X^{[n]}$ is spanned by this divisor and $H^{[n]}$.

For the proof of this statement, we need to describe the image of the Donaldson morphism more precisely. In this case, the vector v is given by $(1, 0, -n)$ and $\mathcal{E} = \mathcal{U}_n \in \mathrm{Coh}(X \times X^{[n]})$ is the universal ideal sheaf for $X^{[n]}$.

Proposition 8.7 *Choose m such that $(1, 0, m) \in v^{\#}$ and for any divisor D on X choose m_D such that $(0, D, m_D) \in v^{\#}$. Then*

$$\lambda_{\mathcal{U}_n}(1, 0, m) = \frac{E}{2},$$
$$\lambda_{\mathcal{U}_n}(0, D, m_D) = -D^{[n]}.$$

Sketch of the proof *Let $x \in X$ be an arbitrary point and $\mathbb{C}(x)$ the corresponding skyscraper sheaf. The Grothendieck-Riemann-Roch Theorem implies*

$$\mathrm{ch}((p_{X^{[n]}})_*(p_X^*\mathbb{C}(x) \otimes \mathcal{U}_n)) = (p_{X^{[n]}})_*(\mathrm{ch}(p_X^*\mathbb{C}(x) \otimes \mathcal{U}_n) \cdot \mathrm{td}(T_{p_{X^{[n]}}})).$$

As a consequence

$$\lambda_{\mathcal{U}_n}(0, 0, 1) = p_X^*\left(-[x] \cdot \frac{K_X}{2}\right) = 0$$

holds. Therefore, the values of m and m_D are irrelevant for the remaining computation. Similarly, we can show that

$$-\lambda_{\mathcal{U}_n}\left(0, D, -\frac{D^2}{2}\right) = (p_{X^{[n]}})_*(p_X^*(-D) \cdot \mathrm{ch}_2(\mathcal{U}_n)).$$

Intuitively, this divisor consists of those schemes Z parameterized in $X^{[n]}$ that intersect the divisor D, i.e., it is $D^{[n]}$ (see [36]).

Finally, we have to compute $\lambda_{\mathcal{U}_n}(1, 0, 0)$. By definition it is given as $\det((p_{X^{[n]}})_*\mathcal{U}_n) = \frac{E}{2}$. □

Corollary 8.8 *Let W be a numerical semicircular wall for $v = (1, 0, -n)$ with center s_W. Assume that all ideal sheaves I_Z on X with $\mathrm{ch}(I_Z) = (1, 0, -n)$ are semistable along W. Then*

$$D_{\alpha,\beta,\mathcal{U}_n} \in \mathbb{R}_{>0}\left(\frac{K_X^{[n]}}{2} - s_W H^{[n]} - \frac{E}{2}\right)$$

for all $\sigma \in W$.

Proof *We fix $\sigma = (\alpha, \beta) \in W$. By Proposition 8.5 there is a class $w_\sigma \in v^{\#}$ such that the divisor $D_{\sigma,\mathcal{U}_n}$ is given by $\lambda_{\mathcal{U}_n}(w_\sigma)$. This class is characterized by the property*

$$\chi(w_\sigma \cdot w') = \Im\left(-\frac{Z_{\alpha,\beta}(w')}{Z_{\alpha,\beta}(v)}\right)$$

for all $w' \in K_{\mathrm{num}}(X)_{\mathbb{R}}$. We define $x, y \in \mathbb{R}$ by

$$-\frac{1}{Z_{\alpha,\beta}(v)} = x + \sqrt{-1}y.$$

The fact that $Z_{\alpha,\beta}$ is a stability function implies $y \geq 0$. We even have $y > 0$ because $y = 0$ holds if and only if (α, β) is on the unique numerical vertical wall contrary to assumption. The strategy of the proof is to determine w_σ by pairing it with some easy to calculate classes and then using the previous proposition.

Let $w_\sigma = (r, C, d)$, where $r \in \mathbb{Z}$, $d \in \frac{1}{2}\mathbb{Z}$ and C is a curve class. We have

$$r = \chi(w_\sigma \cdot (0,0,1)) = \Im((x + \sqrt{-1}y)Z_{\alpha,\beta}(0,0,1)) = -y.$$

A similar computation together with Riemann-Roch shows

$$\begin{aligned}
(x + \beta y)H^2 &= \Im((x + iy) \cdot Z_{\alpha,\beta}(0, H, 0)) \\
&= \chi(w_\sigma \cdot (0, H, 0)) \\
&= \chi(0, -yH, H \cdot C) \\
&= \int_X (0, -yH, H \cdot C) \cdot \left(1, -\frac{K_X}{2}, \chi(\mathcal{O}_X)\right) \\
&= \frac{y}{2} H \cdot K_X + H \cdot C.
\end{aligned}$$

Since $\mathrm{Pic}(X) = \mathbb{Z} \cdot H$, *we obtain*

$$C = (x + \beta y)H - \frac{y}{2}K_X = y s_W H - \frac{y}{2}K_X.$$

The last step used $x + \beta y = y s_W$ which is a straightforward computation. We get

$$D_{\alpha,\beta,\mathcal{U}_n} \in \mathbb{R}_{>0}\left(\lambda_{\mathcal{U}_n}(-1, s_W H - \frac{K_X}{2}, m)\right),$$

where m is uniquely determined by $w_\sigma \in v^\#$. The statement follows now from a direct application of Proposition 8.7. □

Proof of Theorem 8.6 By Proposition 7.1 and the assumption $n \geq a^2H^2$ we know that the largest wall destabilizes those ideal sheaves $\mathcal{I}_Z$ that fit into an exact sequence

$$0 \to \mathcal{O}(-aH) \to \mathcal{I}_Z \to \mathcal{I}_{Z/C} \to 0,$$

where $C \in |aH|$. This wall has center

$$s = -\frac{a}{2} - \frac{n}{aH^2}.$$

By Corollary 8.8 we get that

$$D = \frac{1}{2}K_X^{[n]} + \left(\frac{a}{2} + \frac{n}{aH^2}\right) H^{[n]} - \frac{1}{2}E$$

is nef. We are left to show extremality of the divisor in case $n \geq g + 1$. By part (2) of the Positivity Lemma, we have to construct a one dimensional family of S-equivalent objects. It exists if and only if $\mathrm{ext}^1(\mathcal{I}_{Z/C}, \mathcal{O}(-aH)) \geq 2$. A Riemann-Roch calculation shows

$$\begin{aligned} 1 - g = \chi(\mathcal{O}_C) &= \int_X \left(0, aH, -\frac{a^2H^2}{2}\right) \cdot \left(1, -\frac{K_X}{2}, \chi(\mathcal{O}_X)\right) \\ &= -\frac{a}{2}H \cdot K_X - \frac{a^2H^2}{2}. \end{aligned}$$

Another application of Riemann-Roch for surfaces shows

$$\begin{aligned} \mathrm{ext}^1(\mathcal{I}_{Z/C}, \mathcal{O}(-aH)) &\geq -\chi(\mathcal{I}_{Z/C}, \mathcal{O}(-aH)) \\ &= -\int_X \left(1, -aH, \frac{a^2H^2}{2}\right)\left(0, -aH, -\frac{a^2H^2}{2} - n\right)\left(1, -\frac{K_X}{2}, \chi(\mathcal{O}_X)\right) \\ &= n - \frac{a}{2}H \cdot K_X - \frac{a^2H^2}{2} \\ &= n + 1 - g. \end{aligned}$$

Therefore, $n \geq g + 1$ implies that D is extremal. □

Remark 8.9 In particular cases, Theorem 8.6 can be made more precise and general. In the case of the projective plane [37, 38, 75] and of K3 surfaces [16, 17], given any primitive vector, varying stability conditions corresponds to a directed Minimal Model Program for the corresponding moduli space. This allows to completely describe the nef cone, the movable cone, and the pseudo-effective cone for them. Also, all corresponding birational models appear as moduli spaces of Bridgeland stable objects. This has also deep geometrical applications; for example, see [12].

9 Stability Conditions on Threefolds

In this section we will give a short sketchy introduction to the higher-dimensional case. The main result is the construction of stability condition on $X = \mathbb{P}^3$. By abuse of notation, we will identify $\mathrm{ch}_i^\beta(E) \in \mathbb{Q}$, for any $E \in \mathrm{D}^\mathrm{b}(\mathbb{P}^3)$.

9.1 Tilt Stability and the Second Tilt

The construction of $\sigma_{\alpha,\beta} = (\mathrm{Coh}^{\beta}(X), Z_{\alpha,\beta})$ for $\alpha > 0$ and $\beta \in \mathbb{R}$ can carried out as before. However, this will not be a stability condition, because $Z_{\alpha,\beta}$ maps skyscraper sheaves to the origin. In [18], this (weak) stability condition is called *tilt stability*. The idea is that, by repeating the previous process of tilting another time with $\sigma_{\alpha,\beta}$ instead of slope stability, this might allow to construct a Bridgeland stability condition on $\mathrm{D^b}(X)$. Let

$$\mathcal{T}'_{\alpha,\beta} = \{E \in \mathrm{Coh}^{\beta}(\mathbb{P}^3) : \text{any quotient } E \twoheadrightarrow G \text{ satisfies } \nu_{\alpha,\beta}(G) > 0\},$$

$$\mathcal{F}'_{\alpha,\beta} = \{E \in \mathrm{Coh}^{\beta}(\mathbb{P}^3) : \text{any subobject } F \hookrightarrow E \text{ satisfies } \nu_{\alpha,\beta}(F) \leq 0\}$$

and set $\mathcal{A}^{\alpha,\beta}(\mathbb{P}^3) = \langle \mathcal{F}'_{\alpha,\beta}[1], \mathcal{T}'_{\alpha,\beta} \rangle$. For any $s > 0$, we define

$$Z_{\alpha,\beta,s} = -\mathrm{ch}_3^{\beta} + (s + \tfrac{1}{6})\alpha^2 \mathrm{ch}_1^{\beta} + \sqrt{-1}(\mathrm{ch}_2^{\beta} - \frac{\alpha^2}{2} \mathrm{ch}_0^{\beta}).$$

and the corresponding slope

$$\lambda_{\alpha,\beta,s} = \frac{\mathrm{ch}_3^{\beta} - (s + \frac{1}{6})\alpha^2 \mathrm{ch}_1^{\beta}}{\mathrm{ch}_2^{\beta} - \frac{\alpha^2}{2} \cdot \mathrm{ch}_0^{\beta}}.$$

Recall that the key point in the construction stability conditions on surfaces was the classical Bogomolov inequality for slope stable sheaves. That is the idea behind the next theorem.

Theorem 9.1 ([18, 19, 82]) *The pair $(\mathcal{A}^{\alpha,\beta}(\mathbb{P}^3), \lambda_{\alpha,\beta,s})$ is Bridgeland stability condition for all $s > 0$ if and only if for all tilt stable objects $E \in \mathrm{Coh}^{\beta}(\mathbb{P}^3)$ we have*

$$Q_{\alpha,\beta}(E) = \alpha^2 \Delta(E) + 4(\mathrm{ch}_2^{\beta}(E))^2 - 6\,\mathrm{ch}_1^{\beta}(E)\,\mathrm{ch}_3^{\beta}(E) \geq 0.$$

It was more generally conjectured that similar inequalities are true for all smooth projective threefolds. This turned out to be true in various cases. The first proof was in the case of $\mathbb{P}^3$ in [82] and a very similar proof worked for the smooth quadric hypersurface in $\mathbb{P}^4$ in [101]. These results were generalized with a fundamentally different proof to all Fano threefolds of Picard rank one in [73]. The conjecture is also known to be true for all abelian threefolds with two independent proofs by Maciocia and Piyaratne [80] and Bayer et al. [19]. Unfortunately, it turned out to be problematic in the case of the blow up of $\mathbb{P}^3$ in a point as shown in [103].

Remark 9.2 Many statements about Bridgeland stability on surfaces work almost verbatim in tilt stability. Analogously to Theorem 5.27 walls in tilt stability are locally finite and stability is an open property. The structure of walls from

Proposition 6.22 is the same in tilt stability. The Bogomolov inequality also works, i.e., any tilt semistable object E satisfies $\Delta(E) \geq 0$. The bound for high degree walls in Lemma 7.2 is fine as well. Finally, slope stable sheaves E are $\nu_{\alpha,\beta}$-stable for all $\alpha \gg 0$ and $\beta < \mu(E)$.

We will give the proof of Chunyi Li in the special case of $\mathbb{P}^3$ in these notes. We first recall the Hirzebruch-Riemann-Roch Theorem for $\mathbb{P}^3$.

Theorem 9.3 *Let $E \in \mathrm{D}^{\mathrm{b}}(\mathbb{P}^3)$. Then*

$$\chi(\mathbb{P}^3, E) = \mathrm{ch}_3(E) + 2\,\mathrm{ch}_2(E) + \frac{11}{6}\,\mathrm{ch}_1(E) + \mathrm{ch}_0(E).$$

In order to prove the inequality in Theorem 9.1, we want to reduce the problem to a simpler case.

Definition 9.4 For any object $E \in \mathrm{Coh}^{\beta}(\mathbb{P}^3)$, we define

$$\overline{\beta}(E) = \begin{cases} \frac{\mathrm{ch}_1(E) - \sqrt{\Delta_H(E)}}{\mathrm{ch}_0(E)} & \mathrm{ch}_0(E) \neq 0, \\ \frac{\mathrm{ch}_2(E)}{\mathrm{ch}_1(E)} & \mathrm{ch}_0(E) = 0. \end{cases}$$

The object E is called *$\overline{\beta}$-(semi)stable*, if E (semi)stable in a neighborhood of $(0, \overline{\beta}(E))$.

A straightforward computation shows that $\mathrm{ch}_2^{\overline{\beta}}(E) = 0$.

Lemma 9.5 **([19])** *Proving the inequality in Theorem 9.1 can be reduced to $\overline{\beta}$-stable objects. Moreover, in that case the inequality reduces to $\mathrm{ch}_3^{\overline{\beta}}(E) \leq 0$.*

Proof Let $E \in \mathrm{Coh}^{\beta_0}(\mathbb{P}^3)$ be a ν_{α_0,β_0}-stable object with $\mathrm{ch}(E) = v$. If (α_0, β_0) is on the unique numerical vertical wall for v, then $\mathrm{ch}_1^{\beta_0}(E) = 0$ and therefore,

$$Q_{\alpha_0,\beta_0} = \alpha_0^2 \Delta(E) + 4(\mathrm{ch}_2^{\beta_0}(E))^2 \geq 0.$$

Therefore, (α_0, β_0) lies on a unique numerical semicircular wall W with respect to v. One computes that there are $x, y \in \mathbb{R}$ such that

$$Q_{\alpha,\beta}(E) \geq 0 \Leftrightarrow \Delta(E)\alpha^2 + \Delta(E)\beta^2 + x\beta + y\alpha \geq 0,$$
$$Q_{\alpha,\beta}(E) = 0 \Leftrightarrow \nu_{\alpha,\beta}(E) = \nu_{\alpha,\beta}\left(\mathrm{ch}_1(E), 2\,\mathrm{ch}_2(E), 3\,\mathrm{ch}_3(E), 0\right).$$

In particular, the equation $Q_{\alpha,\beta}(E) \geq 0$ defines the complement of a semi-disc with center on the β-axis or a quadrant to one side of a vertical line in case $\Delta(E) = 0$. Moreover, $Q_{\alpha,\beta}(E) = 0$ is a numerical wall with respect to v.

We will proceed by an induction on $\Delta(E)$ to show that it is enough to prove the inequality for $\overline{\beta}$-stable objects. Assume $\Delta(H) = 0$. Then $Q_{\alpha_0,\beta_0}(E) \geq 0$ is equivalent to $Q_{0,\overline{\beta}}(E) \geq 0$. If E would not be $\overline{\beta}$-stable, then it must destabilize along a wall between W and $(0, \overline{\beta}(E))$. By part (3) of Lemma 5.29 with $Q = \Delta$ all stable factors of E along that wall have $\Delta = 0$. By part (4) of the same lemma this can only happen if at least one of the stable factors satisfies $\mathrm{ch}_{\leq 2}(E) = (0, 0, 0)$ which only happens at the numerical vertical wall.

Assume $\Delta(H) > 0$. If E is $\overline{\beta}$-stable, then it is enough to show $Q_{0,\overline{\beta}(E)} \geq 0$. Assume E is destabilized along a wall between W and $(0, \overline{\beta}(E))$ and let $F_1, \ldots, F_n$ be the stable factors of E along this wall. By Lemma 5.29 (3) we have $\Delta(F_i) < \Delta(E)$ for all $i = 1, \ldots n$. We can then use Lemma 5.29 (3) with $Q = Q_{\alpha,\beta}$ to finish the proof by induction. □

The idea for the following proof is due to [73].

Theorem 9.6 *For all tilt stable objects $E \in \mathrm{Coh}^{\beta}(\mathbb{P}^3)$ we have*

$$Q_{\alpha,\beta}(E) = \alpha^2 \Delta(E) + 4(\mathrm{ch}_2^{\beta}(E))^2 - 6\,\mathrm{ch}_1^{\beta}(E)\,\mathrm{ch}_3^{\beta}(E) \geq 0.$$

Proof As observed in Remark 9.2, tilt semistable objects satisfy $\Delta \geq 0$. Hence, line bundles are stable everywhere in tilt stability in the case $X = \mathbb{P}^3$.

Let $E \in \mathrm{Coh}^{\overline{\beta}}(X)$ be a $\overline{\beta}$-stable object. By tensoring with line bundles, we can assume $\overline{\beta} \in [-1, 0)$. By assumption we have $\nu_{0,\overline{\beta}}(E) = 0 < \nu_{0,\overline{\beta}}(\mathcal{O}_{\mathbb{P}^3})$ which implies $\mathrm{Hom}(\mathcal{O}_{\mathbb{P}^3}, E) = 0$. Moreover, the same argument together with Serre duality shows

$$\mathrm{Ext}^2(\mathcal{O}_{\mathbb{P}^3}, E) = \mathrm{Ext}^1(E, \mathcal{O}_{\mathbb{P}^3}(-4)) = \mathrm{Hom}(E, \mathcal{O}_{\mathbb{P}^3}(-4)[1]) = 0.$$

Therefore,

$$\begin{aligned}
0 \geq \chi(\mathcal{O}, E) &= \mathrm{ch}_3(E) + 2\,\mathrm{ch}_2(E) + \frac{11}{6}\mathrm{ch}_1(E) + \mathrm{ch}_0(E) \\
&= \mathrm{ch}_3^{\overline{\beta}}(E) + (\overline{\beta} + 2)\,\mathrm{ch}_2^{\overline{\beta}}(E) + \frac{1}{6}(3\overline{\beta}^2 + 12\overline{\beta} + 11)\,\mathrm{ch}_1^{\overline{\beta}}(E) \\
&\quad + \frac{1}{6}(\overline{\beta}^3 + 6\overline{\beta}^2 + 11\overline{\beta} + 6)\,\mathrm{ch}_0^{\overline{\beta}}(E) \\
&= \mathrm{ch}_3^{\overline{\beta}}(E) + \frac{1}{6}(3\overline{\beta}^2 + 12\overline{\beta} + 11)\,\mathrm{ch}_1^{\overline{\beta}}(E) + \frac{1}{6}(\overline{\beta}^3 + 6\overline{\beta}^2 + 11\overline{\beta} + 6)\,\mathrm{ch}_0^{\overline{\beta}}(E).
\end{aligned}$$

By construction of $\mathrm{Coh}^{\overline{\beta}}(X)$, we have $\mathrm{ch}_1^{\overline{\beta}}(E) \geq 0$. Due to $\overline{\beta} \in [-1, 0)$ we also have $3\overline{\beta}^2 + 12\overline{\beta} + 11 \geq 0$ and $\overline{\beta}^3 + 6\overline{\beta}^2 + 11\overline{\beta} + 6 \geq 0$. If $\mathrm{ch}_0^{\overline{\beta}}(E) \geq 0$ or $\overline{\beta} = -1$, this finishes the proof. If $\mathrm{ch}_0^{\overline{\beta}}(E) < 0$ and $\overline{\beta} \neq -1$, the same type of argument works with $\chi(\mathcal{O}(3), E) \leq 0$. □

9.2 Castelnuovo's Genus Bound

The next exercise outlines an application of this inequality to a proof of Castelnuovo's classical theorem on the genus of non-degenerate curves in $\mathbb{P}^3$.

Theorem 9.7 ([35]) *Let $C \subset \mathbb{P}^3$ be an integral curve of degree $d \geq 3$ and genus g. If C is not contained in a plane, then*

$$g \leq \frac{1}{4}d^2 - d + 1.$$

Exercise 9.8 Assume there exists an integral curve $C \subset \mathbb{P}^3$ of degree $d \geq 3$ and arithmetic genus $g > \frac{1}{4}d^2 - d + 1$ which is not contained in a plane.

(1) Compute that the ideal sheaf $\mathcal{I}_C$ satisfies

$$\mathrm{ch}(\mathcal{I}_C) = (1, 0, -d, 2d + g - 1).$$

Hint: Grothendieck-Riemann-Roch!

(2) Show that $\mathcal{I}_C$ has to be destabilized by an exact sequence

$$0 \to F \to \mathcal{I}_C \to G \to 0,$$

where $\mathrm{ch}_0(F) = 1$. *Hint: Use Lemma 7.2 to show that any wall of higher rank has to be contained inside $Q_{\alpha,\beta}(\mathcal{I}_C) < 0$.*

(3) Show that $\mathcal{I}_C$ has to be destabilized by a sequence of the form

$$0 \to \mathcal{O}(-a) \to \mathcal{I}_C \to G \to 0,$$

where $a > 0$ is a positive integer. *Hint: Show that if the subobject is an actual ideal sheaf and not a line bundle, then integrality of C implies that $\mathcal{I}_C$ is destabilized by a line bundle giving a bigger or equal wall.*

(4) Show that the wall is inside $Q_{\alpha,\beta}(\mathcal{I}_C) < 0$ if $a > 2$. Moreover, if $d = 3$ or $d = 4$, then the same holds additionally for $a = 2$.

(5) Derive a contradiction in the cases $d = 3$ and $d = 4$.

(6) Let $E \in \mathrm{Coh}^{\beta}(\mathbb{P}^3)$ be a tilt semistable object for some $\alpha > 0$ and $\beta \in \mathbb{R}$ such that $\mathrm{ch}(E) = (0, 2, d, e)$ for some d, e. Show that the inequality

$$e \leq \frac{d^2}{4} + \frac{1}{3}$$

holds. *Hint: Use Theorem 9.6 and Lemma 7.2.*

(7) Obtain a contradiction via the quotient G in $0 \to \mathcal{O}(-2) \to \mathcal{I}_C \to G \to 0$.

Acknowledgements We would very much like to thank Benjamin Bakker, Arend Bayer, Aaron Bertram, Izzet Coskun, Jack Huizenga, Daniel Huybrechts, Martí Lahoz, Ciaran Meachan, Paolo Stellari, Yukinobu Toda, and Xiaolei Zhao for very useful discussions and many explanations on the topics of these notes. We are also grateful to Jack Huizenga for sharing a preliminary version of his survey article [60] with us and to the referee for very useful suggestions which improved the readability of these notes. The first author would also like to thank very much the organizers of the two schools for the kind invitation and the excellent atmosphere, and the audience for many comments, critiques, and suggestions for improvement. The second author would like to thank

Northeastern University for the hospitality during the writing of this article. This work was partially supported by NSF grant DMS-1523496 and a Presidential Fellowship of the Ohio State University.

Appendix: Background on Derived Categories

This section contains definition and important properties of the bounded derived category $\mathrm{D}^{\mathrm{b}}(\mathcal{A})$ for an abelian category $\mathcal{A}$. Most of the time the category $\mathcal{A}$ will be the category $\mathrm{Coh}(X)$ of coherent sheaves on a smooth projective variety X. To simplify notation $\mathrm{D}^{\mathrm{b}}(X)$ will be written for $\mathrm{D}^{\mathrm{b}}(\mathrm{Coh}(X))$. Derived categories were introduced by Verdier in his thesis under the supervision of Grothendieck. The interested reader can find a detailed account of the theory in [49], the first two chapters of [61] or the original source [114].

Definition 1

(1) A complex

$$\ldots \to A^{i-1} \to A^i \to A^{i+1} \to \ldots$$

is called *bounded* if $A^i = 0$ for both $i \gg 0$ and $i \ll 0$.

(2) The objects of the category $\mathrm{Kom}^b(\mathcal{A})$ are bounded complexes over $\mathcal{A}$ and its morphisms are homomorphisms of complexes.

(3) A morphism $f : A \to B$ in $\mathrm{Kom}^b(\mathcal{A})$ is called a *quasi isomorphism* if the induced morphism of cohomology groups $H^i(A) \to H^i(B)$ is an isomorphism for all integers i.

The bounded derived category of $\mathcal{A}$ is the localization of $\mathrm{Kom}^b(\mathcal{A})$ by quasi isomorphisms. The exact meaning of this is the next theorem.

Theorem 2 ([61, Theorem 2.10]) *There is a category* $\mathrm{D}^{\mathrm{b}}(\mathcal{A})$ *together with a functor* $Q : \mathrm{Kom}^b(\mathcal{A}) \to \mathrm{D}^{\mathrm{b}}(\mathcal{A})$ *satisfying two properties.*

(1) *The morphism $Q(f)$ is an isomorphism for any quasi-isomorphism f in the category* $\mathrm{Kom}^b(\mathcal{A})$.

(2) *Any functor $F : \mathrm{Kom}^b(\mathcal{A}) \to \mathcal{D}$ satisfying property (i) factors uniquely through Q, i.e., there is a unique (up to natural isomorphism) functor $G : \mathrm{D}^{\mathrm{b}}(\mathcal{A}) \to \mathcal{D}$ such that F is naturally isomorphic to $G \circ Q$.*

In particular, Q identifies objects in $\mathrm{Kom}^b(\mathcal{A})$ and $\mathrm{D}^{\mathrm{b}}(\mathcal{A})$. By the definition of quasi isomorphisms we still have well defined cohomology groups $H^i(A)$ for any $A \in \mathrm{D}^{\mathrm{b}}(\mathcal{A})$. The category $\mathcal{A}$ is equivalent to the full subcategory of $\mathrm{D}^{\mathrm{b}}(\mathcal{A})$ consisting of those objects $A \in \mathcal{A}$ that satisfy $H^i(A) = 0$ for all $i \neq 0$. In the next section we will learn that this is the simplest example of what is known as the heart of a bounded t-structure.

Notice, there is the automorphism $[1] : \mathrm{D}^{\mathrm{b}}(\mathcal{A}) \to \mathrm{D}^{\mathrm{b}}(\mathcal{A})$, where $E[1]$ is defined by $E[1]^i = E^{i+1}$. It simply changes the grading of a complex. Moreover, we define the shift functor $[n] = [1]^n$ for any integer n. The following lemma will be used to actually compute homomorphisms in the derived category.

Lemma 3 ([61, Proposition 2.56]) *Let $\mathcal{A}$ be either an abelian category with enough injectives or* $\mathrm{Coh}(X)$ *for a smooth projective variety X. For any $A, B \in \mathcal{A}$ and $i \in \mathbb{Z}$ we have the equality*

$$\mathrm{Hom}_{\mathrm{D^b}(\mathcal{A})}(A, B[i]) = \mathrm{Ext}^i(A, B).$$

In contrast to $\mathrm{Kom}^b(\mathcal{A})$ the bounded derived category $\mathrm{D^b}(\mathcal{A})$ is not abelian. This lead Verdier and Grothendieck to the notion of a triangulated category which will be explained in the next theorem.

Definition 4 For any morphism $f : A \to B$ in $\mathrm{Kom}^b(\mathcal{A})$ the cone $C(f)$ is defined by $C(f)^i = A^{i+1} \oplus B^i$. The differential is given by the matrix

$$\begin{pmatrix} -d_A & 0 \\ f & d_B \end{pmatrix}.$$

The inclusion $B^i \hookrightarrow A^{i+1} \oplus B^i$ leads to a morphism $B \to C(f)$ and the projection $B^i \oplus A^{i+1} \twoheadrightarrow A^{i+1}$ leads to a morphism $C(f) \to A[1]$.

Definition 5 A sequence of maps $F \to E \to G \to F[1]$ in $\mathrm{D^b}(\mathcal{A})$ is called a *distinguished triangle* if there is a morphism $f : A \to B$ in $\mathrm{Kom}^b(\mathcal{A})$ and a commutative diagram with vertical isomorphisms in $\mathrm{D^b}(\mathcal{A})$ as follows

$$\begin{array}{ccccccc} F & \longrightarrow & E & \longrightarrow & G & \longrightarrow & F[1] \\ \downarrow & & \downarrow & & \downarrow & & \downarrow \\ A & \longrightarrow & B & \longrightarrow & C(f) & \longrightarrow & A[1]. \end{array}$$

These distinguished triangles should be viewed as the analogue of exact sequences in an abelian category. If $0 \to A \to B \to C \to 0$ is an exact sequence in $\mathcal{A}$, then $A \to B \to C \to A[1]$ is a distinguished triangle where the map $C \to A[1]$ is determined by the element in $\mathrm{Hom}(C, A[1]) = \mathrm{Ext}^1(C, A)$ that determines the extension B. The following properties of the derived category are essentially the defining properties of a *triangulated category*.

Theorem 6 ([49, Chap. IV])

(1) *Any morphism $A \to B$ in $\mathrm{D^b}(\mathcal{A})$ can be completed to a distinguished triangle $A \to B \to C \to A[1]$.*
(2) *A triangle $A \to B \to C \to A[1]$ is distinguished if and only if the induced triangle $B \to C \to A[1] \to B[1]$ is distinguished.*
(3) *Assume we have two distinguished triangles with morphisms f and g making the diagram below commutative.*

$$\begin{array}{ccccccc} A & \longrightarrow & B & \longrightarrow & C & \longrightarrow & A[1] \\ \downarrow{\scriptstyle f} & & \downarrow{\scriptstyle g} & & \downarrow{\scriptstyle \exists h} & & \downarrow{\scriptstyle f[1]} \\ A' & \longrightarrow & B' & \longrightarrow & C' & \longrightarrow & A'[1]. \end{array}$$

Then we can find $h : C \to C'$ making the whole diagram commutative.

(4) *Assume we have two morphisms $A \to B$ and $B \to C$. Then together with (1) and (3) we can get a commutative diagram as follows where all rows and columns are distinguished triangles.*

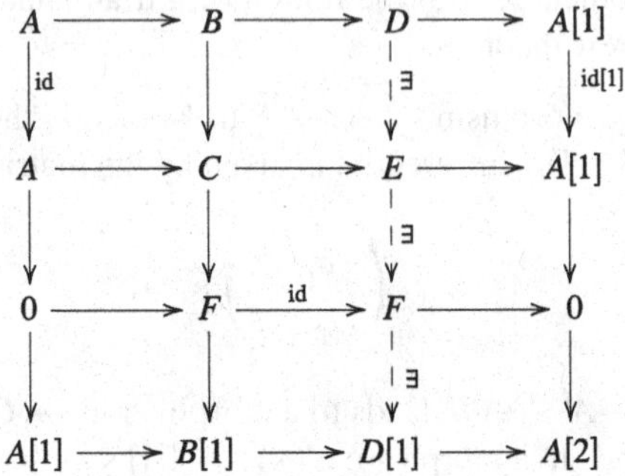

The key in property (4) is that the triangle $D \to E \to F \to D[1]$ is actually distinguished. Be aware that contrary to most definitions in category theory the morphism in (3) is not necessarily unique.

Exercise 7 Let $A \to B \to C \to A[1]$ be a distinguished triangle and $E \in \mathrm{D}^{\mathrm{b}}(\mathcal{A})$ be an arbitrary object. Then there are long exact sequences

$$\ldots \to \mathrm{Hom}(E, C[-1]) \to \mathrm{Hom}(E, A) \to \mathrm{Hom}(E, B) \to \mathrm{Hom}(E, C) \to \mathrm{Hom}(E, A[1]) \to \ldots$$

and

$$\ldots \to \mathrm{Hom}(A[-1], E) \to \mathrm{Hom}(C, E) \to \mathrm{Hom}(B, E) \to \mathrm{Hom}(A, E) \to \mathrm{Hom}(C[-1], E) \to \ldots$$

Show the existence of one of the two long exact sequences (their proofs are almost the same).

Exercise 8 Let $f : A \to B$ be a morphism in $\mathrm{D}^{\mathrm{b}}(\mathcal{A})$. Show that f is an isomorphism if and only if $C = 0$.

Exercise 9 Prove the corresponding statement to the Five Lemma for derived categories: Assume there is a commutative diagram between distinguished triangles

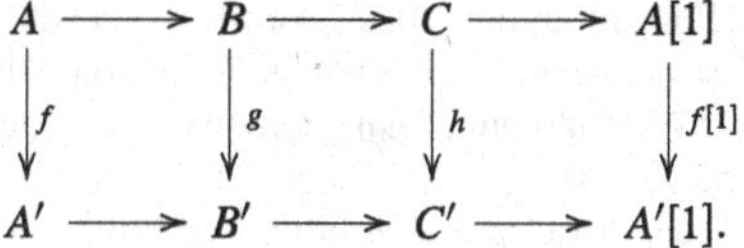

If two of the morphisms f, g, h are isomorphisms, so is the third one.

We will need the following technical statement in the main text.

Proposition 10 ([33, Proposition 5.4]) *Let X be a smooth projective variety and $E \in \mathrm{D}^{\mathrm{b}}(X)$. If $\mathrm{Ext}^i(E, \mathbb{C}(x)) = 0$ for all $x \in X$ and $i < 0$ or $i > s \in \mathbb{Z}$. Then E is isomorphic to a complex $F^\bullet$ of locally free sheaves such that $F^i = 0$ for $i > 0$ and $i < -s$.*

All triangles coming up in this article are distinguished. Therefore, we will simply drop the word distinguished from the notation.

References

1. D. Abramovich, A. Polishchuk, Sheaves of t-structures and valuative criteria for stable complexes. J. Reine Angew. Math. **590**, 89–130 (2006)
2. J. Alper, D.I. Smyth, Existence of good moduli spaces for A_k-stable curves (2012). arXiv:1206.1209
3. J. Alper, J. Hall, D. Rydh, A Luna étale slice theorem for algebraic stacks (2015). arXiv:1504.06467
4. L. Álvarez-Cónsul, A. King, A functorial construction of moduli of sheaves. Invent. Math. **168**(3), 613–666 (2007)
5. R. Anno, R. Bezrukavnikov, I. Mirković, Stability conditions for Slodowy slices and real variations of stability. Mosc. Math. J. **15**(2), 187–203, 403 (2015)
6. D. Arcara, A. Bertram, Bridgeland-stable moduli spaces for K-trivial surfaces. J. Eur. Math. Soc. (JEMS) **15**(1), 1–38 (2013). With an appendix by Max Lieblich
7. D. Arcara, E. Miles, Projectivity of Bridgeland moduli spaces on del Pezzo surfaces of Picard rank 2 (2015). arXiv:1506.08793
8. D. Arcara, A. Bertram, I. Coskun, J. Huizenga, The minimal model program for the Hilbert scheme of points on $\mathbb{P}^2$ and Bridgeland stability. Adv. Math. **235**, 580–626 (2013)
9. P.S. Aspinwall, D-branes on Calabi-Yau manifolds, in *Progress in String Theory* (World Scientific Publishing, Hackensack, NJ, 2005), pp. 1–152
10. M.F. Atiyah, Vector bundles over an elliptic curve. Proc. Lond. Math. Soc. (3) **7**, 414–452 (1957)
11. A. Bayer, A tour to stability conditions on derived categories, http://www.maths.ed.ac.uk/~abayer/dc-lecture-notes.pdf (2011)
12. A. Bayer, Wall-crossing implies Brill-Noether. Applications of stability conditions on surfaces (2016). arXiv:1604.08261
13. A. Bayer, A short proof of the deformation property of Bridgeland stability conditions (2016). arXiv:1606.02169
14. A. Bayer, T. Bridgeland, Derived automorphism groups of *K3* surfaces of Picard rank 1. Duke Math. J. **166**(1), 75–124 (2017)
15. A. Bayer, E. Macrì, The space of stability conditions on the local projective plane. Duke Math. J. **160**(2), 263–322 (2011)

16. A. Bayer, E. Macrì, MMP for moduli of sheaves on K3s via wall-crossing: nef and movable cones, Lagrangian fibrations. Invent. Math. **198**(3), 505–590 (2014)
17. A. Bayer, E. Macrì, Projectivity and birational geometry of Bridgeland moduli spaces. J. Am. Math. Soc. **27**(3), 707–752 (2014)
18. A. Bayer, E. Macrì, Y. Toda, Bridgeland stability conditions on threefolds I: Bogomolov-Gieseker type inequalities. J. Algebraic Geom. **23**(1), 117–163 (2014)
19. A. Bayer, E. Macrì, P. Stellari, The space of stability conditions on abelian threefolds, and on some Calabi-Yau threefolds. Invent. Math. **206**(3), 869–933 (2016)
20. A. Bayer, M. Lahoz, E. Macrì, P. Stellari, Stability conditions on Kuznetsov components (2016). arXiv:1703.10839
21. A.A. Beilinson, Coherent sheaves on $\mathbf{P}^n$ and problems in linear algebra. Funktsional. Anal. i Prilozhen. **12**(3), 68–69 (1978)
22. A.A. Beĭlinson, J. Bernstein, P. Deligne, Faisceaux pervers, in *Analysis and Topology on Singular Spaces, I (Luminy, 1981)*. Astérisque, vol. 100 (Société Mathématique de France, Paris, 1982), pp. 5–171
23. M. Bernardara, E. Macrì, B. Schmidt, X. Zhao, Bridgeland stability conditions on Fano threefolds (2016). arXiv:1607.08199
24. F.A. Bogomolov, Holomorphic tensors and vector bundles on projective manifolds. Izv. Akad. Nauk SSSR Ser. Mat. **42**(6), 1227–1287, 1439 (1978)
25. B. Bolognese, J. Huizenga, Y. Lin, E. Riedl, B. Schmidt, M. Woolf, X. Zhao, Nef cones of Hilbert schemes of points on surfaces, 2015. Algebra Number Theory **10**(4), 907–930 (2016)
26. A. Bondal, D. Orlov, Semiorthogonal decomposition for algebraic varieties (1995). arXiv:alg-geom/9506012
27. A. Bondal, M. van den Bergh, Generators and representability of functors in commutative and noncommutative geometry. Mosc. Math. J. **3**(1), 1–36, 258 (2003)
28. T. Bridgeland, Derived categories of coherent sheaves, in *International Congress of Mathematicians*. vol. II (European Mathematical Society, Zürich, 2006), pp. 563–582
29. T. Bridgeland, Stability conditions on a non-compact Calabi-Yau threefold. Commun. Math. Phys. **266**(3), 715–733 (2006)
30. T. Bridgeland, Stability conditions on triangulated categories. Ann. Math. (2) **166**(2), 317–345 (2007)
31. T. Bridgeland, Stability conditions on $K3$ surfaces. Duke Math. J. **141**(2), 241–291 (2008)
32. T. Bridgeland, Spaces of stability conditions, in *Algebraic Geometry—Seattle 2005. Part 1*. Proceedings of Symposia in Pure Mathematics, vol. 80 (American Mathematical Society, Providence, RI, 2009), pp. 1–21
33. T. Bridgeland, A. Maciocia, Fourier-Mukai transforms for $K3$ and elliptic fibrations. J. Algebraic Geom. **11**(4), 629–657 (2002)
34. T. Bridgeland, I. Smith, Quadratic differentials as stability conditions. Publ. Math. Inst. Hautes Études Sci. **121**, 155–278 (2015)
35. G. Castelnuovo, *Richerche di geometria sulle curve algebriche* (Zanichelli, Bologna, 1937)
36. F. Catanese, L. Göttsche, d-very-ample line bundles and embeddings of Hilbert schemes of 0-cycles. Manuscr. Math. **68**(3), 337–341 (1990)
37. I. Coskun, J. Huizenga, The nef cone of the moduli space of sheaves and strong Bogomolov inequalities (2015). arXiv:1512.02661
38. I. Coskun, J. Huizenga, M. Woolf, The effective cone of the moduli space of sheaves on the plane. J. Eur. Math. Soc. **19**(5), 1421–1467 (2017)
39. U.V. Desale, S. Ramanan, Classification of vector bundles of rank 2 on hyperelliptic curves. Invent. Math. **38**(2), 161–185 (1976/1977)
40. G. Dimitrov, F. Haiden, L. Katzarkov, M. Kontsevich, Dynamical systems and categories, in *The Influence of Solomon Lefschetz in Geometry and Topology*. Contemporary Mathematics, vol. 621 (American Mathematical Society, Providence, RI, 2014), pp. 133–170
41. M.R. Douglas, Dirichlet branes, homological mirror symmetry, and stability, in *Proceedings of the International Congress of Mathematicians, Vol. III (Beijing, 2002)* (Higher Education Press, Beijing, 2002), pp. 395–408

42. J.-M. Drezet, M.S. Narasimhan, Groupe de Picard des variétés de modules de fibrés semi-stables sur les courbes algébriques. Invent. Math. **97**(1), 53–94 (1989)
43. J. Fogarty, Algebraic families on an algebraic surface. Am. J. Math. **90**, 511–521 (1968)
44. J. Fogarty, Algebraic families on an algebraic surface. II. The Picard scheme of the punctual Hilbert scheme. Am. J. Math. **95**, 660–687 (1973)
45. W. Fulton, *Intersection Theory. Ergebnisse der Mathematik und ihrer Grenzgebiete. 3. Folge. A Series of Modern Surveys in Mathematics [Results in Mathematics and Related Areas. 3rd Series. A Series of Modern Surveys in Mathematics]*, vol. 2, 2nd edn. (Springer, Berlin, 1998)
46. D. Gaiotto, G.W. Moore, A. Neitzke, Wall-crossing, Hitchin systems, and the WKB approximation. Adv. Math. **234**, 239–403 (2013)
47. P. Gallardo, C.L. Huerta, B. Schmidt, Families of elliptic curves in $\mathbb{P}^3$ and Bridgeland stability (2016). arXiv:1609.08184
48. F.J. Gallego, B.P. Purnaprajna, Vanishing theorems and syzygies for $K3$ surfaces and Fano varieties. J. Pure Appl. Algebra **146**(3), 251–265 (2000)
49. S.I. Gelfand, Y.I. Manin, *Methods of Homological Algebra*, 2nd edn. Springer Monographs in Mathematics (Springer, Berlin, 2003)
50. D. Gieseker, On the moduli of vector bundles on an algebraic surface. Ann. Math. (2) **106**(1), 45–60 (1977)
51. D. Gieseker, On a theorem of Bogomolov on Chern classes of stable bundles. Am. J. Math. **101**(1), 77–85 (1979)
52. D.R. Grayson, Reduction theory using semistability. Comment. Math. Helv. **59**(4), 600–634 (1984)
53. F. Haiden, L. Katzarkov, M. Kontsevich, Flat surfaces and stability structures (2014). arXiv:1409.8611
54. D. Happel, I. Reiten, S.O. Smalø, Tilting in abelian categories and quasitilted algebras. Mem. Am. Math. Soc. **120**(575), viii+ 88 (1996)
55. G. Harder, M.S. Narasimhan, On the cohomology groups of moduli spaces of vector bundles on curves. Math. Ann. **212**, 215–248 (1974/1975)
56. R. Hartshorne, *Ample Subvarieties of Algebraic Varieties*. Lecture Notes in Mathematics, vol. 156 (Springer, Berlin/New York, 1970). Notes written in collaboration with C. Musili
57. R. Hartshorne, *Algebraic Geometry*. Graduate Texts in Mathematics, vol. 52 (Springer, New York/Heidelberg, 1977)
58. G. Hein, D. Ploog, Postnikov-stability for complexes on curves and surfaces. Int. J. Math. **23**(2), 1250048, 20 pp. (2012)
59. G. Hein, D. Ploog, Postnikov-stability versus semistability of sheaves. Asian J. Math. **18**(2), 247–261 (2014)
60. J. Huizenga, Birational geometry of moduli spaces of sheaves and Bridgeland stability (2016). arXiv:1606.02775
61. D. Huybrechts, *Fourier-Mukai Transforms in Algebraic Geometry*. Oxford Mathematical Monographs (The Clarendon Press/Oxford University Press, Oxford, 2006)
62. D. Huybrechts, Introduction to stability conditions, in *Moduli Spaces*. London Mathematical Society Lecture Note Series, vol. 411 (Cambridge University Press, Cambridge, 2014), pp. 179–229
63. D. Huybrechts, M. Lehn, *The Geometry of Moduli Spaces of Sheaves*. Cambridge Mathematical Library, 2nd edn. (Cambridge University Press, Cambridge, 2010)
64. M. Inaba, Toward a definition of moduli of complexes of coherent sheaves on a projective scheme. J. Math. Kyoto Univ. **42**(2), 317–329 (2002)
65. D. Joyce, Conjectures on Bridgeland stability for Fukaya categories of Calabi-Yau manifolds, special Lagrangians, and Lagrangian mean curvature flow. EMS Surv. Math. Sci. **2**(1), 1–62 (2015)
66. A.D. King, Moduli of representations of finite-dimensional algebras. Quart. J. Math. Oxford Ser. (2) **45**(180), 515–530 (1994)
67. S. Kobayashi, *Differential Geometry of Complex Vector Bundles*. Publications of the Mathematical Society of Japan, vol. 15. (Princeton University Press, Princeton, NJ; Iwanami Shoten, Tokyo, 1987). Kanô Memorial Lectures, 5

68. M. Kontsevich, Y. Soibelman, Stability structures, motivic Donaldson-Thomas invariants and cluster transformations (2008). arXiv:0811.2435
69. A. Kuznetsov, Semiorthogonal decompositions in algebraic geometry (2014). arXiv:1404.3143
70. A. Langer, Semistable sheaves in positive characteristic. Ann. Math. (2) **159**(1), 251–276 (2004)
71. R. Lazarsfeld, A sampling of vector bundle techniques in the study of linear series, in *Lectures on Riemann Surfaces (Trieste, 1987)* (World Scientific Publishing, Teaneck, NJ, 1989), pp. 500–559
72. J. Le Potier, *Lectures on Vector Bundles*. Cambridge Studies in Advanced Mathematics, vol. 54 (Cambridge University Press, Cambridge, 1997). Translated by A. Maciocia
73. C. Li, Stability conditions on Fano threefolds of Picard number one (2015). arXiv:1510.04089
74. C. Li, X. Zhao, The MMP for deformations of Hilbert schemes of points on the projective plane (2013). arXiv:1312.1748
75. C. Li, X. Zhao, Birational models of moduli spaces of coherent sheaves on the projective plane (2016). arXiv:1603.05035.
76. M. Lieblich, Moduli of complexes on a proper morphism. J. Algebraic Geom. **15**(1), 175–206 (2006)
77. A. Maciocia, Computing the walls associated to Bridgeland stability conditions on projective surfaces. Asian J. Math. **18**(2), 263–279 (2014)
78. A. Maciocia, C. Meachan, Rank 1 Bridgeland stable moduli spaces on a principally polarized abelian surface. Int. Math. Res. Not. IMRN **9**, 2054–2077 (2013)
79. A. Maciocia, D. Piyaratne, Fourier-Mukai transforms and Bridgeland stability conditions on abelian threefolds. Algebr. Geom. **2**(3), 270–297 (2015)
80. A. Maciocia, D. Piyaratne, Fourier–Mukai transforms and Bridgeland stability conditions on abelian threefolds II. Int. J. Math. **27**(1), 1650007, 27 (2016)
81. E. Macrì, Stability conditions on curves. Math. Res. Lett. **14**(4), 657–672 (2007)
82. E. Macrì, A generalized Bogomolov-Gieseker inequality for the three-dimensional projective space. Algebra Number Theory **8**(1), 173–190 (2014)
83. M. Maruyama, Moduli of stable sheaves. I. J. Math. Kyoto Univ. **17**(1), 91–126 (1977)
84. M. Maruyama, Moduli of stable sheaves. II. J. Math. Kyoto Univ. **18**(3), 557–614 (1978)
85. K. Matsuki, R. Wentworth, Mumford-Thaddeus principle on the moduli space of vector bundles on an algebraic surface. Int. J. Math. **8**(1), 97–148 (1997)
86. H. Minamide, S. Yanagida, K. Yoshioka, Some moduli spaces of Bridgeland's stability conditions. Int. Math. Res. Not. IMRN **19**, 5264–5327 (2014)
87. M.S. Narasimhan, S. Ramanan, Moduli of vector bundles on a compact Riemann surface. Ann. Math. (2) **89**, 14–51 (1969)
88. M.S. Narasimhan, C.S. Seshadri, Holomorphic vector bundles on a compact Riemann surface. Math. Ann. **155**, 69–80 (1964)
89. M.S. Narasimhan, C.S. Seshadri, Stable and unitary vector bundles on a compact Riemann surface. Ann. Math. (2) **82**, 540–567 (1965)
90. P.E. Newstead, *Introduction to Moduli Problems and Orbit Spaces*. Tata Institute of Fundamental Research Lectures on Mathematics and Physics, vol. 51 (Tata Institute of Fundamental Research, Bombay; Narosa Publishing House, New Delhi, 1978)
91. H. Nuer, Projectivity and birational geometry of Bridgeland moduli spaces on an Enriques surface. Proc. Lond. Math. Soc. **113**(3), 345–386 (2016)
92. S. Okada, Stability manifold of $\mathbb{P}^1$. J. Algebraic Geom. **15**(3), 487–505 (2006)
93. M. Olsson, *Algebraic Spaces and Stacks*. Colloquium Publications, vol. 62 (American Mathematical Society, Providence, RI, 2016)
94. A. Ortega, On the moduli space of rank 3 vector bundles on a genus 2 curve and the Coble cubic. J. Algebraic Geom. **14**(2), 327–356 (2005)
95. D. Piyaratne, Generalized Bogomolov-Gieseker type inequalities on Fano 3-folds (2016). arXiv:1607.07172

96. D. Piyaratne, Y. Toda, Moduli of Bridgeland semistable objects on 3-folds and Donaldson-Thomas invariants (2015). arXiv:1504.01177
97. A. Polishchuk, *Abelian Varieties, Theta Functions and the Fourier Transform*. Cambridge Tracts in Mathematics, vol. 153 (Cambridge University Press, Cambridge, 2003)
98. A. Polishchuk, Constant families of t-structures on derived categories of coherent sheaves. Mosc. Math. J. **7**(1), 109–134, 167 (2007)
99. M. Reid, Bogomolov's theorem $c_1^2 \leq 4c_2$, in *Proceedings of the International Symposium on Algebraic Geometry (Kyoto University, Kyoto, 1977)* (Kinokuniya Book Store, Tokyo, 1978), pp. 623–642
100. I. Reider, Vector bundles of rank 2 and linear systems on algebraic surfaces. Ann. Math. (2) **127**(2), 309–316 (1988)
101. B. Schmidt, A generalized Bogomolov-Gieseker inequality for the smooth quadric threefold. Bull. Lond. Math. Soc. **46**(5), 915–923 (2014)
102. B. Schmidt, Bridgeland stability on threefolds – Some wall crossings (2015). arXiv:1509.04608
103. B. Schmidt, Counterexample to the generalized Bogomolov–Gieseker inequality for three-folds. Int. Math. Res. Not. (8), 2562–2566 (2017)
104. C.S. Seshadri, *Fibrés vectoriels sur les courbes algébriques*. Astérisque, vol. 96 (Société Mathématique de France, Paris, 1982). Notes written by J.-M. Drezet from a course at the École Normale Supérieure, June 1980
105. C.S. Seshadri, Vector bundles on curves, in *Linear Algebraic Groups and Their Representations (Los Angeles, CA, 1992)*. Contemporary Mathematics, vol. 153 (American Mathematical Society, Providence, RI, 1993), pp. 163–200
106. S. Shatz, Degeneration and specialization in algebraic families of vector bundles. Bull. Am. Math. Soc. **82**(4), 560–562 (1976)
107. C.T. Simpson, Moduli of representations of the fundamental group of a smooth projective variety. I. Inst. Hautes Études Sci. Publ. Math. **79**, 47–129 (1994)
108. The Stacks Project Authors, *Stacks Project*, http://stacks.math.columbia.edu (2016)
109. Y. Toda, Moduli stacks and invariants of semistable objects on $K3$ surfaces. Adv. Math. **217**(6), 2736–2781 (2008)
110. Y. Toda, Limit stable objects on Calabi-Yau 3-folds. Duke Math. J. **149**(1), 157–208 (2009)
111. Y. Toda, Introduction and open problems of Donaldson-Thomas theory, in *Derived Categories in Algebraic Geometry*, EMS Series of Congress Reports, pp. 289–318 (European Mathematical Society, Zürich, 2012)
112. Y. Toda, Stability conditions and curve counting invariants on Calabi-Yau 3-folds. Kyoto J. Math. **52**(1), 1–50 (2012)
113. Y. Toda, Derived category of coherent sheaves and counting invariants (2014). arXiv:1404.3814
114. J.-L. Verdier, Des catégories dérivées des catégories abéliennes. Astérisque **239**, xii+253 pp. (1997). 1996 With a preface by Luc Illusie, Edited and with a note by Georges Maltsiniotis
115. M. Woolf, Nef and effective cones on the moduli space of torsion sheaves on the projective plane (2013). arXiv:1305.1465
116. B. Xia, The Hilbert scheme of twisted cubics as simple wall-crossing (2016). arXiv:1608.04609
117. K. Yoshioka, Some remarks on Bridgeland stability conditions on K3 and Enriques surfaces (2016). arXiv:1607.04946
118. S. Yanagida, K. Yoshioka, Bridgeland's stabilities on abelian surfaces. Math. Z. **276**(1-2), 571–610 (2014)

Algebraic Curves and Their Moduli

E. Sernesi

This survey is modeled on the CIMPA introductory lectures I gave in Guanajuato in February 2016 about moduli of curves. Surveys on this topic abound and for this reason I decided to focus on few specific aspects giving some illustration of the relation between local and global properties of moduli. The central theme is the interplay between the various notions of "generality" for a curve of genus g. Of course these notes reflect my own view of the subject. For a recent comprehensive text on moduli of curves I refer to [5].

1 Parameters

All schemes will be defined over $\mathbb{C}$. The category of algebraic schemes will be denoted by (Schemes).

Algebraic Varieties Depend on Parameters This is clear if we define them by means of equations in some (affine or projective) space, because one can vary the coefficients of the equations. For example, by moving the coefficients of their equation we parametrize *nonsingular plane curves* of degree d in $\mathbb{P}^2$ by the points of (an open subset of) a $\mathbb{P}^N$, where $N = \frac{d(d+3)}{2}$.

Less obviously, consider a *nonsingular rational cubic curve* $C \subset \mathbb{P}^3$. Up to choice of coordinates it can be defined by the three quadric equations:

$$X_1X_3 - X_2^2 = X_0X_3 - X_1X_2 = X_0X_2 - X_1^2 = 0 \tag{1}$$

E. Sernesi (✉)
Dipartimento di Matematica e Fisica, Università Roma Tre, L.go S.L. Murialdo 1, 00146 Roma, Italy
e-mail: sernesi@mat.uniroma3.it

L. Brambila Paz et al. (eds.), *Moduli of Curves*, Lecture Notes of the Unione Matematica Italiana 21, DOI 10.1007/978-3-319-59486-6_6

that I will write as

$$\mathrm{rk}\begin{pmatrix} X_0 \, X_1 \, X_2 \\ X_1 \, X_2 \, X_3 \end{pmatrix} \leq 1 \tag{2}$$

If we deform arbitrarily the coefficients of Eq. (1) their intersection will consist of eight distinct points. Bad choice!

Good choice: deform the entries of the matrix (2) to general linear forms in $X_0, \ldots, X_3$. This will correspond to ask that the corresponding family of subvarieties is *flat*, and will guarantee that we obtain twisted cubics again.

2 From Parameters to Moduli

Parameters are a naive notion. Moduli are a more refined notion: they are *parameters of isomorphism classes* of objects.

Typical example: the difference between parametrizing plane conics and parametrizing plane cubics. Conics depend on five parameters but have no moduli, cubics depend on nine parameters and have one modulus.

What does it mean that cubics have one modulus? This has been an important discovery in the nineteenth century. It consists of the following steps [13, Chap. 3]:

- given an ordered 4-tuple of pairwise distinct points $P_i = [a_i, b_i] \in \mathbb{P}^1$, $i = 1, \ldots, 4$, consider their *cross ratio*

$$\lambda = \frac{(a_1b_3 - a_3b_1)(a_2b_4 - a_4b_2)}{(a_1b_4 - a_4b_1)(a_2b_3 - a_3b_2)}$$

It is invariant under linear coordinate changes (direct computation) and takes all values $\neq 0, 1$.

As we permute the points their cross ratio takes the values

$$\lambda, 1 - \lambda, \frac{1}{\lambda}, \frac{1}{1-\lambda}, \frac{\lambda}{1-\lambda}, \frac{1-\lambda}{\lambda}$$

and the expression

$$j(\lambda) := 2^8 \frac{(\lambda^2 - \lambda + 1)^3}{\lambda^2(\lambda - 1)^2}$$

takes the same value if and only if we replace λ by any of the above six expressions. Moreover each $j \in \mathbb{C}$ is of the form $j(\lambda)$ for some $\lambda \neq 0, 1$. Therefore we obtain a 1-1 correspondence between $\mathbb{A}^1_{\mathbb{C}}$ and unordered 4-tuples of distinct points of $\mathbb{P}^1$ up to coordinate changes.

- Given a nonsingular cubic $C \subset \mathbb{P}^2$ and $P \in C$ there are four distinct tangent lines to C passing through P besides the tangent line at P. View them as points of $\mathbb{P}(T_P\mathbb{P}^2) \cong \mathbb{P}^1$, and compute their $j(\lambda)$. Then it can be proved that $j(\lambda)$ is independent of P. Call it $j(C)$.

  For example if C is in *Hesse normal form*

$$X_0^3 + X_1^3 + X_2^3 + 6\alpha X_0X_1X_2 = 0$$

  then $1 + 8\alpha^3 \neq 0$ and $j(C) = \frac{64(\alpha-\alpha^4)^3}{(1+8\alpha^3)^3}$
- For every $j \in \mathbb{C}$ there exists a nonsingular cubic C such that $j = j(C)$.
- (Salmon) Two nonsingular cubics C, C' are isomorphic if and only if $j(C) = j(C')$.

Therefore *in some sense* the set of isomorphism classes of plane cubics is identified with $\mathbb{A}^1_{\mathbb{C}}$. Or we might say that $\mathbb{A}^1_{\mathbb{C}}$ is the *moduli space* of plane cubics.

This is not yet a satisfactory definition, because the relation between $\mathbb{A}^1_{\mathbb{C}}$ and parametrized curves is not fully transparent. We will discuss why in Sect. 4.

It is difficult to distinguish which parameters are moduli.

For example consider the following linear pencil of plane quartics:

$$\lambda F_4(X_0, X_1, X_2) + \mu(X_0^4 + X_1^4 + X_2^4) = 0, \quad (\lambda, \mu) \in \mathbb{P}^1 \tag{3}$$

where $F_4(X_0, X_1, X_2)$ is a general quartic homogeneous polynomial. The two quartics $F_4(X_0, X_1, X_2) = 0$ and $X_0^4 + X_1^4 + X_2^4 = 0$ are not isomorphic because F_4 has 24 ordinary flexes and the other quartic has 12 hyperflexes (i.e. nonsingular points where the tangent lines meets the curve with multiplicity 4).

Can we infer, from the fact that it contains two non-isomorphic members, that the pencil depends continuously on one modulus? Of course our intuition suggests that this should be the case. But in moduli theory there are two pathologies that can appear, the existence of *isotrivial families* and the *jumping phenomenon* (both related to the presence of automorphisms). They are fatal to the existence of a good moduli space, and we cannot exclude just by intuition that they appear in this case. We will discuss later the subtle role of isotriviality, and we now give an example of jumping phenomenon.

Example 1 Consider the graded $\mathbb{C}[t]$-algebra $R = \mathbb{C}[t, X_0, X_1]$ and the graded R-module

$$M = \operatorname{coker}\left[\, R(-1) \xrightarrow{\begin{pmatrix} X_0 \\ X_1 \\ t \end{pmatrix}} R \oplus R \oplus R(-1) \,\right]$$

Then $\mathrm{Proj}(R) = \mathbb{A}^1 \times \mathbb{P}^1$ and $\mathcal{F} := \widetilde{M}$ is a rank-two locally free sheaf on $\mathbb{A}^1 \times \mathbb{P}^1$. Then $\mathbb{P}(\mathcal{F}) \to \mathbb{A}^1 \times \mathbb{P}^1$ is a $\mathbb{P}^1$-bundle and, after composing with the projection, we get a proper smooth family:

$$f : \mathbb{P}(\mathcal{F}) \longrightarrow \mathbb{A}^1$$

whose fibres are rational ruled surfaces. Precisely, one checks that

$$f^{-1}(\alpha) = \mathbb{P}(\mathcal{O}_{\mathbb{P}^1} \oplus \mathcal{O}_{\mathbb{P}^1}), \quad \alpha \neq 0$$
$$f^{-1}(0) = \mathbb{P}(\mathcal{O}_{\mathbb{P}^1}(-1) \oplus \mathcal{O}_{\mathbb{P}^1}(1))$$

Therefore all fibres over $\mathbb{A}^1 \setminus \{0\}$ are pairwise isomorphic, while $f^{-1}(0)$ belongs to a different isomorphism class. There is no continuous variation of isomorphism classes of the fibres because we have a jumping above 0. This excludes the possibility of having a space parametrizing isomorphism classes of ruled surfaces and with reasonable geometric properties. For details about this example we refer to [36, pp. 53–54].

As observed, we cannot a priori exclude that a similar jumping phenomenon takes place in the pencil (3). We will eventually do, but after some hard work has been done. It's now time to move to a more advanced point of view.

3 Families

A *family* of projective nonsingular irreducible curves of genus g parametrized by a scheme B is a projective smooth morphism:

$$f : \mathcal{C} \longrightarrow B$$

whose fibres are projective nonsingular curves of genus g. The fibre over a point $b \in B$ will be denoted by $f^{-1}(b)$ or by $\mathcal{C}(b)$. Recall that the *genus* of a nonsingular curve C is

$$g(C) := \dim(H^1(C, \mathcal{O}_C)) = \dim(H^0(C, \Omega^1_C))$$

More generally, if C is a possibly singular projective curve we define its *arithmetic genus* by

$$p_a(C) := 1 - \chi(\mathcal{O}_C)$$

Clearly $g(C) = p_a(C)$ if C is nonsingular and irreducible.

A *family of deformations of a given projective nonsingular curve C* parametrized by a pointed scheme (B, b) is a pullback diagram:

$$\begin{array}{ccc} C & \hookrightarrow & \mathcal{C} \\ \downarrow & & \downarrow f \\ \mathrm{Spec}(\mathbb{C}) & \xrightarrow{b} & B \end{array} \qquad (4)$$

where f is a family of projective nonsingular curves of genus g. This means that, in addition to the family f, an isomorphism $C \cong \mathcal{C}(b)$ is given.

In the above definitions we can replace smooth curves by possibly singular ones, but then we must require that the family f is flat.

An *isomorphism between two families of curves* $f : \mathcal{C} \to B$ and $\varphi : \mathcal{D} \to B$ is just a B-isomorphism:

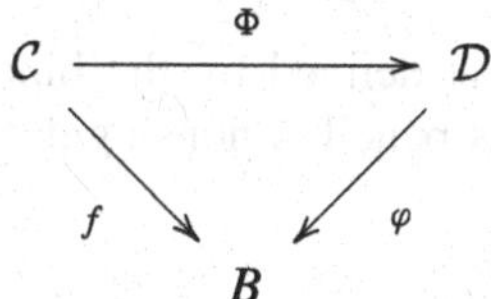

The family $f : \mathcal{C} \to B$ is *trivial* if it is isomorphic to a *product family*

$$p : B \times C \to B$$

for some curve C.

An *isomorphism between two families of deformations of* C, say (4) and

$$\begin{array}{ccc} C & \hookrightarrow & \mathcal{D} \\ \downarrow & & \downarrow \varphi \\ \mathrm{Spec}(\mathbb{C}) & \xrightarrow{b} & B \end{array}$$

is an isomorphism between the families f and φ which commutes with the identifications of C with $\mathcal{C}(b)$ and with $\mathcal{D}(b)$. Similarly one defines the notion of *trivial deformation of C*.

A family of curves *embedded* in a projective variety X is a commutative diagram:

$$\begin{array}{ccc} \mathcal{C} & \overset{j}{\hookrightarrow} & B \times X \\ f \downarrow & \swarrow \pi & \\ B & & \end{array}$$

where f is a family of projective curves of genus g and π is the projection.

Most important case: $X = \mathbb{P}^r$. One may include the case $r = 1$ by replacing the inclusion j by a finite flat morphism. In this case for each closed point $b \in B$ the fibre $j(b) : \mathcal{C}(b) \to \mathbb{P}^1$ will be a ramified cover of degree independent of b.

If B is irreducible then $\dim(B)$ is defined to be the *number of parameters of the family* f.

In practice one often considers families including singular curves among their fibres. One then replaces the smoothness condition by flatness, which guarantees the constancy of the arithmetic genus of the fibres.

For example, the pencil of plane quartics considered before defines a family of curves embedded in $\mathbb{P}^2$:

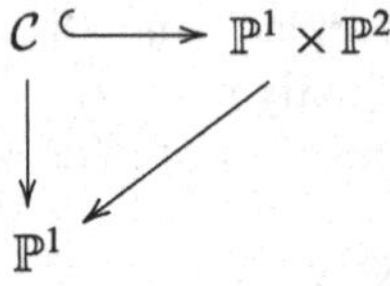

parametrized by $\mathbb{P}^1$, where $\mathcal{C}$ is defined by the bihomogeneous equation of the pencil. The general fibre of this pencil is nonsingular, but there are some singular fibres.

4 Moduli Functors

We would like to construct, for each $g \geq 0$, an algebraic $\mathbb{C}$-scheme M_g, to be called "moduli space of curves", whose closed points are in 1-1 correspondence with the set of isomorphism classes of projective nonsingular irreducible curves (shortly "curves") of a given genus g.

We saw in Sect. 2 that in the case $g = 1$ the affine line $\mathbb{A}^1_{\mathbb{C}}$ does the job, even though we were not completely satisfied by this solution. One question remained unanswered:

Where should the structure of scheme of M_g come from?

We expect that such structure is *natural* in some sense, i.e. that it reflects in a precise way the nature of moduli as parameters. More clearly stated, we want a precise relation between M_g and families of curves of genus g. The functorial point of view comes to help at this point.

To every scheme X there is associated its *functor of points*

$$h_X : (\text{Schemes}) \to (\text{Sets}), \quad h_X(S) = \text{Mor}(S, X)$$

and conversely X can be reconstructed from this functor. So we must look for a functor on the first place and it must be related with families of curves of genus g. Here is one.

Setting

$$\mathcal{M}_g(B) = \left\{ \begin{array}{c} \text{families } C \to B \\ \text{of curves of genus } g \end{array} \right\} \Big/ \cong$$

(where by $\cong$ we mean "isomorphism") we obtain a contravariant functor

$$\mathcal{M}_g : (\text{Schemes}) \to (\text{Sets})$$

called the *moduli functor* of nonsingular curves of genus g.

The optimistic expectation is that $\mathcal{M}_g$ is representable, i.e. that it is the functor of points of a scheme M_g. The representability implies that M_g comes equipped with a *universal family* $\pi : \mathcal{X} \to M_g$ of curves of genus g.

"Universal" means that every other family $f : \mathcal{C} \to B$ of curves of genus g is obtained by pulling back π via a unique morphism $\mu_f : B \to M_g$. This property implies that the closed points of M_g are in 1-1 correspondence with the set of isomorphism classes of curves of genus g (because such isomorphism classes are in turn in 1-1 correspondence with families of the form $C \to \text{Spec}(\mathbb{C})$ up to isomorphism). Therefore the morphism μ_f necessarily maps a $\mathbb{C}$-rational point $b \in B$ to the isomorphism class $[\mathcal{C}(b)]$.

The pair (M_g, π) would then represent the functor $\mathcal{M}_g$. In other words it would imply the existence of an isomorphism of functors

$$\mathcal{M}_g \cong h_{M_g}$$

and it would be fair to call such M_g *the moduli space* (or moduli scheme) of curves of genus g. Actually its name would be *fine moduli space*.

The situation is not that simple though. A universal family of curves of genus g does not exist and this is due to the existence of non-trivial families $f : \mathcal{C} \to B$ whose geometric fibres are pairwise isomorphic. In fact, such a family, like any other, should be induced by pulling back the universal family:

$$\begin{array}{ccc} \mathcal{C} & \longrightarrow & \mathcal{X} \\ {\scriptstyle f}\downarrow & & \downarrow{\scriptstyle \pi} \\ B & \xrightarrow[\mu_f]{} & M_g \end{array}$$

But since all fibres of f are isomorphic μ_f is constant and therefore f would be trivial.

Example 2 The above phenomenon appears already in genus $g = 0$. There is only one isomorphism class of curves of genus zero, namely $[\mathbb{P}^1]$. So if $\mathcal{M}_0$ were representable the universal family would just be $\mathbb{P}^1 \to \text{Spec}(\mathbb{C})$. This would imply, as shown above, that every family $f : \mathcal{C} \to B$ of curves of genus zero is trivial. But this contradicts the existence of non-trivial ruled surfaces $\mathcal{C} \to B$, i.e. ruled surfaces not isomorphic to $B \times \mathbb{P}^1 \to B$.

Example 3 Consider the pencil of nonsingular plane cubics $\mathcal{C} \to \mathbb{A}^1$ given in affine coordinates by:

$$Y^2 = X^3 + t, \quad t \in \mathbb{A}^1 \setminus \{0\} \tag{5}$$

The j-invariant of $\mathcal{C}(t)$ is constant and equal to zero. Therefore all $\mathcal{C}(t)$ are pairwise isomorphic. But it is impossible to give an isomorphism of this family with the constant family $Y^2 = X^3 + 1$ without introducing the irrationality $t^{1/6}$. Therefore (5) is non-trivial.

A systematic way of producing non-trivial families of curves of higher genus whose geometric fibres are pairwise isomorphic is by means of the notion of isotriviality.

Definition 4 Let $f : \mathcal{Z} \to S$ be a flat family of algebraic schemes. Then f is called *isotrivial* if there is a finite surjective and etale morphism $S' \to S$ such that the induced family $f_{S'} : S' \times_S \mathcal{Z} \to S'$ is trivial. If $S' \times_S \mathcal{Z} \cong S' \times Z$ we say that f is *isotrivial with fibre Z*.

Example 5 (An Example of Isotrivial Family) Consider the family

$$f : \mathrm{Spec}(\mathbb{C}[Z, t, t^{-1}]/(Z^2 - t)) \longrightarrow \mathrm{Spec}(\mathbb{C}[t, t^{-1}]) = \mathbb{A}^1 \setminus \{0\}$$

For each $0 \neq a \in \mathbb{A}^1$ the fibre $f^{-1}(a) = \mathrm{Spec}(\mathbb{C}[Z]/(Z^2 - a))$ consists of two distinct reduced points: hence all fibres of f are isomorphic to $X = \mathrm{Spec}(\mathbb{C}[Z]/(Z^2 - 1))$. As in Example 3 one checks that the family is not trivial. Consider the etale morphism:

$$\beta : \mathrm{Spec}(\mathbb{C}[u, u^{-1}]) \longrightarrow \mathrm{Spec}(\mathbb{C}[t, t^{-1}]), \quad t \mapsto u^2$$

Pulling back f by β we obtain the family:

$$\mathrm{Spec}(\mathbb{C}[Z, u, u^{-1}]/(Z^2 - u^2)) \longrightarrow \mathrm{Spec}(\mathbb{C}[u, u^{-1}])$$

and this family is trivial. Therefore f is isotrivial but not trivial.

The existence of isotrivial families is regulated by the following simple result.

Proposition 6 *The following conditions are equivalent on a quasi-projective scheme X:*

- *There exists a non-trivial isotrivial family with fibre X.*
- *The group Aut(X) contains a non-trivial finite subgroup.*

For the proof we refer to [43, Th. 2.6.15]. For each $g \geq 2$ there exist curves of genus g with non-trivial automorphisms, and all such curves have a finite group of automorphisms; therefore the proposition applies to them and implies the existence of non-trivial isotrivial families of curves of any genus $g \geq 2$.

The conclusion is that *a universal family of curves of genus g does not exist,* for all $g \geq 0$. Equivalently, *the moduli functor $\mathcal{M}_g$ is not representable* for all g.

No Panic: despite these discouraging phenomena we still are on the right track because any reasonable structure on the set of genus g curves must be somehow compatible with the moduli functor. All we have to do is to weaken the condition that there is a universal family. There are several ways to do this. The one we choose is via the notion of *coarse moduli space.*

5 The Coarse Moduli Space of Curves

The following definition is due to Mumford [38].

Definition 7 The *coarse moduli space of curves of genus g* is an algebraic scheme M_g such that:

- There is a morphism of functors $\mathcal{M}_g \to h_{M_g}$ which induces a bijection

$$\mathcal{M}_g(\mathbb{C}) \cong M_g(\mathbb{C}) = \{\mathbb{C}\textit{-rational points of } M_g\}$$

- If N is another scheme such that there is a morphism of functors $\mathcal{M}_g \to h_N$ then there is a unique morphism $M_g \to N$ such that the following diagram commutes:

$$\mathcal{M}_g \longrightarrow h_{M_g} \longrightarrow h_N$$

The definition implies that:

- The closed points of M_g are in 1-1 correspondence with the isomorphism classes of (nonsingular projective) curves of genus g.
- for every family $f : C \to B$ of curves of genus g the set theoretic map

$$B(\mathbb{C}) \ni b \mapsto [f^{-1}(b)] \in M_g$$

defines a morphism $\mu_f : B \to M_g$. (this is the *universal property* of M_g).

It is easy to prove that, if it exists, M_g is unique up to isomorphism. In that case we say that $\mathcal{M}_g$ is *coarsely represented* by M_g.

The case $g = 0$ is trivial: $M_0 = \mathrm{Spec}(\mathbb{C})$ because $[\mathbb{P}^1]$ is the unique isomorphism class of curves of genus 0.

The case of genus 1, despite having served us as a useful heuristic introductory example, requires a special treatment because curves of genus 1 have a continuous group of automorphisms. It turns out to be more natural to consider families of *pointed* curves of genus 1. The functor of such curves then admits a coarse moduli space, which is isomorphic to $\mathbb{A}^1_{\mathbb{C}}$. The details of its construction are worked out in [27]. The general case is covered by the following deep result.

Theorem 8 *Let $g \geq 2$. Then:*

(i) *(Mumford [38]) M_g exists and is a quasi-projective normal algebraic scheme of dimension* $3g - 3$.
(ii) *(Deligne-Mumford [11], Fulton [18]) M_g is irreducible.*

The construction of M_g is obtained by means of Geometric Invariant Theory, which will not be considered in these lectures. As explained in the introduction of [11] the irreducibility of M_g had already been proved classically, via a topological analysis of families of branched coverings of $\mathbb{P}^1$. But an algebro-geometric proof was still lacking.

Moduli are local parameters on M_g around a given point $[C]$ and the *number of moduli* on which an abstract curve C depends is the dimension of M_g at $[C]$.

Now it is clear, at least theoretically, *how to distinguish moduli among parameters.* A family of curves $f : \mathcal{C} \to B$, with B an irreducible algebraic scheme, depends on $\dim(B)$ parameters and on $\dim(\mu_f(B))$ moduli.

For example in the *product family*

$$p : B \times C \longrightarrow B$$

all closed fibres are isomorphic to C and therefore $\mu_f(B) = \{[C]\}$: thus the number of moduli of this family is zero. This happens more generally if the family is isotrivial.

On the opposite side, an *effectively parametrized family* is one which depends on $\dim(B)$ moduli and such that μ_f is finite. This means that the fibre $\mathcal{C}(b)$ over any closed point $b \in B$ is isomorphic to only finitely many others.

Because of the universal property, every 1-parameter family of curves which contains two non-isomorphic fibres is effectively parametrized. In particular we can now deduce that the pencil of plane quartics (3) considered in Sect. 2 depends on one modulus.

A family $f : \mathcal{C} \to B$ is said to have *general moduli* if $\mu_f : B \to M_g$ is dominant. If a curve C is given as the general member of a family having general moduli, we say *C has general moduli* or that C is a *general curve of genus g*. This definition can be sometimes misleading because it presupposes that a family containing C has been given before we can say that it is a general curve. Nevertheless it is a classical and ubiquitous terminology.

Variants: Moduli of Pointed Curves

Given $g \geq 0$ and $n \geq 1$ a useful variant of M_g is the coarse moduli space $M_{g,n}$ of *n-pointed curves of genus g*.

It parametrizes pairs $(C; p_1, \ldots, p_n)$ consisting of a curve C of genus g and an ordered n-tuple $(p_1, \ldots, p_n)$ of distinct points of C.

The corresponding moduli functor is

$$\mathcal{M}_{g,n}(B) = \{(f : \mathcal{C} \to B, \sigma_1, \ldots, \sigma_m)\} \,/\text{isomorphism}$$

where $\sigma_1, \ldots, \sigma_n : B \to \mathcal{C}$ are disjoint sections of $f : \mathcal{C} \to B$, and the notion of isomorphism is the obvious one.

6 The Dimension of M_g

Riemann was able to count the number of moduli of curves of genus g, i.e. $\dim(M_g)$, by exhibiting a family of curves of genus g with general moduli in the following way. Assume $g \geq 4$.

Consider the family of all ramified covers of $\mathbb{P}^1$ of genus g and of a fixed degree n such that

$$\frac{g+2}{2} \leq n \leq g-1$$

We can represent it as

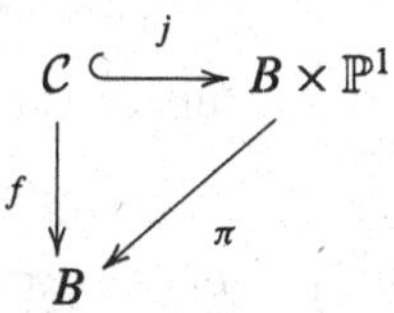

where B is a certain irreducible scheme. Riemann existence theorem implies that, associating to a cover the (ordered) set of its branch points we obtain a *finite* morphism $r : B \to (\mathbb{P}^1)^{2(n+g-1)}$. Therefore $\dim(B) = 2(n + g - 1)$.

Consider $\mu_f : B \to M_g$. We have the following facts:

- each curve of genus g can be expressed as a ramified cover of $\mathbb{P}^1$ defined by a line bundle of degree n provided $n \geq \frac{g+2}{2}$. Therefore μ_f is surjective.
- Composing a cover $f : C \to \mathbb{P}^1$ with a non-trivial automorphism $\alpha : \mathbb{P}^1 \to \mathbb{P}^1$ we obtain another cover $\alpha \cdot f : C \to \mathbb{P}^1$ defined on the same curve by the same line bundle L.
- In the range $\frac{g+2}{2} \leq n \leq g-1$ the line bundles L of degree n with two sections on a given curve C depend on $2n - 2 - g$ parameters.

Therefore the general fibre of μ_f has dimension

$$\dim(\text{PGL}(2)) + 2n - 2 - g = 2n + 1 - g$$

Then we conclude that:

$$\begin{aligned}\dim(M_g) &= \dim(\mathrm{Im}(\mu_f)) \\ &= \dim(B) - (2n+1-g) \\ &= 2(n+g-1) - (2n+1-g) \\ &= 3g-3\end{aligned}$$

This computation depends on several implicit assumptions but is essentially correct.

If $g = 2, 3$ one can take $n = 2, 3$ resp. and get the same result by a similar computation.

The previous computation is an example of *parameter counting*, a method that can be applied in several situations and is useful in computing the dimension of various loci in M_g. In such computations the universal property of M_g is used. There is a better way to perform them and it is by means of *deformation theory* (see Sect. 9).

7 Stable Curves

The moduli space M_g is quasi-projective but it is not projective. The reason is that curves varying in a family may become singular. In that case one speaks of a *degenerating family* of curves.

It is therefore natural to consider the functor $\widetilde{\mathcal{M}}_g$, of which $\mathcal{M}_g$ is a subfunctor, defined as follows:

$$\widetilde{\mathcal{M}}_g(B) := \left\{ \begin{array}{c} \text{isom. classes of flat families} \\ \text{of curves of arithmetic genus } g \end{array} \right\}$$

and ask: is it possible to embed M_g into a projective scheme $\widetilde{M}_g$ which is a coarse moduli scheme for the functor $\widetilde{\mathcal{M}}_g$?

The answer is NO, as the following example shows.

Example 9 Consider a nonsingular curve C and a parameter nonsingular curve B. Let $\beta : S \to B \times C$ be the blow-up at a point $x \in B \times C$. We get a flat family:

$$f : S \xrightarrow{\ \beta\ } B \times C \xrightarrow{\ p\ } B$$

whose fibres over $B \setminus p(x)$ are isomorphic to C while $f^{-1}(p(x))$ is a reducible curve. This is again an example of jumping phenomenon and it implies that the functor $\widetilde{\mathcal{M}}_g$ cannot be coarsely represented.

But there is a nice solution if we modify the question by *allowing only certain singular curves.*

Definition 10 A *stable curve* of genus $g \geq 2$ is a connected reduced curve of arithmetic genus g having at most nodes (i.e. ordinary double points) as singularities and such that every nonsingular rational component meets the rest of the curve in ≥ 3 points.

Define the *moduli functor of stable curves* of genus $g \geq 2$ as follows:

$$\overline{\mathcal{M}}_g(B) := \left\{ \begin{array}{c} \text{isom. classes of flat families} \\ \text{of stable curves of genus } g \end{array} \right\}$$

We obviously have natural inclusions:

$$\mathcal{M}_g(B) \subseteq \overline{\mathcal{M}}_g(B) \subseteq \widetilde{\mathcal{M}}_g(B)$$

Theorem 11 (Deligne-Mumford) *There is a projective scheme $\overline{M}_g$ containing M_g and coarsely representing the functor $\overline{\mathcal{M}}_g$. The complement $\overline{M}_g \setminus M_g$ is a divisor with normal crossings.*

Remark 12 M_g is not projective but not affine either. It is known that a priori it may contain projective subvarieties having up to dimension $g-2$ [12] but the precise bound for their dimension is not known in all genera.

Other variants of M_g are the moduli spaces of *stable pointed curves*. We will not introduce them since they will not appear in our discussion.

8 The Local Structure of a Scheme

Denote by $\widehat{\mathfrak{Loc}}$ the category of local noetherian complete $\mathbb{C}$-algebras with residue field $\mathbb{C}$, and local homomorphisms. Let

$$\mathbb{C}[\epsilon] := \mathbb{C}[t]/(t^2)$$

and $D := \mathrm{Spec}(\mathbb{C}[\epsilon])$. If $(R, \mathfrak{m})$ is in $\widehat{\mathfrak{Loc}}$ then

$$t_R = \left(\mathfrak{m}/\mathfrak{m}^2\right)^{\vee} = \mathrm{Hom}(R, \mathbb{C}[\epsilon])$$

is the (Zariski) tangent space of R.

The strictly local structure of a scheme X around a point $x \in X(\mathbb{C})$ is encoded by the *complete* local ring $\widehat{\mathcal{O}}_{X,x}$, which is by definition the following object of $\widehat{\mathfrak{Loc}}$:

$$\widehat{\mathcal{O}}_{X,x} = \varprojlim \mathcal{O}_{X,x}/\mathfrak{m}_{X,x}^n$$

This ring tells us in particular about the dimension and the singularity of X at x. For example, X is nonsingular of dimension d at x if and only if $\widehat{\mathcal{O}}_{X,x}$ is isomorphic to the formal power series ring $\mathbb{C}[[X_1, \ldots, X_d]]$ where $d = \dim(t_R)$.

If $\mu : B \to M$ is a morphism of reduced and irreducible schemes and $b \in B(\mathbb{C})$ is a nonsingular point then much of the local behaviour of μ at b is encoded by the local homomorphism:

$$\widehat{\mu}^\sharp : \widehat{\mathcal{O}}_{M,\mu(b)} \to \widehat{\mathcal{O}}_{B,b}$$

induced by μ. For example if

$$\mathrm{Spec}(\widehat{\mathcal{O}}_{B,b}) \to \mathrm{Spec}(\widehat{\mathcal{O}}_{M,\mu(b)})$$

is dominant then μ is dominant. The smoothness of μ at b is also encoded by the local homomorphism $\widehat{\mu}^\sharp$ because it is equivalent to the surjectivity of the differential:

$$d\mu_b : t_{\widehat{\mathcal{O}}_{B,b}} = T_b B \to T_{\mu(b)} M = t_{\widehat{\mathcal{O}}_{M,\mu(b)}}$$

To a ring R in $\widehat{\mathfrak{Loc}}$ one may associate a covariant functor defined on the category $\mathfrak{Art}$ of local artinian $\mathbb{C}$-algebras with residue field $\mathbb{C}$:

$$\widehat{h}_R : \mathfrak{Art} \longrightarrow (\mathrm{Sets})\ .$$

defined by:

$$\widehat{h}_R(A) = \mathrm{Hom}_{\mathbb{C}-alg}(R, A)$$

for all objects A of $\mathfrak{Art}$. Covariant functors $F : \mathfrak{Art} \longrightarrow (\mathrm{Sets})$ are called *functors of Artin rings* and one of the type $\widehat{h}_R$ is said to be *prorepresented* by R. Note that $\widehat{h}_R$ is the restriction to $\mathfrak{Art}$ of the representable functor

$$h_R : \widehat{\mathfrak{Loc}} \longrightarrow (\mathrm{Sets})\,, \quad h_R(S) = \mathrm{Hom}_{\mathbb{C}-alg}(R, S)$$

on the larger category $\widehat{\mathfrak{Loc}}$.

A morphism $\varphi : R \to S$ in $\widehat{\mathfrak{Loc}}$ induces a natural transformation $\widehat{h}_S \to \widehat{h}_R$ which, among other information, encodes the differential

$$d\varphi : t_S = \widehat{h}_S(\mathbb{C}[\epsilon]) \to \widehat{h}_R(\mathbb{C}[\epsilon]) = t_R$$

For an arbitrary functor of Artin rings F one can consider $t_F := F(\mathbb{C}[\epsilon])$. Under some conditions t_F ha a structure of $\mathbb{C}$-vector space and we are authorized to call it the *tangent space to the functor* F.

It is interesting that a ring R in $\widehat{\mathfrak{Loc}}$ can be recovered if we know the functor $\widehat{h}_R$ [43, Prop. 2.3.1]. Therefore the prorepresentable functors play a role with respect to strictly local properties of schemes analogous to (the role of) functors of points in characterizing schemes globally. One can start from a functor of

Artin rings and try to find conditions for its prorepresentability. This problem arises naturally for example in the formalization of deformation theory due to A. Grothendieck. He introduced the functors of Artin rings and gave a characterization of the prorepresentable ones (see [22, 23]). His results have been later improved by Schlessinger [41].

9 The Local Structure of M_g

We want to apply the remarks made in the previous section to the study of the local structure of M_g at a given point $[C]$. For doing this some of the technicalities of deformation theory are needed. Since they are not appropriate for a survey article of this kind we will only briefly outline the main steps.

Using the universal property it is natural to consider families of the form $f : \mathcal{C} \to \mathrm{Spec}(A)$, where $(A, \mathfrak{m})$ is in $\mathfrak{Art}$, such that there exists an isomorphism $C \cong f^{-1}([\mathfrak{m}])$. We call them *infinitesimal families at* $[C]$ parametrized by A (or by $\mathrm{Spec}(A)$).

We can now define a functor of Artin rings[1]

$$\mathrm{Inf}_{[C]} : \mathfrak{Art} \longrightarrow (\mathrm{Sets})$$

by setting:

$$\mathrm{Inf}_{[C]}(A) := \begin{pmatrix} \text{infinitesimal families at } [C] \\ \text{parametrized by } A \end{pmatrix} \Big/ \cong$$

Since M_g represents the moduli functor $\mathcal{M}_g$ only coarsely, we cannot expect that $\mathrm{Inf}_{[C]}$ is prorepresentable. All we can get from the universal property of M_g is a natural transformation:

$$\mu_C : \mathrm{Inf}_{[C]} \longrightarrow \widehat{h}_{\widehat{\mathcal{O}}}$$

where $\widehat{\mathcal{O}} = \widehat{\mathcal{O}}_{M_g,[C]}$. But this is not very useful. On the other hand an infinitesimal family at $[C]$

$$f : \mathcal{C} \to \mathrm{Spec}(A)$$

is very close to being a deformation of C; we only need to further specify an isomorphism $C \cong f^{-1}([\mathfrak{m}])$. Once we do this we call the resulting deformation:

$$\begin{array}{ccc} C & \longrightarrow & \mathcal{C} \\ \downarrow & & \downarrow f \\ \mathrm{Spec}(\mathbb{C}) & \longrightarrow & \mathrm{Spec}(A) \end{array}$$

[1] This functor is called the *crude local functor* in [27, § 18].

an *infinitesimal deformation* of C parametrized by A and we define a functor of Artin rings Def_C by:

$$\mathrm{Def}_C(A) := \left(\begin{matrix} \text{infinitesimal deformations of } C \\ \text{parametrized by } A \end{matrix} \right) \Big/ \cong$$

This functor is more interesting. For example its tangent space is easy to describe.

Definition 13 A *first order deformation* of C is a family of deformations of C parametrized by $\mathbb{C}[\epsilon]$:

$$\begin{array}{ccc} C & \hookrightarrow & \mathcal{C} \\ \downarrow & & \downarrow f \\ \mathrm{Spec}(\mathbb{C}) & \xrightarrow{(\epsilon)} & D \end{array} \tag{6}$$

Therefore $t_{\mathrm{Def}_C} = \mathrm{Def}_C(\mathbb{C}[\epsilon])$ is the set of isomorphism classes of first order deformations of C.

Proposition 14 *There is a natural identification:*

$$\kappa : \mathrm{Def}_C(\mathbb{C}[\epsilon]) \cong H^1(C, T_C)$$

The cohomology class $\kappa(f)$ associated to a first order deformation (6) *is called the* Kodaira-Spencer class *of f (KS class).*

For a proof of Proposition 14 we refer to [43, Prop. 1.2.9], where it is proved, more generally, for the deformation functor of a nonsingular projective variety of any dimension. The KS class can be used to introduce an important map associated to any family of deformations of C:

$$\begin{array}{ccc} C & \hookrightarrow & \mathcal{C} \\ \downarrow & & \downarrow f \\ \mathrm{Spec}(\mathbb{C}) & \xrightarrow{b} & B \end{array}$$

called *Kodaira-Spencer map* of f. It is the map:

$$\kappa : T_bB \longrightarrow H^1(C, T_C)$$

which associates to a tangent vector $v : D \to B$ at b the KS class $\kappa(f_v)$ of the first order deformation of C:

$$\begin{array}{ccccc} C & \hookrightarrow & \mathcal{C}_D & \longrightarrow & \mathcal{C} \\ \downarrow & & \downarrow f_v & & \downarrow f \\ \mathrm{Spec}(\mathbb{C}) & \xrightarrow{(\epsilon)} & D & \xrightarrow{v} & B \end{array}$$

obtained by pulling back f to D. The KS map is linear. As the name suggests, it has been introduced by Kodaira and Spencer in [33], and will play an important role in what follows.

Since two isomorphic infinitesimal deformations of C are supported on isomorphic infinitesimal families at $[C]$, we get a natural "forgetful" transformation of functors

$$\mathrm{Def}_C \longrightarrow \mathrm{Inf}_{[C]}$$

obtained by forgetting the isomorphism $C \cong f^{-1}([\mathfrak{m}])$. By composition with the transformation μ_C we obtain a natural transformation:

$$\Phi : \mathrm{Def}_C \longrightarrow \widehat{h}_{\widehat{\mathcal{O}}}$$

The precise relation between these functors is given by the following:

Theorem 15

(i) Def_C *is prorepresentable. Precisely, if* $g \geq 2$ Def_C *is prorepresented by* $R = \mathbb{C}[[X_1, \ldots, X_{3g-3}]]$, *if* $g = 1$ *by* $\mathbb{C}[[X]]$ *and if* $g = 0$ *by* $\mathbb{C}$.

(ii) Φ *corresponds to a local homomorphism* $\widehat{\mathcal{O}} \to R$ *such that the induced morphism* $\widetilde{\Phi} : \mathrm{Spec}(R) \to \mathrm{Spec}(\widehat{\mathcal{O}})$ *is dominant with finite fibres.*

(iii) *The following conditions are equivalent:*

- C *has no non-trivial automorphisms.*
- $\widetilde{\Phi}$ *is an isomorphism.*
- M_g *is nonsingular at* $[C]$.

Proof See [43] and [27]. □

As far as the study of families of curves is concerned, it is not so important to dig into the structure of the local ring $\widehat{\mathcal{O}}$ anymore. It suffices to record that the GIT construction shows that, for $g \geq 2$, M_g is locally the quotient of a nonsingular variety of dimension $3g - 3$ by the finite group $\mathrm{Aut}(C)$. Theorem 15 is the formal analogue of this fact.

10 Families with General Moduli

Suppose given a family of curves of genus g:

$$f : \mathcal{C} \to B$$

parametrized by an algebraic variety B. Let $b \in B$ be a closed nonsingular point and $C = f^{-1}(b)$. We would like to have a criterion to decide whether $\mu_f : B \to M_g$ is dominant, i.e. if f has general moduli, using only local information about f at b. If the differential:

$$d\mu_{f,b} : T_b B \longrightarrow T_{[C]} M_g$$

is surjective then μ_f is smooth at b and therefore the family f has general moduli. This criterion can possibly work only if M_g is nonsingular at $[C]$. A more efficient result is the following.

Proposition 16 *Let $f : \mathcal{C} \to B$ be a family of curves of genus g parametrized by an algebraic variety B, let $b \in B$ be a closed nonsingular point and $C = f^{-1}(b)$. If the KS map:*

$$\kappa : T_b B \longrightarrow H^1(C, T_C)$$

is surjective then f is a family with general moduli.

Proof Let $S = \widehat{\mathcal{O}}_{B,b}$. The family f defines a natural transformation of functors of Artin rings:

$$\widehat{h}_S \longrightarrow \mathrm{Def}_C$$

by associating to every A in $\mathfrak{A}rt$ and $(\alpha : S \to A) \in \widehat{h}_S(A)$ the infinitesimal deformation:

$$\begin{array}{ccc} C & \longrightarrow & \mathrm{Spec}(A) \times_B \mathcal{C} \\ \downarrow & & \downarrow \\ \mathrm{Spec}(\mathbb{C}) & \longrightarrow & \mathrm{Spec}(A) \end{array}$$

obtained by pulling f via α. This natural transformation corresponds to a homomorphism: $\Psi : R \to S$ where R is the ring prorepresenting Def_C (Theorem 15(i)). We have a commutative diagram:

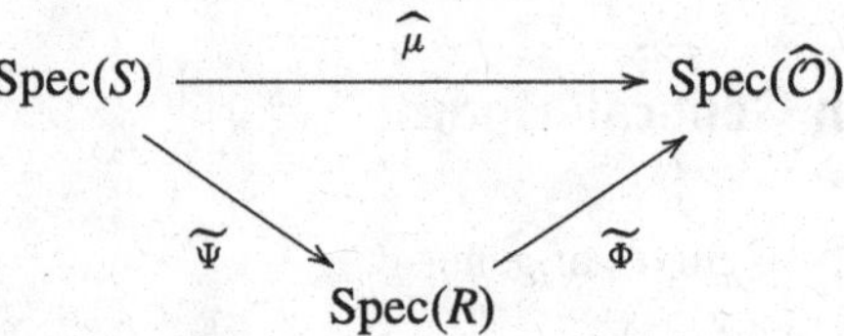

where $\widehat{\mu}$ is induced by the functorial morphism $\mu_f : B \to M_g$. Since $\kappa = d\Psi$ is surjective $\widetilde{\Psi}$ is smooth, and by Theorem 15(ii) $\widetilde{\Phi}$ is dominant. Thus $\widehat{\mu}$ is dominant and therefore also μ_f is dominant. □

Note that the criterion of Proposition 16 applies regardless of whether $\mathrm{Aut}(C)$ is trivial or not.

11 The Hilbert Scheme

The next step is to have a good description of families of curves in a given projective space, which are the most important families appearing in nature. This is done by introducing new objects, the Hilbert schemes. For details on this section we refer to [5, 43].

We will introduce first the functors that are represented by the Hilbert schemes. Fix a polynomial $p(t) \in \mathbb{Q}[t]$ and define a contravariant functor:

$$Hilb^r_{p(t)} : (\text{Schemes}) \longrightarrow (\text{Sets})$$

setting

$$Hilb^r_{p(t)}(B) = \left\{ \begin{array}{c} \text{families of closed subschemes of } \mathbb{P}^r \\ \text{param. by } B \text{ and with Hilbert polyn. } p(t) \end{array} \right\}$$

This is the *Hilbert functor* for the polynomial $p(t)$.

Theorem 17 (Grothendieck [24]) *For every $r \geq 2$ and $p(t)$ there is a projective scheme* $\mathrm{Hilb}^r_{p(t)}$ *and a family*

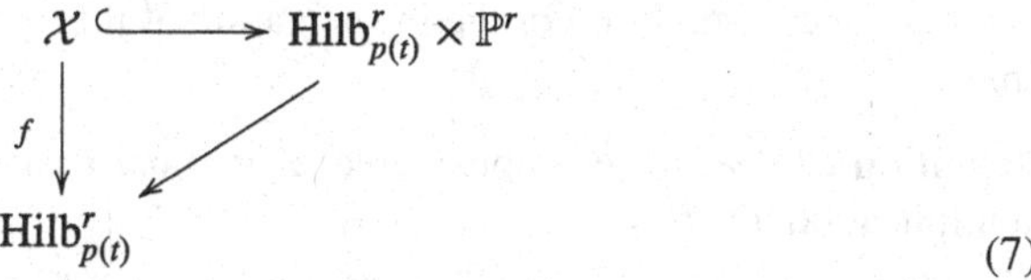

$$(7)$$

which is universal for the functor $Hilb^r_{p(t)}$. In particular $Hilb^r_{p(t)}$ is representable.

$\mathrm{Hilb}^r_{p(t)}$ is called *Hilbert scheme* of $\mathbb{P}^r$ relative to the Hilbert polynomial $p(t)$.

It is a very complicated object, highly reducible and singular, in general non-reduced, but connected. Its local properties at a point $[X \subset \mathbb{P}^r]$ depend only on the geometry of the embedding $X \subset \mathbb{P}^r$, as the following shows.

Theorem 18 (Grothendieck [24]) *Let $X \subset \mathbb{P}^r$ be a local complete intersection with Hilbert polynomial $p(t)$. Let $\mathcal{I} \subset \mathcal{O}_{\mathbb{P}^r}$ be its ideal sheaf and $N = N_{X/\mathbb{P}^r} := Hom(\mathcal{I}/\mathcal{I}^2, \mathcal{O}_X)$ its normal bundle. Then:*

- *$H^0(X,N)$ is the Zariski tangent space to $\mathrm{Hilb}^r_{p(t)}$ at $[X]$.*
- *$h^0(X,N) - h^1(X,N) \leq \dim_{[X]}(\mathrm{Hilb}^r_{p(t)}) \leq h^0(X,N)$.*
- *If $H^1(X,N) = 0$ then $\mathrm{Hilb}^r_{p(t)}$ is nonsingular of dimension $h^0(X,N)$ at $[X]$.*

If $C \subset \mathbb{P}^r$ is a nonsingular curve of degree d and genus g then the Hilbert polynomial of C is $p(t) = dt + 1 - g$ and we write $\mathrm{Hilb}^r_{d,g}$ instead of $\mathrm{Hilb}^r_{p(t)}$. Then:

$$h^0(C,N) - h^1(C,N) = \chi(C,N) = (r+1)d + (r-3)(1-g)$$

For example, if C is a nonsingular plane curve of degree d its genus is $g = \binom{d-1}{2}$ and $N = \mathcal{O}_C(d)$. Then $H^1(C,N) = 0$ and

$$h^0(C,N) = 3d + g - 1 = \frac{d(d+3)}{2} = \binom{d+2}{2} - 1$$

If $C \subset \mathbb{P}^3$ is a nonsingular curve of degree d and genus g then $\chi(N) = 4d$ does not depend on g. $\mathrm{Hilb}^3_{d,g}$ can be singular and/or of dimension $> 4d$.

If $C \subset \mathbb{P}^r$ is nonsingular of degree d and genus g the KS map of the universal family (7) at the point $[C]$ is a map:

$$\kappa_C : T_{[C]}\mathrm{Hilb}^r_{d,g} = H^0(C, N_{C/\mathbb{P}^r}) \longrightarrow H^1(C, T_C) \tag{8}$$

Proposition 19 *κ_C is the coboundary map of the* normal sequence

$$0 \longrightarrow T_C \longrightarrow T_{\mathbb{P}^r|C} \longrightarrow N_{C/\mathbb{P}^r} \longrightarrow 0$$

Proof An easy diagram chasing. □

It is of primary importance to decide for which d, g, r there is an irreducible component of the Hilbert scheme $\mathrm{Hilb}^r_{d,g}$ such that the universal family restricted to it has general moduli. For the investigation of this condition we introduce a new object.

Definition 20 Let C be a nonsingular curve and L an invertible sheaf on C. The multiplication map:

$$\mu_0(L) : H^0(L) \otimes H^0(\omega_C L^{-1}) \longrightarrow H^0(\omega_C)$$

is the *Petri map* relative to L.

We have:

Proposition 21 *Let $C \subset \mathbb{P}^r$ be of degree d and genus g, $L = \mathcal{O}_C(1)$, and assume that $\mu_0(L)$ is injective. Then $[C] \in \mathrm{Hilb}^r_{d,g}$ is a nonsingular point and the map* (8) *is surjective. Therefore the universal family* (7) *has general moduli around* $[C]$ *(shortly, the curves parametrized by* $\mathrm{Hilb}^r_{d,g}$ *nearby* $[C]$ *have general moduli).*

Proof We can reduce to the case $r + 1 = h^0(L)$. Then we use the *restricted Euler sequence*

$$0 \longrightarrow \mathcal{O}_C \longrightarrow H^0(L)^\vee \otimes L \longrightarrow T_{\mathbb{P}^r|C} \longrightarrow 0$$

to show that the injectivity of $\mu_0(L)$ is equivalent to $H^1(C, T_{\mathbb{P}^r|C}) = 0$. Then we apply Proposition 16, Theorem 18 and Proposition 19 to conclude. □

If L is an invertible sheaf on C such that $\deg(L) = d$ and $h^0(L) = r + 1$ then the *expected corank* of $\mu_0(L)$ is

$$\rho(g, r, n) := g - (r + 1)(g - d + r)$$

It is called the *Brill-Noether number* relative to g, r, d. Of course for $\mu_0(L)$ to be injective it is necessary that $\rho(g, r, n) \geq 0$. The condition $\rho(g, r, d) \geq 0$ is equivalent to $d \geq \frac{1}{2}g + 1$ if $r = 1$, and to $d \geq \frac{2}{3}g + 2$ if $r = 2$, etc.

On the other hand $\mu_0(L)$ can have a non-zero kernel even if $\rho(g, r, n) \geq 0$, as shown by plenty of examples. Nevertheless the following important result holds.

Theorem 22 *Let $r \geq 3$ and $d, g \geq 0$ such that $\rho(g, r, d) \geq 0$. Then* $\mathrm{Hilb}^r_{d,g}$ *has a unique irreducible component $\mathbf{I}^r_{d,g}$ whose general point parametrizes a nonsingular irreducible curve $C \subset \mathbb{P}^r$ such that $h^0(\mathcal{O}_C(1)) = r + 1$ and $\mu_0(\mathcal{O}_C(1))$ is injective. In particular an open non-empty subset of $\mathbf{I}^r_{d,g}$ parametrizes a family of curves of genus g with general moduli.*

If $r = 2$ and $d, g \geq 0$ are such that $d \geq \frac{2}{3}g + 2$ then there is an irreducible locally closed $\mathcal{V}_{d,g} \subset |H^0(\mathbb{P}^2, \mathcal{O}(d))|$ parametrizing irreducible plane curves of degree d and geometric genus g such that:

(a) *Every $\Gamma \subset \mathbb{P}^2$ parametrized by a point of $\mathcal{V}_{d,g}$ has only nodes as singularities and the sheaf $L = \nu^*\mathcal{O}_\Gamma(1)$ on the normalization $\nu : C \to \Gamma$ satisfies $h^0(L) = 3$ and $\mu_0(L)$ injective.*
(b) *The restriction to $\mathcal{V}_{d,g}$ of the universal family can be simultaneously desingularized and the resulting family of nonsingular curves of genus g has general moduli.*
(c) *There is a unique $\mathcal{V}_{d,g}$ maximal with respect to properties (a) and (b).*

This theorem has been foreseen since the beginning of curve theory, starting from Brill-Noether [6] and Severi [45]. It is due to the concentrated efforts of several mathematicians during the 1970s and 1980s: Kleiman-Laksov [32], Kempf [31], Griffiths-Harris [21], Arbarello-Cornalba [2], Eisenbud-Harris [14], Gieseker [20], Fulton-Lazarsfeld [19], Harris [25].

12 Petri General Curves

The Petri map is a central object in curve theory.

Definition 23 A curve C is called *Petri general* if the Petri map $\mu_0(L)$ is injective for all invertible sheaves $L \in \mathrm{Pic}(C)$.

Note that if $\deg(L) < 0$ or $\deg(L) > 2g - 2$ then $\mu_0(L)$ is clearly injective. Moreover $\mu_0(L) = \mu_0(\omega_C L^{-1})$. Therefore the condition that C is Petri general has to be checked only on invertible sheaves such that $0 \leq \deg(L) \leq g - 1$.

A Petri general curve has the property that any embedding $C \subset \mathbb{P}^r$, $r \geq 3$, by a complete linear system corresponds to a nonsingular point of one of the components $\mathbf{I}^r_{d,g}$ of the Hilbert scheme described by Theorem 22.

The definition of Petri general curve is completely intrinsic, i.e. it does not make use of families. In a footnote to [40] K. Petri, a student of M. Noether, stated as a fact what has been subsequently considered as

Petri's Conjecture For every g a general curve of genus g is Petri general.

This conjecture asserts that for a curve C the condition of being Petri general does not impose any closed condition on its moduli. According to the conjecture Petri general curves should be the most natural ones available in nature. But in fact this is not the case.

For example nonsingular plane curves of degree $d \geq 5$ are not Petri general.

In fact if $C \subset \mathbb{P}^2$ is of degree d then $\omega_C = \mathcal{O}(d-3)$ and therefore

$$H^0(\omega_C L^{-2}) = H^1(\mathcal{O}_C(2))^\vee \neq 0$$

if $d \geq 5$. Then use a simple remark which shows that $H^0(\omega_C L^{-2}) \subset \ker(\mu_0(L))$ for any L on any curve. A similar argument holds for complete intersections of multidegree $(d_1, \ldots, d_{r-1})$ such that $\sum d_j \geq r+3$.

It is very difficult to produce explicit examples of Petri general curves (see the final section for more about this). So the challenge of Petri's conjecture, if true, is to modify our intuitive idea of a general curve.

The conjecture is in fact true. It has been proved for the first time by Gieseker [20], and subsequently it has been given simpler proofs by Eisenbud and Harris [14] and by Lazarsfeld [34]. Special cases of the conjecture had been proved before by Arbarello and Cornalba [2].

13 Unirationality

To describe explicitly a general curve of genus g is the most elusive part of the theory. To write down equations of a general curve with coefficients depending on parameters requires solving a mixture of abstract and concrete problems that are very difficult to concile.

For instance, it is very difficult to give an explicit description/construction of the family parametrized by $\mathbf{I}^r_{d,g}$ described in Theorem 22, even if we know that it exists. The attempts to construct such explicit families have a long history and have motivated a large amount of work on M_g. The classical geometers succeeded for the first values of g and observed that for the families they got the parameter variety is rational or unirational. For example, a canonical curve of genus 3 is just a nonsingular plane quartic: it moves in the linear system $|H^0(\mathbb{P}^2, \mathcal{O}(4))| \cong \mathbb{P}^{14}$. A similar remark can be made for genus 4 and 5 since canonical curves of genus 4 and 5 are complete intersections in $\mathbb{P}^3$ (resp. $\mathbb{P}^4$).

One can prove, because of Theorem 22, that the general curve C of genus 6 can be realized as a plane sextic curve with 4 double points (see the table below). The linear system of adjoints to C of degree 3 maps $\mathbb{P}^2$ birationally to a Del

Pezzo surface $S \subset \mathbb{P}^5$ containing the canonical model of C. Now $C \subset \mathbb{P}^5$ is the complete intersection of S with a quadric, therefore it varies in the linear system $|\mathcal{O}_S(2)|$, which is 15-dimensional. Since any two general 4-tuples of points in $\mathbb{P}^2$ are projectively equivalent, all the Del Pezzo surfaces constructed in this way are isomorphic. Therefore the general curve C of genus 6 is parametrized by a point varying in a $(15 = 3 \cdot 6 - 3)$-dimensional linear system on a fixed surface S, and M_6 is thus unirational. A more delicate argument, due to Shepherd-Barron [46], shows that M_6 is even rational.

One may ask whether an analogous statement is true for higher values of g, namely whether it is possible to produce a family of projective curves with general moduli parametrized by a rational variety, say an open subset of a projective space. For such a family $f : C \to B$ the functorial morphism $\mu_f : B \to M_g$, being dominant, proves that M_g is unirational.

To my knowledge M. Noether was the first to ask such a question. He extended the above constructions to the more difficult case of genus 7 proving the unirationality of M_7 by explicit describing the canonical curves [39].

Subsequently Severi [44] extended the result up to genus 10. The proof given by Severi is quite simple and can be easily understood by looking at the following table which lists degree and number of nodes of plane curves of minimal degree and non-negative Brill-Noether number.

Families of plane nodal curves with general moduli

Genus	Degree	δ	$\frac{d(d+3)}{2} - 3\delta$	ρ
0	1	0	2	0
1	3	0	9	1
2	4	1	11	2
3	4	0	14	0
4	5	2	14	1
5	6	5	12	2
6	6	4	15	0
7	7	8	11	1
8	8	13	5	2
9	8	12	8	0
10	9	18	0	1
11	10	25	−10	2
12	10	24	−7	0
⋮	⋮	⋮	⋮	⋮

In outline it goes as follows. The table shows that if $g \leq 10$ it is possible to realize a general curve of genus g as a plane curve of degree d with δ singular points in such a way that

$$3\delta \leq \frac{d(d+3)}{2}$$

This implies, modulo a careful argument, that we can assign the singular points of such a curve in general position. Then the parameter space of the corresponding family $f : \mathcal{C} \to \mathcal{V}_{d,g}$ is fibered over an open subset of $(\mathbb{P}^2)^{(\delta)}$ with fibres linear systems, and therefore $\mathcal{V}_{d,g}$ is rational.

Based on such limited evidence Severi conjectured the unirationality of M_g for all g [44]. This conjecture resisted for 67 years, even though it was certainly considered with great interest (see e.g. [37]), until it was finally disproved by Harris and Mumford in 1982 [26]. One year earlier the unirationality of M_{12} had been proved [42] and other results followed promptly [8, 9, 15, 28–30, 35, 46]. Other results have been proved in more recent years [7, 16, 17, 47]. Now we have a quite precise information about the Kodaira dimension of M_g for almost all g, summarized in the table below. The problem is not yet closed though because as of today (June 2017) we have no idea about the Kodaira dimension of M_g for $17 \leq g \leq 21$. This is a challenge for younger generations!

For a detailed discussion of the vast topic considered in this section I refer the reader to the survey article of Verra [48].

State of the art about the Kodaira dimension κ of M_g

Genus	K-dim	Credit
$1, 2, 3, 4, 6$	Rational	Weierstrass-Salmon (1), Igusa (2) Katsylo (3, 4, 5), Shepherd-Barron (6)
11	Uniruled	Mori-Mukai
≤ 14	Unirational	Noether (≤ 7), Severi (≤ 10), Sernesi (12), Chang-Ran (11, 13), Verra (14)
15	rat.lly connected	Chang-Ran ($\kappa = -\infty$), Bruno-Verra
16	Uniruled	Chang-Ran ($\kappa = -\infty$), Farkas
$22, \geq 24$	gen. type	Farkas (22), Harris-Mumford (odd), Eisenbud-Harris (even)
23	≥ 2	Farkas

14 Construction of Petri General Curves

The existence of Petri general curves has been proved originally [20] by degeneration, thus in a non-effective way. The following has been a breakthrough:

Theorem 24 (Lazarsfeld [34]) *If S is a K3 surface with* $\mathrm{Pic}(S) = \mathbb{Z}[H]$ *then a general curve $C \in |H|$ is Petri general.*

This result is still non-effective, but it brings very clearly on the forefront the fact that *Petri general curves are not necessarily appearing in families with general moduli.* In fact the specific classes of Petri general curves described by

Theorem 24 (we will call them *K3-curves*) are an illustration of this fact. Let's count parameters.

- Pairs (S, H) depend on 19 moduli.
- The linear system $|H|$ on a given (S, H) has dimension $g = \frac{1}{2}(H \cdot H) + 1$.

Therefore K3-curves depend on $\leq g + 19$ moduli. If $g \geq 12$ then $3g - 3 > g + 19$ and therefore a K3-curve of genus g is not a general curve if $g \geq 12$, even though it is Petri general.

K3-curves of genus $g \geq 12$ have been characterized recently among the Petri general ones by means of a cohomological condition by Arbarello et al. [4]. The condition is that the so-called *Wahl map*

$$\bigwedge^2 H^0(\omega_C) \longrightarrow H^0(\omega^3) \tag{9}$$

is not surjective. Precisely, the map (9) is known to be surjective on general curves [10] and therefore its non-surjectivity defines a closed locus $W \subset M_g$. It has also been known since some time [49] that the locus of K3-curves is contained in W. In [4] it is proved that W intersects the open set of Petri general curves precisely along the closure of the locus of K3-curves.

This argument leaves open the possibility that there exist Petri general curves in W that are limits of K3-curves without being K3-curves, namely Petri general curves not contained in a K3 surface but only on a (singular) limit of K3 surfaces. This question has been considered recently by Arbarello, Bruno, Farkas and Saccà in [3]. In their work the authors give examples of Petri general curves C of every genus $g \geq 1$ on certain rational surfaces that are limits of K3 surfaces. These surfaces are obtained by contracting the exceptional elliptic curve on the blow-up of $\mathbb{P}^2$ at 10 points $p_1, \ldots, p_{10} \in \mathbb{P}^2$ conveniently chosen on a cubic. The curves C, called *Du Val curves*, are the images of the proper transforms of curves of degree $3g$ passing through the points $p_1, \ldots, p_9, p_{10}$ with multiplicity $(g, \ldots, g, g-1, 1)$. In [1] Arbarello and Bruno have finally shown that there exist Du Val curves which are not K3-curves even though they are limits of K3-curves, thereby proving that the result of Arbarello et al. [4] is sharp.

Acknowledgements I want to warmly thank the Organizing Committee of the CIMPA School for inviting me to lecture in Guanajuato. Special additional thanks go to Leticia Brambila Paz for her perfect organizing job and for her enthusiastic presence during the school. I also thank the other speakers and the participants for creating a friendly and stimulating atmosphere. I am grateful to the referee for a careful reading and for useful remarks.

References

1. E. Arbarello, A. Bruno, Rank two vector bundles on polarized Halphen surfaces and the Gauss-Wahl map for du Val curves. arXiv:1609.09256
2. E. Arbarello, M. Cornalba, Su una congettura di Petri. Comment. Math. Helv. **56**, 1–38 (1981)

3. E. Arbarello, A. Bruno, G. Farkas, G. Saccà, Explicit Brill-Noether-Petri general curves. arXiv:1511.07321
4. E. Arbarello, A. Bruno, E. Sernesi, On two conjectures by J. Wahl. arXiv:1507.05002
5. E. Arbarello, M. Cormalba, P. Griffiths, *Geometry of Algebraic Curves*, vol. II. Grundlehren der mathematischen Wissenschaften, vol. 268 (Springer, Berlin, 2011)
6. A. Brill, M. Noether, Ueber die algebraischen Functionen und ihre Anwendung in der Geometrie. Math. Ann. **7**, 269–310 (1873)
7. A. Bruno, A. Verra, M_{15} is rationally connected, in *Projective Varieties with Unexpected Properties*, pp. 51–65 (Walter de Gruyter, Berlin, 2005)
8. M.C. Chang, Z. Ran, Unirationality of the moduli spaces of curves of genus 11, 13 (and 12). Invent. Math. **76**, 41–54 (1984)
9. M.C. Chang, Z. Ran, On the slope and Kodaira dimension of $\overline{M}_g$ for small g. J. Differ. Geom. **34**, 267–274 (1991)
10. C. Ciliberto, J. Harris, R. Miranda, On the surjectivity of the Wahl map. Duke Math. J. **57**, 829–858 (1988)
11. P. Deligne, D. Mumford, The irreducibility of the space of curves of given genus. Publ. Math. IHES **36**, 75–110 (1969)
12. S. Diaz, *Exceptional Weierstrass Points and the Divisor on Moduli Space that They Define*, vol. 56 (Memoirs of the American Mathematical Society, Providence, RI, 1985)
13. I. Dolgachev, *Classical Algebraic Geometry - A Modern View* (Cambridge University Press, Cambridge, 2012)
14. D. Eisenbud, J. Harris, A simpler proof of the Gieseker-Petri theorem on special divisors. Invent. Math. **74**, 269–280 (1983)
15. D. Eisenbud, J. Harris, The Kodaira dimension of the moduli space of curves of genus ≥ 23. Invent. Math. **90**, 359–387 (1987)
16. G. Farkas, The geometry of the moduli space of curves of genus 23. Math. Ann. **318**, 43–65 (2000)
17. G. Farkas, $\overline{M}_{22}$ is of general type. Manuscript (2005)
18. W. Fulton, On the irreducibility of the moduli space of curves. Appendix to a paper of Harris-Mumford. Invent. Math. **67**, 87–88 (1982)
19. W. Fulton, R. Lazarsfeld, On the connectedness of degeneracy loci and special divisors. Acta Math. **146**, 271–283 (1981)
20. D. Gieseker, Stable curves and special divisors: Petri's conjecture. Invent. Math. **66**, 251–275 (1982)
21. P. Griffiths, J. Harris, The dimension of the variety of special linear systems on a general curve. Duke Math. J. **47**, 233–272 (1980)
22. A. Grothendieck, Géométrie formelle et géométrie algébrique. Seminaire Bourbaki, exp. 182 (1959)
23. A. Grothendieck, Le theoreme d'existence en theorie formelle des modules. Seminaire Bourbaki, exp. 195 (1960)
24. A. Grothendieck, Les schemas de Hilbert. Seminaire Bourbaki, exp. 221 (1960)
25. J. Harris, On the Severi problem. Invent. Math. **84**, 445–461 (1986)
26. J. Harris, D. Mumford, On the Kodaira dimension of the moduli space of curves. Invent. Math. 67, 23–86 (1982)
27. R. Hartshorne, *Deformation Theory*. Graduate text in Mathematics, vol. 257 (Springer, Berlin, 2010)
28. P.I. Katsylo, The moduli variety of curves of genus 4 is rational. Dokl. Akad. Nauk SSSR **290**, 1292–1294 (1986)
29. P.I. Katsylo, Rationality of the moduli variety of curves of genus 5. Math. USSR Sbornik **72**, 439–445 (1992)
30. P.I. Katsylo, Rationality of the moduli variety of curves of genus 3. Comment. Math. Helv. **71**, 507–524 (1996)
31. G. Kempf, *Schubert Methods with an Application to Algebraic Curves* (Publication of Mathematisch Centrum, Amsterdam, 1972)

32. S. Kleiman, D. Laksov, On the existence of special divisors. Am. Math. J. **94**, 431–436 (1972)
33. K. Kodaira, D.C. Spencer, On deformations of complex analytic structures I, II. Ann. Math. **67**, 328–466 (1958)
34. R. Lazarsfeld, Brill-Noether-Petri without degenerations. J. Differ. Geom. **23**, 299–307 (1986)
35. S. Mori, S. Mukai, The uniruledness of the moduli space of curves of genus 11, in *Springer Lecture Notes in Mathematics*, vol. 1016 (Springer, Berlin, 1983), pp. 334–353
36. D. Mumford, *Lectures on Curves on an Algebraic Surface* (Princeton University Press, Princeton, 1966)
37. D. Mumford, Problems of present day mathematics, VI: algebraic geometry, in *Mathematical Developments Arising from Hilbert Problems*. Proceedings of Symposia in Pure Mathematics, Part 1, vol. 28 (American Mathematical Society, Providence, 1976), pp. 44–45
38. D. Mumford, J. Fogarty, *Geometric Invariant Theory*, 2nd enlarged edn. Ergebnisse b. vol. 34 (Springer, Berlin, 1982)
39. M. Noether, Note über die Normalcurven für $p = 5, 6, 7$. Math. Ann. **26**, 143–150 (1886)
40. K. Petri, Über Spezialkurven I. Math. Ann. **93**, 182–209 (1925)
41. M. Schlessinger, Functors of Artin rings. Trans. Am. Math. Soc. **130**, 208–222 (1968)
42. E. Sernesi, L'unirazionalità della varietà dei moduli delle curve di genere dodici. Ann. Sc. Norm. Super. Pisa (4) **8**, 405–439 (1981)
43. E. Sernesi, *Deformations of Algebraic Schemes*. Grundlehren der Mathematischen Wissenschaften, vol. 334 (Springer, Berlin, 2006)
44. F. Severi, Sulla classificazione delle curve algebriche e sul teorema di esistenza di Riemann. Rend. Accademia Naz. Lincei (5) **241**, 877–888/1011–1020 (1915)
45. F. Severi, *Vorlesungen uber Algebraische Geometrie* (Taubner, Leipzig, 1921)
46. N.I. Shepherd-Barron, Invariant theory of S_5 and the rationality of M_6. Compos. Math. **70**, 13–25 (1989)
47. A. Verra, The unirationality of the moduli spaces of curves of genus 14 or lower. Compos. Math. **141**, 1425–1444 (2005)
48. A. Verra, Rational parametrizations of moduli spaces of curves, in *Handbook of Moduli III* (International Press, Boston, 2008–2012), pp. 431–507
49. J. Wahl, The Jacobian algebra of a graded Gorenstein singularity. Duke Math. J. **55**, 843–871 (1987)

LECTURE NOTES IN MATHEMATICS Springer

Editorial Policy

1. Lecture Notes aim to report new developments in all areas of mathematics and their applications – quickly, informally and at a high level. Mathematical texts analysing new developments in modelling and numerical simulation are welcome.

   Manuscripts should be reasonably self-contained and rounded off. Thus they may, and often will, present not only results of the author but also related work by other people. They may be based on specialised lecture courses. Furthermore, the manuscripts should provide sufficient motivation, examples and applications. This clearly distinguishes Lecture Notes from journal articles or technical reports which normally are very concise. Articles intended for a journal but too long to be accepted by most journals, usually do not have this "lecture notes" character. For similar reasons it is unusual for doctoral theses to be accepted for the Lecture Notes series, though habilitation theses may be appropriate.

2. Besides monographs, multi-author manuscripts resulting from SUMMER SCHOOLS or similar INTENSIVE COURSES are welcome, provided their objective was held to present an active mathematical topic to an audience at the beginning or intermediate graduate level (a list of participants should be provided).

   The resulting manuscript should not be just a collection of course notes, but should require advance planning and coordination among the main lecturers. The subject matter should dictate the structure of the book. This structure should be motivated and explained in a scientific introduction, and the notation, references, index and formulation of results should be, if possible, unified by the editors. Each contribution should have an abstract and an introduction referring to the other contributions. In other words, more preparatory work must go into a multi-authored volume than simply assembling a disparate collection of papers, communicated at the event.

3. Manuscripts should be submitted either online at www.editorialmanager.com/lnm to Springer's mathematics editorial in Heidelberg, or electronically to one of the series editors. Authors should be aware that incomplete or insufficiently close-to-final manuscripts almost always result in longer refereeing times and nevertheless unclear referees' recommendations, making further refereeing of a final draft necessary. The strict minimum amount of material that will be considered should include a detailed outline describing the planned contents of each chapter, a bibliography and several sample chapters. Parallel submission of a manuscript to another publisher while under consideration for LNM is not acceptable and can lead to rejection.

4. In general, **monographs** will be sent out to at least 2 external referees for evaluation.

   A final decision to publish can be made only on the basis of the complete manuscript, however a refereeing process leading to a preliminary decision can be based on a pre-final or incomplete manuscript.

   Volume Editors of **multi-author works** are expected to arrange for the refereeing, to the usual scientific standards, of the individual contributions. If the resulting reports can be forwarded to the LNM Editorial Board, this is very helpful. If no reports are forwarded or if other questions remain unclear in respect of homogeneity etc, the series editors may wish to consult external referees for an overall evaluation of the volume.

5. Manuscripts should in general be submitted in English. Final manuscripts should contain at least 100 pages of mathematical text and should always include

   – a table of contents;
   – an informative introduction, with adequate motivation and perhaps some historical remarks: it should be accessible to a reader not intimately familiar with the topic treated;
   – a subject index: as a rule this is genuinely helpful for the reader.
   – For evaluation purposes, manuscripts should be submitted as pdf files.

6. Careful preparation of the manuscripts will help keep production time short besides ensuring satisfactory appearance of the finished book in print and online. After acceptance of the manuscript authors will be asked to prepare the final LaTeX source files (see LaTeX templates online: https://www.springer.com/gb/authors-editors/book-authors-editors/manuscriptpreparation/5636) plus the corresponding pdf- or zipped ps-file. The LaTeX source files are essential for producing the full-text online version of the book, see http://link.springer.com/bookseries/304 for the existing online volumes of LNM). The technical production of a Lecture Notes volume takes approximately 12 weeks. Additional instructions, if necessary, are available on request from; lnm@springer.com.

7. Authors receive a total of 30 free copies of their volume and free access to their book on SpringerLink, but no royalties. They are entitled to a discount of 33.3 % on the price of Springer books purchased for their personal use, if ordering directly from Springer.

8. Commitment to publish is made by a *Publishing Agreement*; contributing authors of multiauthor books are requested to sign a *Consent to Publish form*. Springer-Verlag registers the copyright for each volume. Authors are free to reuse material contained in their LNM volumes in later publications: a brief written (or e-mail) request for formal permission is sufficient.

Addresses:
Professor Jean-Michel Morel, CMLA, École Normale Supérieure de Cachan, France
E-mail: moreljeanmichel@gmail.com

Professor Bernard Teissier, Equipe Géométrie et Dynamique,
Institut de Mathématiques de Jussieu – Paris Rive Gauche, Paris, France
E-mail: bernard.teissier@imj-prg.fr

Springer: Ute McCrory, Mathematics, Heidelberg, Germany,
E-mail: lnm@springer.com

Printed in the United States
By Bookmasters